POWER ENGINEERING
Advances and Challenges
Part B: Electrical Power

Editors

Viorel Badescu

Candida Oancea Institute
Polytechnic University of Bucharest
Spl. Independentei 313, Bucharest 060042, Romania

George Cristian Lazaroiu

Department of Power Systems
University Politehnica of Bucharest
Spl. Independentei 313, Bucharest 060042, Romania

Linda Barelli

Department of Engineering
University of Perugia
via G. Duranti 93, 06125 Perugia, Italy

CRC Press
Taylor & Francis Group
Boca Raton London New York

CRC Press is an imprint of the
Taylor & Francis Group, an **informa** business

A SCIENCE PUBLISHERS BOOK

Cover credit: Maurizio Luigi Cumo and *Renato Gatto*

CRC Press
Taylor & Francis Group
6000 Broken Sound Parkway NW, Suite 300
Boca Raton, FL 33487-2742

First issued in paperback 2020

© 2019 by Taylor & Francis Group, LLC
CRC Press is an imprint of Taylor & Francis Group, an Informa business

No claim to original U.S. Government works

ISBN-13: 978-1-138-31987-5 (hbk)
ISBN-13: 978-0-367-78058-6 (pbk)

Visit the Taylor & Francis Web site at
http://www.taylorandfrancis.com

and the CRC Press Web site at
http://www.crcpress.com

Preface

Power engineering is a subfield of energy engineering and electrical engineering. It deals with the generation, transmission, distribution and utilization of electric power and the electrical devices connected to such systems including generators, motors and transformers. This perception is associated with the generation of power by large plants and distributed for consumption. In the last few decades mankind has faced the climate change phenomena and seen changes in social attitudes including interest in environment protection, and the depletion of classical energy resources. These have had bearings on the power production sector, and resulted in extensive changes and the need to adapt.

Future energy systems must take advantage of the changes and advances in technologies like improvements in natural gas combined cycles and clean coal technologies, carbon dioxide capture and storage, advancements in nuclear reactors and hydropower, renewable energy engineering, power to gas conversion and fuel cells, energy crops, new energy vectors biomass hydrogen, thermal energy storage, new storage systems diffusion, modern substations, high voltage engineering equipment and compatibility, HVDC transmission with FACTS, advanced optimization in a liberalized market environment, active grids and smart grids, power system resilience, power quality and cost of supply, plugin electric vehicles, smart metering, control and communication technologies, new key actors as prosumers, smart cities. These advances will enhance the security of power systems, safety in operation, protection of the environment, high energy efficiency, reliability and sustainability.

The book is a source of information for specialists involved in power engineering related activities and a good starting point for young researchers. The content is structured along logical lines of progressive thought. It presents the current developments and active technological advances in the main aspects of energy engineering, both in thermal and electrical engineering, with contributions from highly qualified experts in each field of study.

The principal audience consists of researchers, engineers, educators involved with the curriculum and research strategies in the field of power engineering. The book is useful for industry and developers interested in joining national or international power development programs. Finally, the book can be used for undergraduate, postgraduate and doctoral teaching in faculties of engineering sciences.

Viorel Badescu
George Cristian Lazaroiu
Linda Barelli

Contents

Chapter 1

New Developments of Electrical Machines

Qingsong Wang[1] and *Shuangxia Niu*[2]

1. Introduction

Electrical machines can be divided into DC machines and AC machines, which are fed with DC current and AC current, respectively. Since brushes or slip rings are needed in DC machines, their relatively low reliability makes the DC machines less competitive. The AC machines can be further divided into induction machines and synchronous machines. Although induction machines are cheap and easy to manufacture, synchronous machines are more attractive due to their high efficiency and precise speed control. Currently, synchronous machines have a wide range of applications and have been extensively investigated by both academics and within the industry. In a synchronous machine, there are two rotary magnetic fields, namely the excitation field and the armature field. The electromagnetic torque is generated through the interaction of these two fields. The armature field can only be excited with the armature winding current, while the excitation field can be produced in various ways. When field winding is used to generate the excitation field, which is referred to as electrically excited machine (EEM), the air-gap field can be easily

[1] HJ810, The Hong Kong Polytechnic University.
   Email: q.s.wang@connect.polyu.hk
[2] CF623, The Hong Kong Polytechnic University.
   Email: eesxniu@polyu.edu.hk

regulated by controlling the field current and therefore the EEM can operate over a wide speed range. However, since the field current will inevitably introduce additional copper loss, the efficiency of the EEM is reduced and the heat generated by the copper loss may cause a problem of heat dissipation. Permanent magnet (PM) machines can solve the aforementioned problems, in which the excitation field is generated by the PMs. High torque density and high efficiency can be achieved in PM machines when high-magnetic-energy-density PM materials are used. The major drawback of the PM machines is that the air-gap flux is difficult to control due to their fixed excitation and their speed range is limited accordingly. In order to achieve high torque density and high efficiency while still maintaining good flux regulating capability, hybrid excited machines (HEMs) are proposed these can be regarded as combinations of EEMs and PM machines (Wang and Niu 2017). The excitation field in the HEM is provided by a primary PM excitation and a secondary field coil excitation source. Since HEMs theoretically have good overall performances, they are being widely studied by researchers.

Although electrical machines were invented more than 100 years ago, their design, manufacturing, testing, condition monitoring, and control techniques are still in permanent evolution. One of the hot research topics in the field of electrical machines is to develop novel machine concepts with better performances. The purpose of this chapter is to present the latest developments of electrical machine concepts. Each of them has some advantages in certain aspects. It should be based on the real application requirements to determine which kind of machine to be used.

2. Magnetic-geared Machine

The co-axial magnetic gear (MG) was first proposed in (Atallah and Howe 2001) and shown in Fig. 1, which has three components, namely the inner rotor and the outer rotor, both of which are surface mounted with PMs, and the middle modulation ring comprises of ferromagnetic segments. The middle modulation ring is the key component of the MG, which can provide a flux modulation effect with the uneven magnetic reluctance distribution. The number of modulation segments N_s, the pole-pair number (PPN) of inner rotor PMs p_i and the PPN of outer rotor PMs p_o are governed by

$$N_s = p_i + p_o$$

Keeping either one of the three components stationary, the other two components can work well as a MG. If we keep the middle modulation

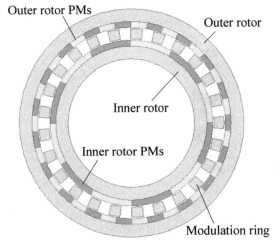

Fig. 1: Configuration of the co-axial MG (Atallah and Howe 2001).

ring stationary, and let the inner rotor be the driving input while letting the outer rotor be the driven output, the gear ratio G_r can be expressed as

$$G_r = \frac{p_o}{p_i}$$

Since all the PMs are involved in the torque transmission, the MG can achieve high torque density. Meanwhile, as there is no mechanical contact between the input and the output, the MG can operate with reduced acoustic noise, less vibration, and free lubrication except with the ball bearings. By working in a slip mode, the MG also features an intrinsic overload protection, which can avoid further damage to the transmission system.

By directly connecting the outer rotor of a PM machine and the inner rotor of a MG, a MG integrated PM machine is invented (Chau et al. 2002), as shown in Fig. 2(a). The electromagnetic torque generated by the PM machine can be amplified, and the inner cave space of the MG can be fully utilized, and therefore this MG integrated PM machine can achieve a very high torque density. The applications in EV propulsion and wind power generation are separately reported in (Chau et al. 2002) and (Jian et al. 2009), respectively. The major drawback of this MG integrated PM machine is its complicated mechanical structure. As can be seen from Fig. 2(a), there are three air-gaps in this MG integrated PM machine, which makes it very difficult to assemble. Figure 2(b) shows another configuration of magnetic-geared machine (Wang et al. 2009), in which the outer rotor of PM machine and inner rotor of MG in Fig. 2(a) are eliminated. The armature field is directly coupled with the PM excitation field and

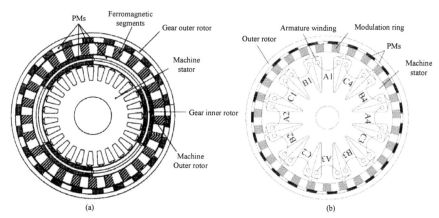

Fig. 2: Configurations of magnetic-geared machines. (a) MG integrated PM machine (Chau et al. 2002). (b) SL-MGPM machine (Wang et al. 2009).

modulated by the modulation ring. This magnetic-geared machine has just one PM layer, and is therefore referred to as a single-layer magnetic-geared PM (SL-MGPM) machine, and it can also be regarded as replacing the inner rotor of MG with a stator. Compared with the MG integrated PM machine shown in Fig. 2(a), the SL-MGPM machine in Fig. 2(b) has much simpler structure and is easy to manufacture.

The concept of MG can be further extended to linear machines. Through integrating a tubular linear PM (TLPM) machine and a tubular linear MG (TLMG), a TLMG integrated PM machine (Du et al. 2010) is constructed as shown in Fig. 3(a). The principle of this TLMG integrated PM machine is very similar to that of the magnetic-geared machine shown in Fig. 2(a), and can be used in tidal energy conversion. The TLPM machine can be designed with high power density, and its thrust force can be amplified by the TLMG. The major drawback of this TLMG integrated PM machine is its complicated structure. Figure 3(b) shows another configuration of the tubular linear magnetic-geared machine with sandwiched stator, which is referred to as a sandwiched stator tubular linear magnetic-geared (SS-TLMG) machine (Ho et al. 2015). The stator housed with windings is sandwiched between the high-speed mover and low-speed mover. The high-speed mover and the stator work as a linear PM machine, while the high-speed mover, the stator and the low-speed mover function like a linear MG. Compared with the TLMG integrated PM machine shown in Fig. 3(a), the structure of the SS-TLMG machine shown in Fig. 3(b) is more compact and simple, and thus improved force density.

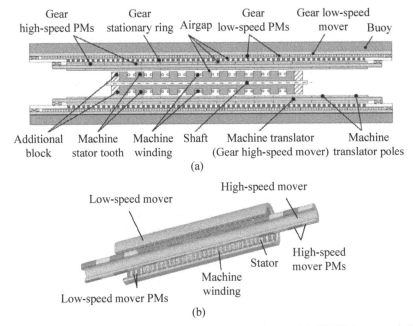

Fig. 3: Configurations of linear magnetic-geared machines. (a) TLMG integrated PM machine (Du et al. 2010). (b) SS-TLMG machine (Ho et al. 2015).

3. Stator-PM Brushless Machine

PM machines are predominant in the industry applications due to their high torque density and high efficiency features. Currently, most of the PM machines locate the PMs on the rotor, which are referred to as rotor-PM machines, that can achieve good overall performances. However, the rotor-PM machines may suffer from thermal problems since the heat generated by the rotor is difficult to dissipate, and cause demagnetization of PMs. Meanwhile, when the rotor-PM machines run at high speeds, the PMs should be protected from the centrifugal force by employing a retaining sleeve. Stator-PM machines can solve these problems, in which the PMs are located on the stator and the rotor is robust with only salient poles. The PMs will not suffer from demagnetization risk since the stator is easy to cool. Basically, stator-PM machines can be categorized into three types, namely, doubly salient PM (DSPM) machine, flux switching PM (FSPM) machine and flux reversal PM (FRPM) machine.

A. DSPM Machine

DSPM machine is oriented from the switched reluctance machine, in which the PMs are inserted into the stator back iron and the armature coils are concentrated wound on the stator teeth. One typical configuration of the DSPM machine is shown in Fig. 4 (Lin et al. 2003). The stator has 12 slots and the rotor has 8 poles. The PM flux linkage increases when the rotor pole rotates close to the corresponding stator tooth, and decreases when the rotor pole rotates away. By specifically design of the arc of stator teeth and rotor poles, the PM flux linkage can linearly varies with the rotor position, as shown in Fig. 5(a). In this case, trapezoidal back-EMF waveforms can be obtained and the DSPM machine is suitable for brushless DC control. A rotor-skewing method is reported in (Ming et al. 2003), which can obtain quasi-sinusoidal back-EMF waveforms as shown in Fig. 5(b) and the DSPM machine can operate in brushless AC mode.

As the PMs are located on the stator, DSPM machines can easily achieve hybrid excitation function. Figure 6 shows several hybrid excited

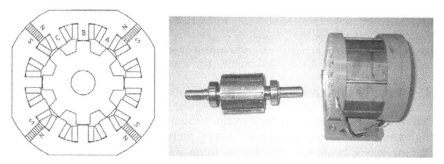

Fig. 4: Configuration of the 12/8 DSPM machine (Lin et al. 2003).

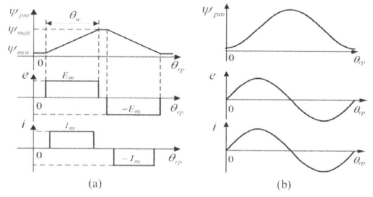

(a)   (b)

Fig. 5: Operating principles of the DSPM machines (Cheng et al. 2011). (a) Unskewed rotor. (b) Skewed rotor.

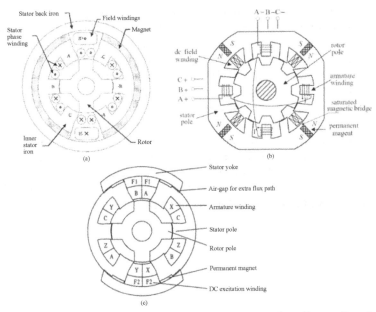

Fig. 6: Configurations of HE-DSPM machines. (a) Primary topology (Leonardi et al. 1996). (b) HE-DSPM machine with magnetic bridges (Xiaoyong et al. 2006). (c) HE-DSPM machine with air bridges (Xiaoyong et al. 2007).

DSPM (HE-DSPM) machines, in which the flux excited by the PMs and field coils are in series. The PMs are radially magnetized in the primary topology shown in Fig. 6(a) (Leonardi et al. 1996). A large field current is needed to realize a good flux regulation, because the flux generated by the field current should pass directly through the large reluctance PMs, which may also result in demagnetization of the PMs. HE-DSPM machine with magnetic bridges is developed as shown in Fig. 6(b) (Xiaoyong et al. 2006), and the magnetic bridges can act as low reluctance paths for the flux of field current, which can greatly reduce the required field current ampere-turns. However, the reluctance of the magnetic bridges will be influenced by magnetic saturation. Figure 6(c) shows another HE-DSPM machine topology (Xiaoyong et al. 2007), in which air bridges are used to achieve the same functions of the magnetic bridges. The improvement is that the air bridge reluctance will no longer be influenced by magnet saturation. Comparative study in (Zhihui et al. 2014) shows that HE-DSPM machine with air bridges can well reduce the demagnetization risk of PMs.

B. FSPM Machine

FSPM machine is another hot topology of stator-PM machines, and the typical configuration is given in Fig. 7. The PMs are inserted into the stator teeth and tangentially magnetized in opposite directions. Concentrated coils are employed and wound on the stator teeth. The stator lamination consists of several U-shaped segments and the rotor only has salient poles, which are mechanically robust and can be used in high-speed applications. The operating principle of FSPM machine can be illustrated based on Fig. 8. At position A, the stator teeth is aligned with the rotor pole, and the PM flux linkage reaches the maximum value because the magnetic resistance is the smallest in this case. When the rotor rotates from position A to position B, the PM flux linkage reduces and theoretically becomes zero at position B. The reason behind this is that the rotor pole is aligned with the stator slots at position B, and the magnetic resistance is the largest. When the rotor rotates from position B to position C, the direction of the PM is flux reversed and maximum PM flux linkage can be obtained at position C

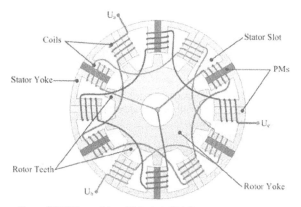

Fig. 7: Configuration of FSPM machine (Shi et al. 2016).

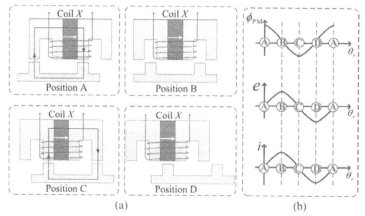

Fig. 8: Operating principle of FSPM machine (Shi et al. 2016). (a) PM flux at different rotor positions. (b) Ideal PM flux-linkage, back-EMF, and phase-current waveforms.

when the stator teeth are aligned with the rotor pole again. The reversion of the PM flux direction is called "flux switching". Finally, when the rotor rotates to position D, the PM flux linkage reduces again and theoretically becomes zero at position D when the rotor pole is aligned with the stator PMs. One can find that the PM flux linkage varies when the rotor rotates, and sinusoidal back-EMF can be induced through proper design of the stator teeth and rotor poles.

Based on the conventional configuration shown in Fig. 7, some other novel topologies of FSPM machines are developed as shown in Fig. 9. The topologies shown in Fig. 9(a)–(c) have alternate poles wound winding, which enables the FSPM machine a good fault tolerant capability because the coils of each phase are isolated and the coupling between phases is minimized. In the first topology, all the stator teeth are inserted with PMs, which is the same as the conventional FSPM machine. However, in the second topology and the third topology, the PMs are removed from the unwound poles and the stator lamination consists of several E-core segments. The PMs in these two machines can be either tangentially magnetized in the same direction or in the opposite directions, as shown

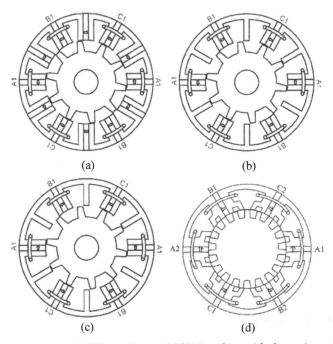

(a)

(b)

(c)

(d)

Fig. 9: Novel topologies of FSPM machines. (a) FSPM machine with alternative poles wound (Chen and Zhu 2010). (b) E-core FSPM machine with the same magnetizing directions (Chen et al. 2011). (c) E-core FSPM machine with opposite magnetizing directions (Chen et al. 2011). (d) Multi-tooth FSPM machine (Chen et al. 2008).

in Fig. 9(b) and Fig. 9(c), respectively. A multi-tooth FSPM machine is also proposed and investigated in (Chen et al. 2008), the configuration for it is given in Fig. 9(d). The PM usage is significantly reduced, but its torque capability is also reduced due to magnetic saturation caused by the higher armature reaction.

C. FRPM Machine

As same as a DSPM machine and FSPM machine, FRPM machine also has concentrated coils but the PMs are surface mounted on the stator teeth, as shown in Fig. 10. It is called "flux reversal" because the PM flux linkage reverses its polarity when the rotor rotates by a pole pitch. The operating principle of FRPM machine is based on the flux modulating effect of the rotor. The FRPM machine is believed to have good potential to be used for low-speed large-torque applications, as reported in (Boldea et al. 2002). However, as the PMs are surface mounted on the stator teeth, the PMs will suffer from demagnetization risk and the effective air-gap length is also increased, which will limit the improvement of torque density. The rotor-skewing method can also be used to improve the sinusoidal feature of EMF waveform and to reduce the cogging torque, which has been reported in (Cheng and Zhou 2001). Compared with DSPM machine and FSPM machine, FRPM machine has simple structure, but its low power factor and high risk of PM demagnetization still need to be improved.

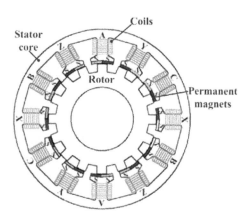

Fig. 10: Configuration of the FRPM machine (Kim and Lee 2004).

4. Vernier Machine

Vernier PM (VPM) machines have attracted much attention in direct-drive applications and are recognized as good candidates to achieve high torque density due to the magnetic gearing effect. The half-cross-sectional view of the original VPM is shown in Fig. 11, which has a simple structure. The PMs are surface mounted on the rotor, and the windings are housed in the stator slots. The stator is designed with open slots, which can function as the modulation segments in a MG and provide a magnetic gearing effect. The PPN of armature winding p_a, PPN of rotor PMs p_o and stator slot number N_s are governed by

$$N_s = p_a + p_o$$

which is similar to the working principle of MG. The design and operating principle of the VPM machine are analytically investigated in Kim and Lipo (2014). One of the major drawbacks of the VPM machine is the low power factor caused by heavy magnet flux leakage and low magnet utilization. Therefore, a large-capacity converter is needed to drive the low-power-factor VPM machine, which will result in high cost. A new VPM topology which can overcome the shortcoming of low power factor of conventional VPM machines is proposed (Li et al. 2014), as shown in Fig. 12. The novelty of this structure is that the inner stator teeth and outer stator teeth are displaced with half teeth pitch, and therefore the inner and outer teeth can compensate with each other. The leakage flux can be greatly reduced and the power factor is improved accordingly.

Besides the VPM machine, stator DC excited Vernier (DCV) machines are also investigated. There is no PM used and the rotor only has salient poles. Two windings are employed, namely, armature winding and field winding, both of which are housed in the stator slots. The configuration of DCV machine is shown in Fig. 13. Compared with VPM machine, DCV machine has improved rotor robustness and reduced cost with the sacrificing of some torque capability. The air-gap flux of a DCV machine can also be easily regulated by controlling the DC field current.

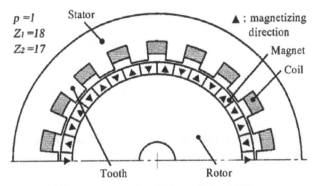

Fig. 11: Configuration of the original VPM (Toba and Lipo 2000).

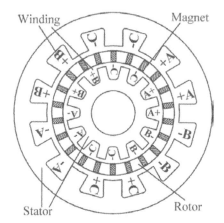

Fig. 12: Configuration of VPM machine with high power factor (Li et al. 2014).

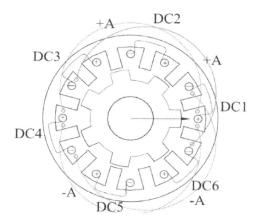

Fig. 13: Configuration of DCV machine (Jia et al. 2015).

5. Dual-PM Flux Bi-directional Modulated Machine

All the aforementioned PM machines employ one set of PMs, which are located either on the stator or the rotor. In this section, a new class of machines which has both stator and rotor PM excitations is studied, which are referred to as a dual-PM machine (Wang and Niu 2016, 2017, Wang et al. 2017, W.Q. et al. 2017). Both the stator and the rotor are employed with PM-iron sequences, in which the magnets are separated by iron segments one by one, as shown in Fig. 14. The PM-iron sequence can provide flux modulation and PM excitation simultaneously. Bi-directional flux modulating effect generated by both the stator and rotor can be achieved, and the magnetic fields excited by the two sets of PMs and armature currents can be coupled effectively. Moreover, the PM arrangement is more flexible because the magnets can be employed on the stator and rotor simultaneously, which can produce various machine configurations and rich the category of PM machines.

One typical configuration of the dual-PM machine is shown in Fig. 15, in which the rotor and the stationary modulation ring are employed with PM-iron sequences. All the magnets are radially outward magnetized, and each magnet and its adjacent modulation segment form a magnetic pole-pair. This machine is oriented from the magnetic-geared machine shown in Fig. 2(b), by moving half of the PMs on the outer rotor to the middle modulation ring. Comparative studies reported in Wang et al. 2017 show that the dual-PM machine in Fig. 15 can achieve a 20% larger output torque than the SL-MGPM machine, due to the bi-directional flux modulating effect. As part of the PMs are stationary, it is easy for the dual-PM machine to achieve hybrid excitation. Figure 16 shows the configuration and mechanical structure of a hybrid excited dual-PM machine, which is developed based on the dual-PM machine in Fig. 15. The field windings are housed in the outer stator slots, and the modulation ring will act as the output rotor. Since the rotor is sandwiched between two stators, the

Fig. 14: Configuration of the PM-iron sequence.

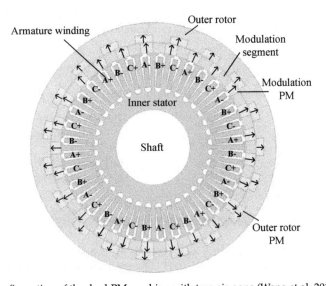

Fig. 15: Configuration of the dual-PM machine with two air-gaps (Wang et al. 2017).

cup rotor is designed to facilitate the assembly. This hybrid excited dual-PM machine can achieve high torque density and a good flux regulating capability, and hence is very suitable for EV propulsion. However, the mechanical structure is relatively complicated since there are two air-gaps and two stators.

Figure 17 shows the configuration and winding connections of an new hybrid excited dual-PM machine with single air-gap, which incorporates the two stators of the machine shown in Fig. 16 into one stator. The AC windings used to produce rotating magnetic field and the DC windings used for flux regulation are both housed in the stator slots, and the stator PMs are alternately surface mounted on the stator teeth. The stator PMs and rotor PMs can generate electromagnetic torque separately, and the

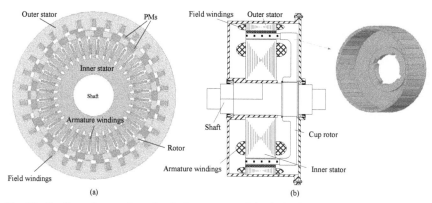

Fig. 16: Configuration and mechanical structure of the hybrid excited dual-PM machine with two air-gaps (Wang and Niu 2016). (a) Configuration. (b) Mechanical structure.

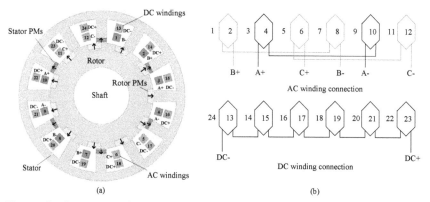

Fig. 17: Configuration and winding connections of the hybrid excited dual-PM machine with single air-gap (Wang and Niu 2017). (a) Configuration. (b) Winding connections.

torque capability of this machine is larger than its hybrid excited machine counterparts. Both flux strengthening and flux weakening can be achieved by controlling the field current. Compared with the hybrid excited dual-PM machine with the two air-gaps shown in Fig. 16. This improved topology can maintain high torque density and good flux regulating capability features with a much simpler structure.

6. Electrical Continuously Variable Transmission System

The Electrical (ECVT) system is an important component and core technology in HEVs, which is a power-splitting unit. The energy flow among the internal combustion engine (ICE) and other energy sources, which are usually electrical machines, and can be flexibly combined or split according to different operational requirements. Therefore, the ICE can operate at optimal conditions when the vehicle speed varies, the efficiency of the ICE is increased and the fuel economy of HEVs is improved consequently. One of the most popular and commercialized ECVT is the one used in the Toyota Prius, which is based on the planetary gear and the configuration is shown in Fig. 18 below. Two sets of electrical machine systems are used, which can be separately controlled and are connected to the sun gear and ring gear, respectively. The crank shaft of ICE is linked to the planetary carrier. During start-up or climbing, the electrical machine will work as a motor to help the ICE drive the vehicle. While during braking or downhill, when the output power of ICE exceeds the requirement of vehicle, the surplus power of ICE can be stored in the battery and the electrical machines are worked in generator mode in this case.

The Toyota Prius ECVT system can optimize the efficiency of ICE well and reduce the fuel consumption, but the planetary gear still suffers from some mechanical drawbacks, such as friction, audible noise, and high maintenance. Moreover, the planetary gear based ECVT system is also bulky and heavy due to its complex structure. Figure 19 shows a novel doubly-fed dual-rotor PM (DFDR-PM) machine based ECVT system. The two electrical machines in the planetary gear based ECVT system are combined into a single machine, namely, the novel DFDR-PM machine. Two rotors are employed, both of which consist of PMs and modulation segments. The inner rotor is linked to the crank shaft of ICE, and the outer rotor is connected to the drive shaft. There are two sets of windings housed in the stator slots, one is the primary winding and the other is the secondary winding. The primary winding interacts with the two rotors through the flux modulating effect and form a dual-rotor flux modulated machine, and the secondary winding interacts with the outer rotor and forms a normal

PM machine. This DFDR-PM based ECVT system can realize both a power split and combination effectively. Compared with the planetary gear based ECVT, the DFDR-PM based ECVT has a compact structure and no mechanical frictions. Besides the application in HEVs as reported in (Wang et al. 2016), the DFDR-PM based ECVT can also be used in variable speed wind power generation (Liu et al. 2016, Luo and Niu 2016).

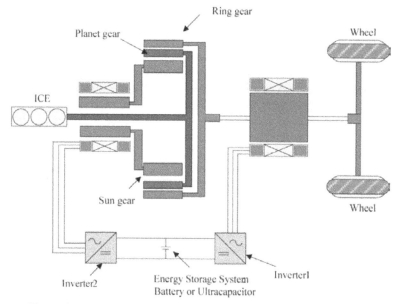

Fig. 18: The configuration of Toyota Prius ECVT system (Niu et al. 2013).

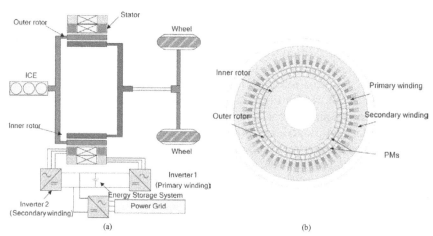

Fig. 19: Configuration of the DFDR-PM machine based ECVT (Niu et al. 2013). (a) ECVT system. (b) DFDR-PM machine.

7. Memory Machine

The concept of memory machine (MM) was firstly proposed in (Ostovic 2001), the novelty is to use PMs with nonlinear hysteresis loops and low coercivities, among which the AlNiCo magnet is the representative. The recoil line never superposes on the demagnetization curve due to the nonlinear demagnetization characteristic. During flux weakening, a demagnetization current is applied and the working point of AlNiCo magnet will move along the recoil line and settle at a lower magnetization level even when the demagnetization current is removed, which is the so-called "memory effect". The same goes for the process of magnetizing. A simplified linear hysteresis model of AlNiCo magnet is shown in Fig. 20 below, in which the intersection of recoil line and load line uniquely determines the operating points of AlNiCo magnet. The initial magnetizing stage is given in process OAMP. When flux strengthening is needed, a magnetizing current pulse is applied and the working point will move along PMBQ and settle at Q. Otherwise, the working point of AlNiCo magnet will move along PNCR and settle at R when flux weakening is needed, by applying demagnetizing current pulse. Since

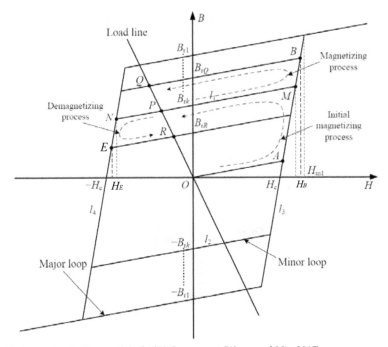

Fig. 20: Linear hysteresis model of AlNiCo magnet (Wang and Niu 2015).

continuous excitation current is not needed due to the memory effect, MM can achieve high efficiency during flux regulation because the copper loss generated by the magnetizing/demagnetizing current pulse is almost negligible.

According to the magnetizing/demagnetizing method, MMs can be divided into two kinds, namely, APR-MMs and DPR-MMs when the magnetizing/demagnetizing is realized by applying AC current pulses and DC current pulses, respectively. Figure 21 below shows the configuration of a primary APR-MM, the PMs are sandwiched in the rotor. In order to achieve regulation, d-axis current pulses are generated by controlling the armature currents. This kind of machine has simple structure and easy to fabricate, but is difficult to control. The rotor position should be accurately detected so as to realize precise flux regulation. DPR-MMs can achieve flux regulation more easily, because the PMs are located on the stator and can be magnetized/demagnetized using DC current pulses. One typical structure of DPR-MMs is shown in Fig. 22 below, in which the stator is divided into two layers. The inner layer stator is used to locate field windings and AlNiCo magnets, while the outer layer stator is used to locate armature windings. With this design, the PMs can only be magnetized/demagnetized by the field currents and not influenced by armature currents. Besides MMs using only AlNiCo magnets, some other MMs with both AlNiCo and NdFeB PM excitations are reported in (Qingsong et al. 2015). These machines can realize better overall performances, the torque density can be improved and good flux regulating capability can still be maintained, but the structure of these machines becomes complicated.

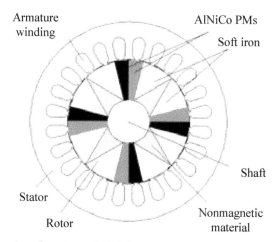

Fig. 21: Configuration of a primary APR-MM (Ostovic 2003).

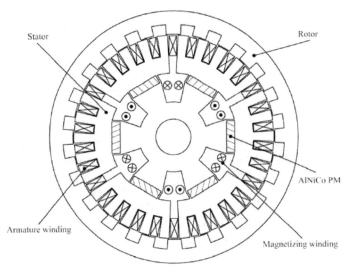

Fig. 22: Configuration of a typical DPR-MM (Chuang and Chau 2011).

8. Conclusion

In this chapter, new developments of electrical machines are presented. The latest proposed machine concepts with distinct features are discussed in detail, which include magnetic-geared machine, stator-PM machine, Vernier machine, dual-PM flux bidirectional modulated machine, ECVT system and memory machine. The aim of developing these novel machine concepts are to achieve better electromagnetic performances, such as high torque density, high power density, good flux regulating capability and high efficiency. Mechanical issues are also being taken into consideration. Generally, PM machines have a larger torque capability and higher efficiency than EE machines, but EE machines have lower cost.

Stator-PM machines can achieve robust rotor structure, and can be used in high-speed applications. The temperature rise of PMs can be easily controlled because the heat on the stator can be directly dissipated outside. However, the stator structure of stator-PM machine is complicated. Meanwhile, the spaces used for armature windings and PM location are both reduced, which will reduce the torque density. In FRPM machines, the PMs also have high demagnetization risks.

Both the magnetic-geared machine and Vernier machine have high torque density due to the magnetic-gearing effect, but the magnetic-geared machine has a relatively complicated structure and Vernier machine is more competitive. Compared with magnetic-geared machine with single-layer PMs, the dual-PM machine with two air-gaps can realize a higher torque density and it can easily achieve hybrid excitation by employing field windings on the stator. When designing the dual-PM machines with a single air-gap, the structure is also simplified.

ECVT system is an attractive power-splitting unit, which can be used in HEVs and variable speed wind power generation. Compared with planetary geared ECVT system, the DFDR-PM machine based ECVT system shows better potential because of its compact structure and free contact transmission.

The most prominent feature of MMs is realizing magnetizing/demagnetizing through current pulse, and therefore online flux regulation can be achieved. Both APR-MMs and DPR-MMs are proposed, and the former have simple structure while the latter have easy control strategy. No matter the APR-MMs or the DPR-MMs, hybrid PM excitations with both the NdFeB magnet and AlNiCo magnet can be used to balance the flux regulating capability and output torque performance.

9. References

Atallah, K. and D. Howe. 2001. A novel high-performance magnetic gear. IEEE Transactions on Magnetics 37: 2844–2846.
Boldea, I., J.C. Zhang and S.A. Nasar. 2002. Theoretical characterization of flux reversal machine in low-speed servo drives—The pole-PM configuration. IEEE Transactions on Industry Applications 38: 1549–1557.
Chau, K.T., D. Zhang, J.Z. Jiang, C. Liu and Y. Zhang. 2007. Design of a magnetic-geared outer-rotor permanent-magnet brushless motor for electric vehicles. IEEE Transactions on Magnetics 43: 2504–2506.
Chen, J.T., Z.Q. Zhu and D. Howe. 2008. Stator and rotor pole combinations for multi-tooth flux-switching permanent-magnet brushless AC machines. IEEE Transactions on Magnetics 44: 4659–4667.
Chen, J.T. and Z.Q. Zhu. 2010. Comparison of all- and alternate-poles-wound flux-switching PM machines having different stator and rotor pole numbers. IEEE Transactions on Industry Applications 46: 1406–1415.
Chen, J.T., Z.Q. Zhu, S. Iwasaki and R.P. Deodhar. 2011. A novel e-core switched-flux PM brushless AC machine. IEEE Transactions on Industry Applications 47: 1273–1282.
Cheng, M. and E. Zhou. 2001. Analysis and control of novel split-winding doubly salient permanent magnet motor for adjustable speed drive. Science in China Series E: Technological Sciences 44: 353–364.
Cheng, M., W. Hua, J. Zhang and W. Zhao. 2011. Overview of stator-permanent magnet brushless machines. IEEE Transactions on Industrial Electronics 58: 5087–5101.
Chuang, Y. and K.T. Chau. 2011. Design, analysis, and control of DC-excited memory motors. IEEE Transactions on Energy Conversion 26: 479–489.
Du, Y., K.T. Chau, M. Cheng and Y. Wang. 2010. A linear magnetic-geared permanent magnet machine for wave energy generation. In 2010 International Conference on Electrical Machines and Systems, pp. 1538–1541.
Ho, S., Q. Wang, S. Niu and W. Fu. 2015. A novel magnetic-geared tubular linear machine with halbach permanent-magnet arrays for tidal energy conversion. IEEE Transactions on Magnetics 51: 1–4.
Jia, S., R. Qu and J. Li. 2015. Analysis of the power factor of stator DC-excited vernier reluctance machines. IEEE Transactions on Magnetics 51: 1–4.

Jian, L., K.T. Chau and J.Z. Jiang. 2009. A magnetic-geared outer-rotor permanent-magnet brushless machine for wind power generation. IEEE Transactions on Industry Applications 45: 954–962.

Kim, B. and T.A. Lipo. 2014. Operation and design principles of a PM vernier motor. IEEE Transactions on Industry Applications 50: 3656–3663.

Kim, T.H. and J. Lee. 2004. A study of the design for the flux reversal machine. IEEE Transactions on Magnetics 40: 2053–2055.

Leonardi, F., T. Matsuo, Y. Li, T.A. Lipo and P. McCleer. 1996. Design considerations and test results for a doubly salient PM motor with flux control. In Industry Applications Conference, 1996. Thirty-First IAS Annual Meeting, IAS '96, Conference Record of the IEEE 1: 458–463.

Li, D., R. Qu and T.A. Lipo. 2014. High-power-factor vernier permanent-magnet machines. IEEE Transactions on Industry Applications 50: 3664–3674.

Lin, M., M. Cheng and E. Zhou. 2003. Design and performance analysis of new 12/8-pole doubly salient permanent-magnet motor. In Sixth International Conference on Electrical Machines and Systems. ICEMS 1: 21–25.

Liu, Y., S. Niu and W. Fu. 2016. Design of an electrical continuously variable transmission based wind energy conversion system. IEEE Transactions on Industrial Electronics 63: 6745–6755.

Luo, X. and S. Niu. 2016. A novel contra-rotating power split transmission system for wind power generation and its dual MPPT control strategy. IEEE Transactions on Power Electronics.

Ming, C., K.T. Chau, C.C. Chan and S. Qiang. 2003. Control and operation of a new 8/6-pole doubly salient permanent-magnet motor drive. IEEE Transactions on Industry Applications 39: 1363–1371.

Niu, S., S.L. Ho and W.N. Fu. 2013. Design of a novel electrical continuously variable transmission system based on harmonic spectra analysis of magnetic field. IEEE Transactions on Magnetics 49: 2161–2164.

Ostovic, V. 2001. Memory motors—a new class of controllable flux PM machines for a true wide speed operation. In Industry Applications Conference, 2001. Thirty-Sixth IAS Annual Meeting. Conference Record of the 2001 IEEE, pp. 2577–2584.

Ostovic, V. 2003. Memory motors. Industry Applications Magazine, IEEE 9: 52–61.

Q, W., S. Niu and X. Luo. 2017. A novel hybrid dual-PM machine excited by AC with DC bias for electric vehicle propulsion. IEEE Transactions on Industrial Electronics, pp. 1–1.

Qingsong, W., N. Shuangxia, H. Siu Lau, F. Weinong and Z. Shuguang. 2015. Design and analysis of novel magnetic flux-modulated mnemonic machines. Electric Power Applications, IET 9: 469–477.

Shi, Y., L. Jian, J. Wei, Z. Shao, W. Li and C.C. Chan. 2016. A new perspective on the operating principle of flux-switching permanent-magnet machines. IEEE Transactions on Industrial Electronics 63: 1425–1437.

Toba, A. and T. Lipo. 2000. Generic torque-maximizing design methodology of surface permanent-magnet vernier machine. IEEE Transactions on Industry Applications 36: 1539–1546.

Wang, L., J.-X. Shen, P.C.-K. Luk, W.-Z. Fei, C. Wang and H. Hao. 2009. Development of a magnetic-geared permanent-magnet brushless motor. IEEE Transactions on Magnetics 45: 4578–4581.

Wang, Q. and S. Niu. 2015. Electromagnetic design and analysis of a novel fault-tolerant flux-modulated memory machine. Energies 8: 8069–8085.

Wang, Q. and S. Niu. 2016. A novel hybrid-excited flux bidirectional modulated machine for electric vehicle propulsion. In Vehicle Power and Propulsion Conference (VPPC), IEEE, pp. 1–6.

Wang, Q. and S. Niu. 2017. A novel hybrid-excited dual-PM machine with bi-directional flux modulation. IEEE Transactions on Energy Conversion.

Wang, Q. and S. Niu. 2017. Overview of flux-controllable machines: Electrically excited machines, hybrid excited machines and memory machines. Renewable and Sustainable Energy Reviews 68: 475–491.

Wang, Q., S. Niu and S. Yang. 2017. Design optimization and comparative study of novel magnetic-geared permanent magnet machines. IEEE Transactions on Magnetics.

Wang, Y., S. Niu and W. Fu. 2016. An electrical-continuously variable transmission system based on doubly-fed flux-bidirectional modulation. IEEE Transactions on Industrial Electronics, pp. 1–1.

Xiaoyong, Z., C. Ming, H. Wei, Z. Jianzhong and Z. Wenxiang. 2006. Design and analysis of a new hybrid excited doubly salient machine capable of field control. In Industry Applications Conference, 2006. 41st IAS Annual Meeting. Conference Record of the IEEE, pp. 2382–2389.

Xiaoyong, Z., C. Ming, Z. Wenxiang, L. Chunhua and K.T. Chau. 2007. A transient cosimulation approach to performance analysis of hybrid excited doubly salient machine considering indirect field-circuit coupling. IEEE Transactions on Magnetics 43: 2558–2560.

Zhihui, C., W. Bo, C. Zhe and Y. Yangguang. 2014. Comparison of flux regulation ability of the hybrid excitation doubly salient machines. IEEE Transactions on Industrial Electronics 61: 3155–3166.

CHAPTER 2

Onshore and Offshore Wind Energy Applications

*J.K. Kaldellis** and *D. Apostolou*

1. Introduction

During the second half of the last century, electricity demand increased dramatically resulting in a rapid growth of fossil fuel-based power capacity. Although the majority of human societies at the end of the 20th century had access to constant electricity supply, which suggests a significant aspect of contemporary life quality, the negative impacts associated with the profligate use of fossil fuels in electricity generation comprise one of the major factors in environmental pollution. Carbon dioxide (CO_2) emissions from the energy sector have been increasing in the last 40 years by 52% reaching at the end of 2014, 32.4 billion metric tons (IEA 2016b).

With this in mind, since the 1970s, governments and policy-makers worldwide enacted measures to promote the development of new sustainable technologies that are able to contribute in the energy mix without the negative effects found in conventional electricity production stations (Yue et al. 2001). However, the transition from traditional methods of producing electricity to renewable energy systems (RES) presents even nowadays drawbacks associated with the penetration limits of RES due to their stochastic and intermittent nature of operation (Kaldellis 2008, 2015). The most mature RES technologies include wind and solar power

Soft Energy Applications and Environmental Protection Laboratory, Piraeus University of Applied Sciences, P.O. Box 41046, Athens, 12201, Greece; j.apostolou@puas.gr
* Corresponding author: jkald@puas.gr

generating plants which in the past 25 years have presented significant growth and are considered to be effective for large scale deployment (Kumar et al. 2016).

Wind energy comprises a fraction of total energy that reaches the earth from the sun. Actually, solar irradiation variation induces near the planet surface cold air flow from the northern and southern hemispheres to the equator of the earth. Furthermore, the Coriolis' effect generated from the rotation of the earth, shifts the air to the east and west concerning the northern and southern hemispheres respectively (Kumar et al. 2016). In this context, Fig. 1 indicates the average wind speed atlas at height of 200 m for both onshore and 30 km offshore areas. Based on the specific map, the regions where wind speed is generally higher comprise northern-southern and high altitude territories.

Wind energy devices date back many centuries ago, as has been observed from the vertical axis windmills that have been discovered at the Persian-Afghan borders. However, it was not until the beginning of the 20th century when the first wind power converters were developed to overcome electrical power shortages (Kaldellis and Zafirakis 2011). Modern wind turbines can be divided in to two major types, the horizontal axis (HAWT) and the vertical axis (VAWT) wind turbines. The vast majority of wind applications both onshore and offshore are based on HAWTs due to their greater efficiency, minimized mechanical failures, and higher energy output compared to VAWTs (Kumar et al. 2016). The main components of a HAWT are depicted in Fig. 2, and include:

- Tower: Ranges from 40 m for large scale machines up to more than 100 m.
- Rotor blades: Length varies up to more than 60 m. They are used to capture the wind energy.
- Rotor hub: It holds the blades in position.
- Nacelle: It covers the turbine's drive train.
- Main shaft: It transfers the rotational torque/power to gearbox.
- Gearbox: Used to increase the rotational speed of the rotor to higher speeds for the generator. In some wind turbine concepts there is no gear-box and instead power electronics are included.
- Generator: Converts mechanical into electrical energy.

In addition, a wind turbine comprises also other primary systems such as a yaw system, an anemometer, a pitch system, one or more brake systems, a power converter, and other auxiliary components such as cables, screws and bearings (IRENA 2015).

Fig. 1: Global mean wind speed atlas. Based on (IRENA 2016a).

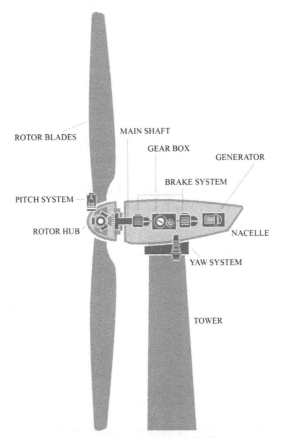

Fig. 2: Main components of a HAWT. Based on (IRENA 2015).

2. Wind Power Evolution

Over the last 30 years, wind energy technology has developed to a spectacular degree, contributing significantly to the energy mix of many countries. Specifically, European Union's policy to support RES, in conjunction with the high wind resource found in many European regions (see also Fig. 1) constitute the main factor for the high number of investments in the wind energy sector. However, wind power capacity in the rest of the world has also increased significantly after the end of the 1990s, as a result of the rapid growth of Asian markets.

Figure 3 indicates the evolution of the annual and cumulative global wind capacity, where according to the latest official data (EPI 2014, GWEC 2017), new annual installations presented a significant increasing rate after 2006 and during the last three years has exceeded the 50 GW, resulting in a cumulative capacity of almost 500 GW at the end of 2016.

Following the above, Fig. 4 suggests that although Europe's total installations present an increasing trend during the last 20 years, its contribution in total capacity has dropped from around 75% at the beginning of the new millennium to below 35% in 2016 (EWEA 2017b, GWEC 2017).

In this context, this significant growth observed in cumulative wind power capacity since 2000, contributed to almost 720 TWh of global wind-generated electricity in 2014. As it is obvious from Fig. 5, although wind produced electricity in Europe presents the higher value compared to the other top wind energy production regions, its share to the worldwide

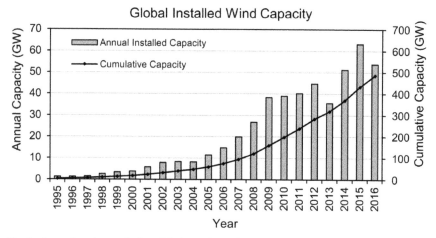

Fig. 3: Annual and cumulative global wind installed capacity. Based on (EPI 2014, GWEC 2017).

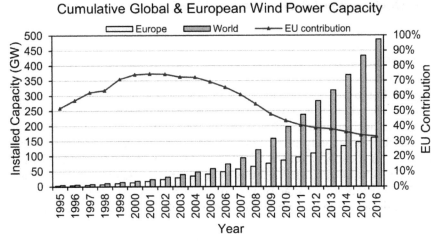

Fig. 4: Evolution of global and Europe's wind power capacity. Based on (EWEA 2017b, GWEC 2017).

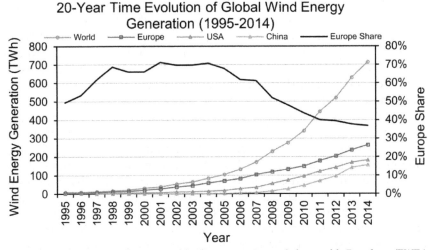

Fig. 5: Wind electricity generation of leading countries and the world. Based on (EWEA 2017b, Energy Information Administration (EIA) 2017).

produced electricity from wind has been reduced below 40% reaching in 2014, approximately 270 TWh$_e$, mostly due to the increase of the wind industry market in China during the last decade (EWEA 2017b, Energy Information Administration (EIA) 2017a).

2.1 Onshore Wind Energy in Global Market

As previously mentioned, the cumulative wind power reached in 2016 487 GW from which the highest share concerns onshore applications (97%). Since mid-1990s, more than 467 GW of onshore wind capacity were installed reaching 472.4 GW in 2016 (EPI 2014, GWEC 2017). In this regard, Fig. 6 presents the evolution of onshore wind capacity by region since 2005, where it is obvious that, Europe comprised the major market in global onshore wind installations in 2005 with a share above 68% and a cumulative capacity of 40.2 GW. North America ranked in the second place with a capacity of 9.8 GW, while Asian installations ranked third with 6.9 GW (GWEC 2017). At the end of 2016, the region with the highest onshore cumulative capacity was Asia, with approximately 202 GW of installations, while Europe ranked second with 161.3 GW of total onshore capacity.

In terms of onshore wind-based electricity generation, several markets presented a significant increase, enhancing the share of RES technologies in gross electricity production (see Fig. 7). This is pretty obvious in the case of the Asian market where its share in global wind electricity increased from 11% in 2006 to 29% in 2014. A similar growth is observed in the case of N. America with a share of 31% in 2014 compared to 23% in 2006. In contrast, Europe's share in global onshore wind generation reduced significantly during the last 10 years from 63% to 34%. However, Europe still holds the biggest share in total electricity production from onshore wind energy (IRENA 2016b).

Acknowledging the above, Fig. 8 indicates the cumulative and added capacity between 2015 and 2016 (a), along with the market share of annual onshore installations for 11 markets (b). Based on the results, China's total onshore capacity reached 151 GW with annual installations above 22.5 GW in 2016 which comprised the major market globally of onshore wind energy. USA ranked second both at cumulative and annual installations while Germany was the third market with cumulative capacity of 46 GW and 2016 additions of 4.6 GW. Concerning other Asian markets, India's new onshore installations were 3.6 GW and its cumulative capacity reached 25 GW, ranking at the fourth place globally. In Brazil on the other hand, new onshore installations for 2016 were 2 GW which comprised the 25% of its total onshore capacity. In Europe, apart from Germany, France added more than 1.5 GW of onshore wind capacity in 2016 reaching 10.3 GW, while the UK's new installations were almost similar with Poland at around 0.7 GW. It is worthy also to mention that in 2016, Turkey invested significantly in wind onshore applications by adding 1.4 GW, exceeding totally 4.6 GW (IRENA 2016b, GWEC 2017).

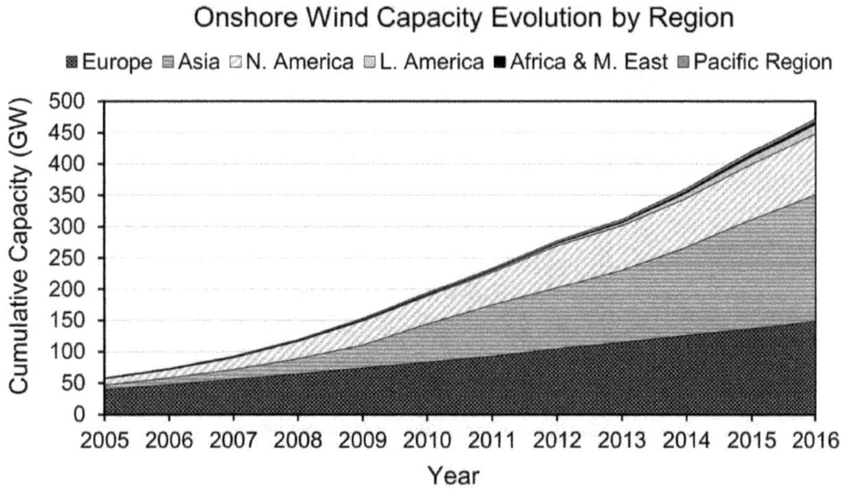

Onshore Wind Capacity Evolution by Region
Europe
Asia
N. America
L. America
Africa & M. East
Pacific Region
Cumulative Capacity (GW)
500
450
400
350
300
250
200
150
100
50
0
2005
2006
2007
2008
2009
2010
2011
2012
2013
2014
2015
2016
Year

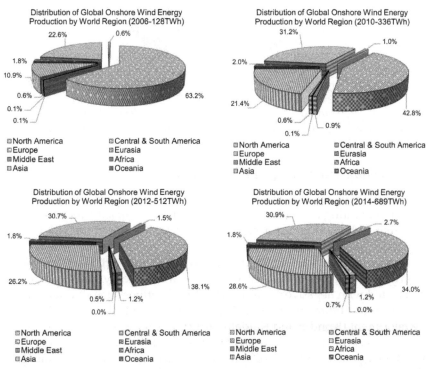

Fig. 6: Cumulative onshore wind capacity by region 2005–2016. Based on (GWEC 2017).

Distribution of Global Onshore Wind Energy Production by World Region (2006-128TWh)
22.6%
0.6%
1.8%
10.9%
0.6%
0.1%
0.1%
63.2%
North America
Central & South America
Europe
Eurasia
Middle East
Africa
Asia
Oceania
Distribution of Global Onshore Wind Energy Production by World Region (2010-336TWh)
31.2%
1.0%
2.0%
21.4%
42.8%
0.6%
0.9%
0.1%
North America
Central & South America
Europe
Eurasia
Middle East
Africa
Asia
Oceania
Distribution of Global Onshore Wind Energy Production by World Region (2012-512TWh)
30.7%
1.5%
1.8%
26.2%
38.1%
0.5%
1.2%
0.0%
North America
Central & South America
Europe
Eurasia
Middle East
Africa
Asia
Oceania
Distribution of Global Onshore Wind Energy Production by World Region (2014-689TWh)
30.9%
2.7%
1.8%
28.6%
34.0%
1.2%
0.7%
0.0%
North America
Central & South America
Europe
Eurasia
Middle East
Africa
Asia
Oceania

Fig. 7: Evolution of onshore wind generated electricity between world regions. Based on (IRENA 2016b).

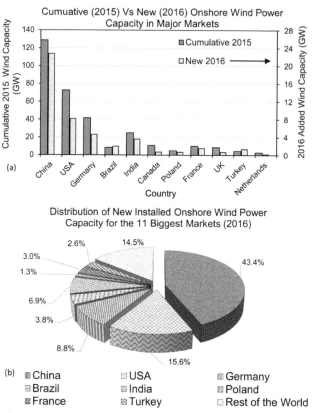

Fig. 8: (a) Onshore wind capacity evolution in 2015–2016, (b) Market share of onshore wind energy in 2016. Based on (IRENA 2016b, GWEC 2017).

2.2 Offshore Wind Energy in Global Market

Nowadays wind power is considered to be one of the most competitive power generating technologies with a cumulative capacity above 486 GW. Apparently, by taking into account the share of 7% (based on 2015 values) of wind energy among cumulative global power capacity, the wind power sector is established as a mainstream power generation technology (World Energy Council 2016). However, onshore wind energy faces constraints associated with negative environmental implications such as visual, noise, land use and other types of ecological impacts, which hinder its future development (Kondili and Kaldellis 2012).

In this context, during the last decade, the point of interest concerning wind energy has moved to offshore locations by also taking advantage of the higher wind speeds and greater resource potential found at open sea which considerably increases the energy yield (Kaldellis and Kapsali 2013). Hence, offshore wind power capacity presented an increasing trend since 2007 reaching at the end of 2016, 14.4 GW (3% of total wind installations). Europe is by far the world leader in offshore applications totalling 12.6 GW in 2016, while the second region with most installations is Asia with 1.7 GW (see Fig. 9) (GWEC 2017). However, in 2016, annual capacity of non-European offshore investments increased significantly compared to previous years, exceeding 650 MW (30% of 2016 added capacity), while 2016 new offshore installations in Europe have been reduced to 1.6 GW compared to 3 GW implemented in 2015 (GWEC 2017).

Since 2009, the offshore wind industry market has presented interesting information (Fig. 10) concerning the future trends of offshore applications. As observed, the European markets comprise the highest share of all annual installations. UK was the most significant market globally until 2015 with almost similar annual installed capacity between 460 and 850 MW, while Denmark in 2009, 2010 and 2013 installed 230, 207, and 350 MW respectively. Of great importance is the evolution of Germany's annual offshore wind installations which has presented an increasing trend since 2010 with a remarkable peak in 2015 of 2.3 GW. Netherlands, on the other hand, in 2015 and 2016 invested significantly in offshore wind applications with 180 and 691 MW respectively. The most significant market outside Europe, is China where annual offshore installations since 2010, with an exception of 2013, presented an increasing trend, reaching in 2016 592 MW of new grid connected capacity (EWEA 2017b, GWEC 2017).

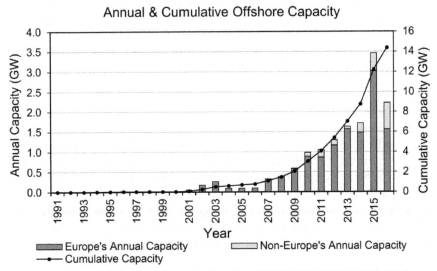

Fig. 9: Evolution of offshore wind capacity. Based on (GWEC 2017, EWEA 2017b, 2011).

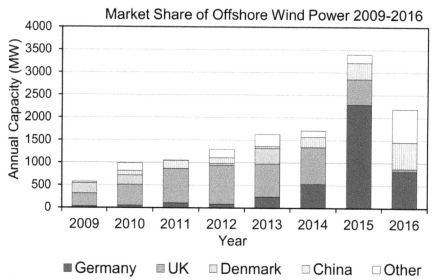

Fig. 10: Annual offshore installations by country 2009–2016. Based on (GWEC 2017).

2.3 *Wind Energy Facts in EU*

The evolution of electrical power capacity in Europe since 2000 is characterized from the adoption of environmentally friendlier technologies compared to the dominated fossil fuels based electricity production of the 20th century. As it is obvious from Fig. 11, in 2000, fossil fuel based power stations comprised above 57% of total capacity, while the rest 43% included RES technologies (23%) and nuclear power plants (20%). At the end of 2015, conventional and nuclear stations share has dropped to 44% and 11% respectively, while RES based stations increased their share to 45%. The highest share between RES technologies belongs to the wind power sector which represents 14% of total EU power capacity (Eurostat 2017, GWEC 2017).

Since 2000, wind energy investments in the European Union (EU), both onshore and offshore, have been significantly expanded. Cumulative capacity in 2016 reached 161.3 GW with many countries being among the biggest markets globally. Figure 12 indicates the evolution of total wind power capacity by country where it is obvious that Germany has by far the highest grid connected wind power capacity, exceeding 50 GW in 2016. Spain, on the other hand, although its new wind investments since 2012 are almost absent, ranks second with total capacity of around 23 GW. 2016 installations in the UK's wind energy sector were 736 MW, increasing

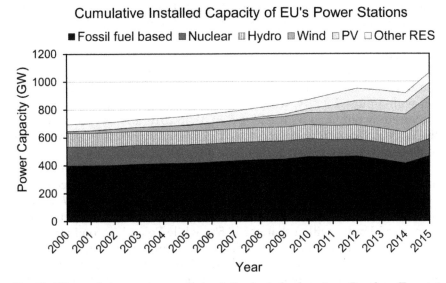

Fig. 11: EU cumulative power capacity evolution by technology type. Based on (Eurostat 2017, GWEC 2017).

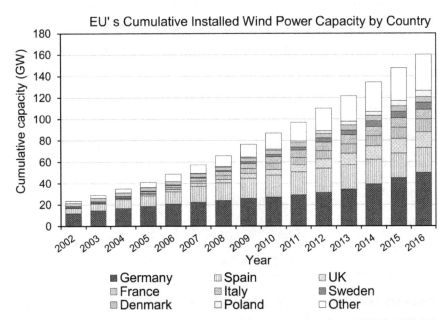

Fig. 12: Evolution of EU's wind power capacity by country. Based on (GWEC 2017, IEA 2016a).

the cumulative capacity up to 14.5 GW, while France is following at 12 GW by adding in 2016, 1.5 GW of wind power. The fifth highest cumulative wind capacity in EU belongs to Italy with 9.2 GW installed power at the end of 2016, from which only 0.3 GW were installed during the same year (GWEC 2017). Denmark on the other hand although at the beginning of the millennium was the country with the third highest capacity in the EU with almost 2.9 GW, wind installations during the last 15 years has increased by only 2.3 GW totalling 5.2 GW in 2016. To this end, it ranks at the eighth place below Poland which presented significant development in the wind energy sector, particularly after 2010 (GWEC 2017).

Wind power market share in Europe for 2016 presented similarities and differences compared to the one of 2015 (see Fig. 13). Specifically, Germany again dominated in the EU market with 5.4 GW annual installations and a share of 43.6%. The second largest EU market in 2016 comprised France with a share of 12.5%, while Poland (second market in 2015) ranks forth with a share of 5.5% and annual installations 682 MW. On the other hand, the Netherlands ranked in the third place compared to the sixth in 2015, by having annual installations of 887 MW. At this point it is necessary to mention that although Spain ranks second below Germany in the cumulative EU's wind capacity, investments during 2015 and 2016 were almost absent (EWEA 2017b).

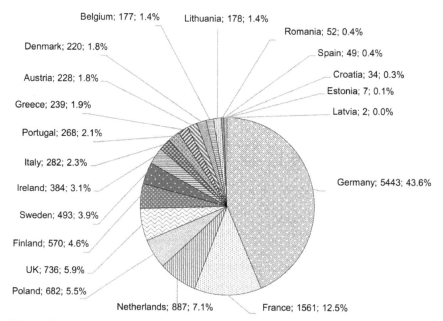

Fig. 13: Wind energy market share of EU-28 countries in 2016. Based on (EWEA 2017b).

Promotion policies in the EU concerning the exploitation of wind energy results nowadays to a share of wind generation technologies in gross electricity production above 7.4% compared to 10 years ago where wind-based electricity comprised only 2% of total electricity production (see Fig. 14a). In this context, Denmark's wind energy share in electricity production for 2014 exceeded 40%, while in Portugal, Ireland, and Spain,

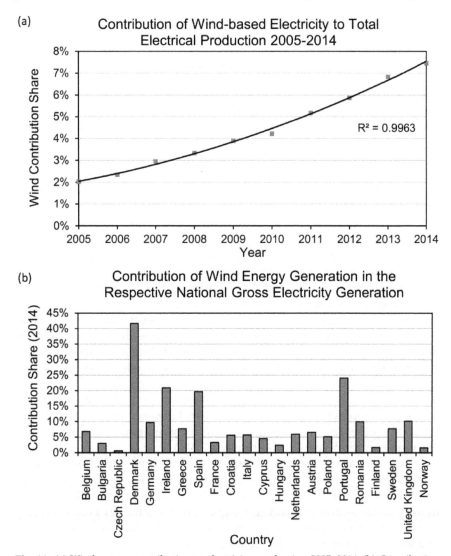

Fig. 14: (a) Wind energy contribution to electricity production 2005–2014, (b) Contribution of wind energy to electricity production by country in 2014. Based on (Energy Information Administration (EIA) 2017b).

wind energy contributed more than 20% in their national electricity generation. By analysing Fig. 14b, one may also notice that in many European countries the wind energy sector covers a significant portion of the electrical demand with a share between 5 and 10% of total electricity production. However, in the EU, there are still countries where wind energy is almost absent and consequently its contribution to gross electricity production equals to zero. These countries include Malta, Slovenia, and Slovakia (Energy Information Administration (EIA) 2017b).

3. Technology Innovations in the Wind Energy Sector

Since the beginning of 1980, contemporary wind turbines development can be correlated with the gradual upscale of wind converters promoted by the necessity of better land exploitation, reduced maintenance and operation (M&O) costs, and numerous development funding programmes (Kaldellis and Zafirakis 2011). According to Fig. 15, the rotor diameter evolution presented an increasing trend until 2004, where machines in the order of 2 MW power rating have been deployed and commercialisation of larger turbines have been stalled. Nevertheless, development of offshore wind technologies along with advances in new materials resulted in a new increase of rotor diameter above 160 m (e.g., Vestas V164). In general, at present the development of wind turbines of 10 MW rated power with rotor diameter above 190 m, is underway by at least three major manufacturers (i.e., AMSC "SeaTitan", Vestas V200, Siemens).

In terms of wind turbines' tower concepts, the most widely used technology comprises tubular steel towers. However, the growing trend in rotor dimensions and larger machines induces the development of new alternative designs able to withstand additional structural loads. Additionally, due to increased height of modern wind turbines, larger diameter of towers poses drawbacks in transportation of onshore components concerning mostly legal regulations of permissible height. In this regard, deployment of concrete or hybrid towers consists of a solution for tackling the disadvantages observed in contemporary tower demand (Serrano-González and Lacal-Arántegui 2016).

Concrete towers are considered to present higher structure strength, low maintenance and high design and construction flexibility (i.e., no height restrictions, adaptable mixture design for strength, stiffness and density optimisation). However, weight of this technology is significantly higher than steel-based towers which causes difficulties in transportation and assembly processes (Tricklebank et al. 2007).

On the other hand, hybrid towers made from steel and concrete present a growing trend supported also by relatively lower costs particularly for high height and markets where steel costs are considered high. The main drawback of this kind of constructions is that assembly may be complicated and consequently this may result to higher installation costs (Tricklebank et al. 2007).

Concerning offshore wind energy, foundation and tower substructures include several technologies that either are considered mature or are at the stage of research and development (R&D). Figure 16 presents the technologies associated with the support structures of offshore wind turbines along with the cumulative share of each technology in 2016 for EU's installations. Monopile-based foundations for 2016 in EU, comprised 88% of total installations, while the remaining 12% was based on Jacket substructures. To this end, Monopile constructions are used nowadays for 3,354 wind turbines in EU (80.8%), gravity based substructures share

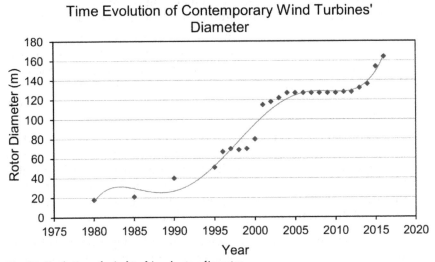

Fig. 15: Evolution of wind turbines' rotor diameter.

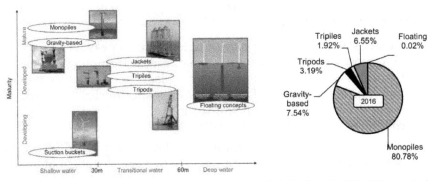

Fig. 16: Offshore wind turbines' support structures and their share in EU offshore wind farms. Based on (Kaldellis and Kapsali 2013, EWEA 2017a).

comprise 7.5% (313 turbines), and Jacket foundations increased to 272 substructures reaching 6.6% of total offshore installations in EU. Tripod and Tripile foundation technologies comprise 3.2 and 1.9% (132 and 80 substructures) respectively (EWEA 2017a).

One of the main parameters of contemporary wind turbines is the drive train configuration which converts the rotor's mechanical power to electric power. Classification of wind turbines based on different configurations of drive train includes three basic categories (Serrano-González and Lacal-Arántegui 2016):

- Turbines with a gearbox and a doubly fed induction generator. Gearbox is used to increase the rotational speed of the main shaft to appropriate speed for the generator to operate. Its main disadvantage results from the faults arising from the demanding operating conditions of gearboxes which contribute significantly in long downtime periods.

- Gearless turbines, where a synchronous generator using permanent magnets or electrically excited is coupled directly to the shaft. The main drawback of this technology includes the weight of machines due to larger generator diameters arising from the large number of poles of the electric generator along with high cost associated with the use of permanent magnets.

- A hybrid configuration being a compromise solution where a gearbox is used for adjusting the rotational speed of the shaft to medium speed for a synchronous generator with lower number of poles. Additionally, the use of a full power converter able to decouple the generator from the grid, contributes to better control of the rotational speed of the rotor and thus to the frequency of the generator's output.

The captured power from the wind is mainly dependant on the tip speed ratio which comprises the ratio between the incoming wind speed and the linear speed of the tip blade. Power regulation of the wind converter can be accomplished via four main technologies (Kaldellis and Zafirakis 2011, Serrano-González and Lacal-Arántegui 2016):

- Passive stall control (PSC) is the simpler form of power control and is used mostly in older machines or small scale contemporary turbines. Control is based on the design of the blades which induce the air flow to stall at higher wind speeds. Its main drawback consists of lower production and high stresses in the case of extreme weather conditions.

- Active stall control (ASC) approach is based on pitched blades in case of higher wind speeds in order to increase the angle of attack

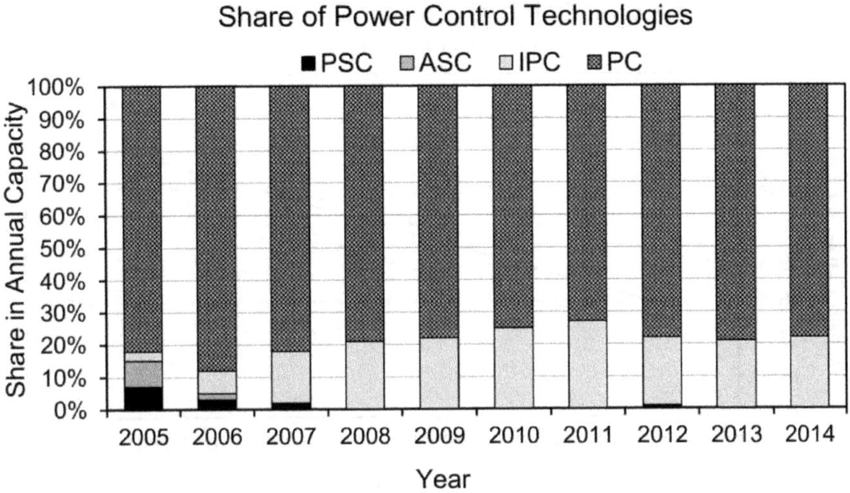

Share of Power Control Technologies

Fig. 17: Wind power control technologies share for EU's annual installations. Based on (Serrano-González and Lacal-Arántegui 2016).

and achieve stall. Although this technology is more complex its main advantage is that power output is smoother and start up is assisted.

- Pitch control (PC) is used in most modern wind turbines and comprises a method where blades turn to opposite angle during ASC. In this way the angle of attack reduces and the captured power is limited. It presents advantages over ASC, due to the fact that power output can be maintained for a higher range of wind speeds.

- Individual pitch control (IPC) is gaining higher share of power regulation technologies during the last years particularly in the European market (see Fig. 17). IPC is based on pitching each blade separately during the movement of the rotor, achieving in this way reduction of asymmetric loads and therefore increase in the lifetime of wind turbines.

Recapitulating, technological innovation in onshore and offshore wind energy during the last two decades has resulted in higher energy yield of wind turbines and reduction of the downtime periods observed more frequently in the past due to lower technical availability of older turbine models (Kaldellis and Zafirakis 2013). This is also reflected by the increase of the mean annual capacity factor of wind projects both in Europe and globally (see Fig. 18), where it can be observed that CF exceeded 23% in 2014 compared to 18% in 1995.

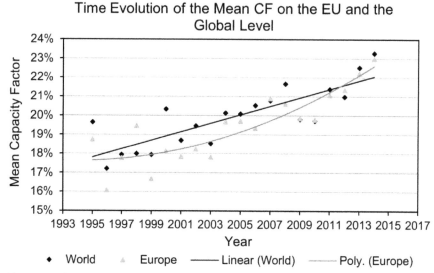

Fig. 18: Evolution of mean annual CF in EU and World's wind power installations. Based on (EIA 2017a, Energy Information Administration (EIA) 2017a).

4. Economic Issues

One of the most important factors, concerning wind energy development, comprises the economic aspects which can be analysed from two perspectives. Firstly, from the investor's point of view concerning mostly expenditures and profits, and secondly from the view of employment opportunities. As it was stated previously, improvements in wind turbine technology have led to increased efficiencies and better energy yield. Furthermore, the trend dominating the market concerning increased rated power of wind converters and subsidised R&D projects (i.e., $1.8 billion for 2014), resulted in reduction of the specific investment cost of wind projects (REN21 2016).

On the other hand, the hidden cost of energy which arises due to the stochastic character of wind energy is minimised. This is achieved by improved wind potential forecasting techniques and geographical distribution of wind farms in suitable regions where grid related effects including stability and power quality issues are limited (Kaldellis 2008).

Acknowledging the above, during the last 30 years there has been a significant decrease of the average investment cost of onshore wind installations to 1,500 €/kW, compared to 1980 where specific cost of wind power was around 3,700 €/kW (see Fig. 19). According to Fig. 19, average investment cost of onshore applications has been reduced below 1,300€/kW at the beginning of the new millennium, where it was stabilised until 2008. Since then, the specific cost presented an increase mostly due to

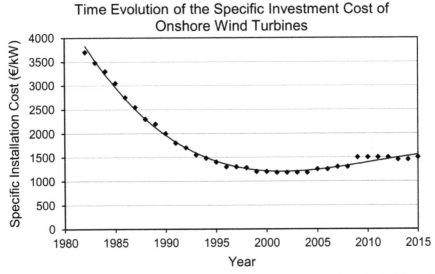

Fig. 19: Investment cost evolution of onshore wind projects. Based on (Kaldellis and Zafirakis 2011, IEA 2016a).

increased sophistication and upscaling of new machines. Countries with high investment cost in 2015 included Japan and Austria with 2,280 and 1,850 €/kW respectively, while countries where specific cost was relatively low included Norway, Spain, Germany, and China with a range between 1,200 and 1,400 €/kW (Kaldellis and Zafirakis 2011, IEA 2016a).

Specific investment cost of offshore wind installations is considerably higher than onshore counterparts because of high installation cost associated with transportation and assembly of the components, construction activities and connection with the grid via undersea cables. By observing Fig. 20, one may notice that cost at the beginning of 1990 was approximately 3,000 €/kW, while between 2000 and 2005 it decreased to around 1,500 €/kW. However, since 2007 specific average cost of offshore installations increased drastically reaching in 2015 4,640 €/kW approximately. Moving to larger machines, and higher depths and distances postpones the reduction of the capital cost of offshore projects causing a significant challenge that hinders offshore wind deployment (The Crown Estate 2012, IEA 2015).

Apart from the investment cost of onshore and offshore wind applications, the levelised cost of energy (LCOE) which includes all the lifecycle stages of a project such as capital investment, M&O costs, and decommissioning expenditures, is of great importance. In this regard, it is essential to analyse LCOE in comparison with other RES and conventional power generating technologies. Figure 21 indicates the LCOE range of most power plant technologies globally where it can be observed that

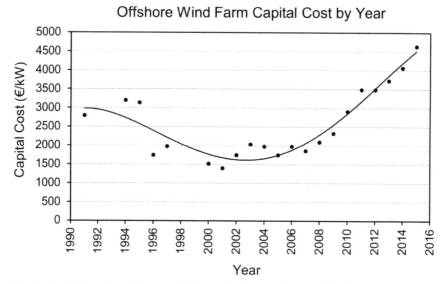

Fig. 20: Evolution of specific cost of offshore wind farms. Based on (The Crown Estate 2012, IEA 2015).

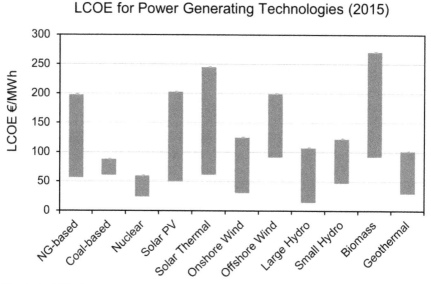

Fig. 21: LCOE of power generating technologies in 2015 with 3% discount rate. Based on (IEA 2015).

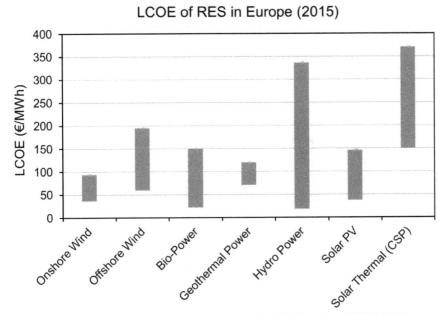

Fig. 22: LCOE of Europe's RES based power stations in 2015. Based on (REN21 2016).

onshore wind energy ranges between 30 and 125 €/MWh, while offshore applications' LCOE is between 90 and 200 €/MWh. It is clear that both wind energy technologies and especially the onshore ones are directly comparable with conventional power plants where in some cases LCOE is lower (i.e., NG open cycle-OCGT fired stations) (IEA 2015).

Furthermore, the LCOE of wind energy applications in Europe is limited further below 100 and 195 €/MWh for onshore and offshore projects respectively (see Fig. 22). In this way, although offshore installations in Europe are considered to be more costly than most conventional technologies, its weighted average is just over the one of geothermal power. In fact, compared to solar thermal and hydroelectric power stations, offshore investments present lower lifecycle costs. In this context, LCOE of offshore wind projects is expected to reduce further in the following years due to a joint declaration signed by DONG Energy, Vestas and Siemens setting a goal to drive levelised cost below 85 €/MWh by 2020 (REN21 2016).

An indicative example of what LCOE—for different power technologies—includes is shown in Fig. 23, where cost of investment, O&M costs, and fuel and carbon costs are depicted. It is observed that

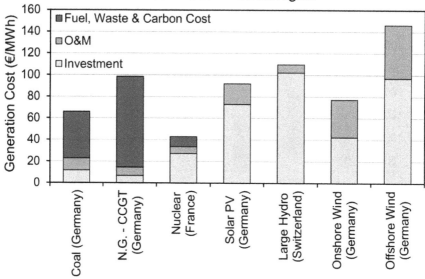

Fig. 23: LCOE breakdown for different power generating technologies. Based on (IEA 2015).

RES-based technologies comprise only investment and O&M costs, while conventional plants include also fuel and carbon emission penalties. Wind energy applications compared to other RES present higher O&M costs, while in conventional technologies the main cost arises from fuel requirements in the case of N.G., and carbon emissions in the case of coal-based plants (IEA 2015).

Concerning support mechanisms for wind energy applications, many countries have established since 1980, incentive programmes for promoting wind energy. An indicative example comprises the Californian outbreak in 1979, where the USA government encouraged renewable investments for coping with a possible similar oil crisis to the one of 1973 (Kaldellis and Zafirakis 2011). Nowadays, several countries are changing their energy policy in order to integrate larger amounts of RES produced electricity in their electrical networks (IEA 2016a).

The main support mechanisms applied to wind energy applications for 2015–2016 include (IEA 2016a):

- Carbon Tax. RES based installations are not subjected to this tax due to zero carbon emissions. It is implemented in 31 European countries through the EU ETS system and in three provinces of Canada.
- Feed-in Tariff (FIT). It is the most prevalent support scheme since the 1980s resulting to a significant increase of the wind power sector.

Nowadays, there are 15 countries that have an established FIT system including major markets such as China, Denmark (for offshore), Germany, USA, and Belgium.

- Renewable Portfolio Standards (RPS). Electricity utilities are obliged to use a portion of their electricity supplies from RES. Eight countries are using this incentive including China, Belgium, the UK, and USA.
- Green Certificates (GCs). Power plants receive certificates for RES generated electricity. Electricity and certificates are sold in a determined price according to the demand set by the consumer's obligation to purchase an electricity share from RES. Seven countries have implemented this measure.
- Net Metering or Net Billing. It comprises a retail value given to a producer for feeding the electrical grid with excess electricity recorded by a bi-directional electricity meter. Seven countries have implemented this support scheme at the moment including Denmark, Canada, the UK, USA, the Netherlands, Italy, and Portugal.
- Some or all expenses are deducted from taxable income of wind energy installations (i.e., Tax Incentives). Seven countries have implemented this mechanism.
- Special incentives for small scale wind energy projects. Connection costs are reduced, capital allowances for corporations are accelerated and value added tax (VAT) for small farmers is rebated. Denmark, the UK, USA, Canada, Italy, and Portugal support this measure.
- Financing and Capital incentives. In the case of wealth creation and business success associated with wind energy, a share in investment may be offered. Additionally, a capital incentive may include a subsidy which aims to cope with the initial cost barrier of a wind power system.

As mentioned above, some of the support mechanisms are deeply associated with promoting social prosperity. In this regard, new opportunities concerning employment in the wind energy sector are established. At the end of 2015, more than 1.08 Million people were employed either directly or indirectly in the wind energy sector. Almost half of global employment jobs were in the Chinese wind sector ($\approx 502,400$ jobs) followed by Germany and USA with 150,000, and 88,000 positions respectively (IEA 2016a).

Figure 24 illustrates the wind energy employers per installed power capacity in major countries. Of the leading countries Denmark's wind power sector employees more than six employers/MW, while Finland ranks second with five employers/MW. Although China ranks first in terms of absolute numbers of employment, wind energy jobs per MW are

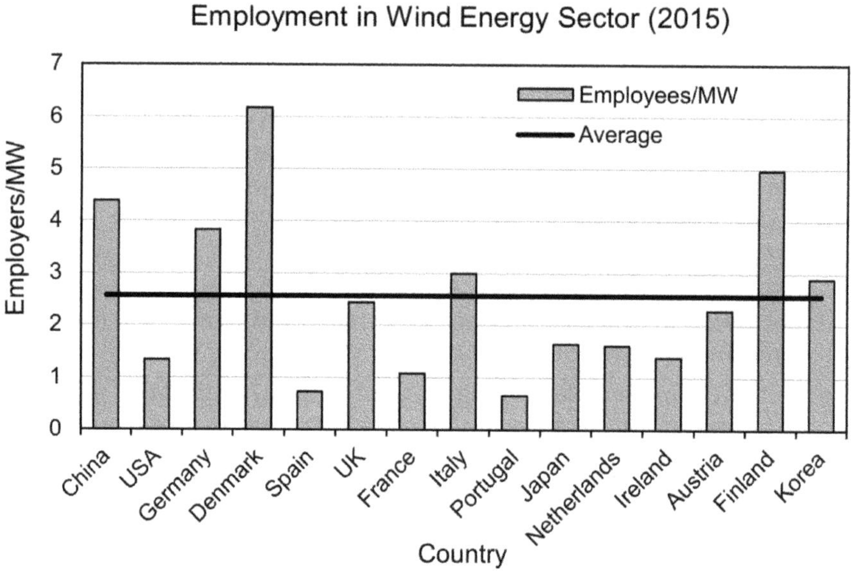

Employment in Wind Energy Sector (2015)

Fig. 24: Wind energy sector employees per installed power capacity in 2015. Based on (GWEC 2017, IEA 2016a, DWIA 2017).

in the order of four employees/MW. However, it is estimated that this number may increase in the following years up to 15 jobs per installed MW including also the manufacturing sector. The countries with the lowest number of employees/MW comprise Spain and Portugal where one employee corresponds to multi-MW of installations (GWEC 2017, IEA 2016a, DWIA 2017).

5. Environmental Issues

Among renewable energy sources, wind energy is considered to be one of the most competitive technologies. However, its sustainable nature is sometimes compromised by possible effects on certain environmental and social parameters which include noise, visual, land use, LC emissions (carbon footprint), ship collisions and other impacts on birds, mammals and fish. Both onshore and offshore wind applications present similar environmental impacts although significant differences are associated with the inherent nature of each environment and the alternative methods used for construction, O&M, and decommissioning activities.

Table 1 indicates the impacts occurred in both onshore and offshore projects distinguished by the three main phases in the lifecycle of a project.

Table 1: Wind energy applications main environmental impacts. Based on (Kaldellis et al. 2016).

Environmental parameter	Impact	Construction phase		Operation phase		Decommissioning	
		Onshore	Offshore	Onshore	Offshore	Onshore	Offshore
Avian Species	Fatalities, barrier to migration, nest displacement, breeding disturbance due to noise	√	√	√	√		
Wildlife	Disturbance & emigration	√		√		√	
Fish	Loss of habitat, disturbance of breeding due to noise-vibrations, reduction of abundance		√		√		√
Mammals	Reduction of habitat, potential hearing damage, displacement of habitat	√	√	√	√	√	√
Zoobenthos	Loss of habitat, reduction of abundance		√		√		√
Hydrography - Land	Contamination of pollutant chemicals, alteration of current flows, land required	√	√			√	√
Humans	Stress symptoms, sleep disturbances due to noise and flickering, aesthetic disruption	√		√		√	
Air Quality	LC emissions	√	√	√	√	√	√

Most developers that are associated with wind energy infrastructure have faced serious oppositions during the development of a wind project due to false impressions generated from the general idea that wind turbines harm the environment. However, several studies indicate that impacts caused by wind-based energy can be perceived as "myths" and actually most oppositions stem from the "Not in my Backyard" syndrome (Kaldellis et al. 2016, Jones and Richard Eiser 2010).

Regarding impacts on avian species associated with wind energy applications, it may include nest displacement, flight routes avoidance and risk of collision fatalities. As one can observe from Figure 25a, wind turbines are responsible for less than one out of 10,000 bird fatalities, while buildings comprise the main cause of collisions exceeding 5,500 out of 10,000 (Kaldellis and Zafirakis 2011, Kaldellis et al. 2016). In terms of avian mortalities caused by other power generating sources (see Fig. 25b), wind energy is responsible for 1 fatality per four GWh, while fossil fuel plants are directly and indirectly responsible for at least 5 fatalities per produced GWh.

Concerning noise impact on local communities, wind energy applications and particularly onshore based ones, are often subjected to serious social oppositions. Keeping in mind the severity of aerodynamic noise, contemporary onshore projects are erected at a distance of at least 300 to 400 m away from the nearest inhabited area. According to a study carried out by Kaldellis et al. 2012, the noise level 300 m far from a wind farm with wind speed 6 m/s at 10 m height is in the range of 45 dB(A) (see Fig. 26). Compared to other sound sources, this value is almost similar to an unoccupied air-conditioned office, while a busy general office presents sound levels of around 60 dB(A) (Kaldellis et al. 2012).

Power generating technologies are greatly associated with primary energy consumption and LC greenhouse gas emissions (GHG), provided they are based on the exploitation of fossil fuels. In this context, although onshore and offshore applications are considered to be low emission technologies, their lifecycle phases enhance energy intensity and consequently carbon footprint. Energy consumption of wind energy applications can be divided into four stages including raw material extraction, manufacturing, O&M, and final disposal processes (Kaldellis and Apostolou 2017).

In this regard, Fig. 27 presents a comparison between electricity production technologies where it is obvious that wind-based electricity requires much lower primary energy in the lifecycle perspective than coal-based power plants and solar-based technologies. Energy intensity for onshore applications ranges between 0.014 and 0.082 kWh_{pr}/kWh_e with a mean value of 0.037 kWh_{pr}/kWh_e, while for offshore energy applications it ranges between 0.029 to 0.054 kWh_{pr}/kWh_e with an average of 0.041

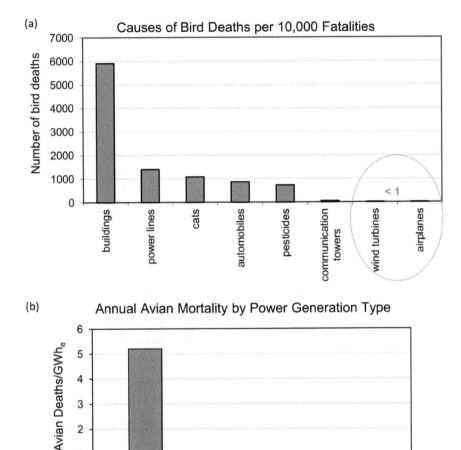

Fig. 25: (a) Bird fatalities caused by different factors, (b) Bird fatalities caused from the power generation industry. Based on (Kaldellis and Zafirakis 2011, Sovacool 2009).

kWh_{pr}/kWh_e. By comparing both wind technologies, offshore applications seem to be more energy intensive throughout their lifecycle due to the higher demand in construction, O&M, and decommissioning activities (Kaldellis and Apostolou 2017).

To this end, by comparing the LC GHG emissions of wind energy with other power generating technologies (Fig. 28), it is observed that wind energy comprises a small fraction in the total GHG emissions of electricity production systems. In fact, the global warming potential of wind energy is much lower than other RES technologies as well. For example, solar PV technologies LC emissions' maximum value has been estimated from several

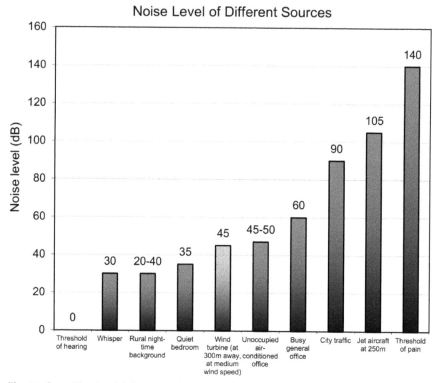

Fig. 26: Sound levels of different sources. Based on (Kaldellis et al. 2012).

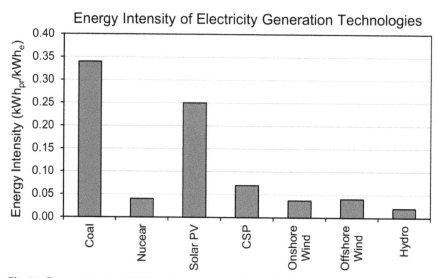

Fig. 27: Energy intensity of different power generating technologies in a lifecycle perspective. Based on (Kaldellis and Apostolou 2017, Aden et al. 2010).

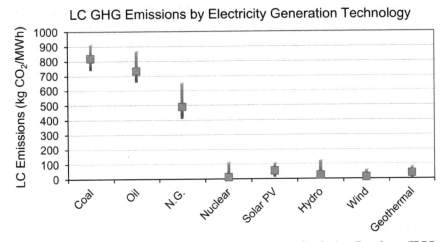

Fig. 28: Lifecycle GHG emissions of power generating technologies. Based on (IPCC, Intergovernmental Panel on Climate Change and Edenhofer 2014, World Energy Council 2004).

studies to around 104 kg/MWh while the corresponding emissions of wind energy does not exceed 56 kg/MWh of produced electricity (IPCC 2014, World Energy Council 2004).

6. Conclusions

This chapter presents a review analysis of onshore and offshore wind applications in the global and EU market. Current situation along with time evolution of power capacity and generated electricity from wind energy have been investigated, proving that significant growth occurred during the last 25 years. Recapitulating, the role of EU concerning wind energy applications and particularly offshore development in the last decade increased substantially, while a new emerging market in Asia since 2010 is represented by China which is at the moment the biggest market by far with relation to onshore wind installations. Additionally, since the last decade USA entered dynamically in the global market of wind energy and in fact at present considered the second largest market behind China. Moreover, R&D advances in wind energy contributed to overcome the technological barriers of the past which played a significant role in hindering the broader exploitation of wind power applications during the previous years. To this end, this research presents the most significant technological innovations applied to wind-based installations towards a more effective deployment of the wind energy sector.

Besides the above, this study focusses also on financial and environmental aspects of wind energy and provides data concerning the lifecycle and economic comparison of wind-based projects with other RES and conventional power generating technologies. In this regard, LCOE of wind energy is greatly comparable with other RES technologies, while onshore applications present a LCOE similar to the one of conventional technologies. On the other hand, of great concern is the environmental profile of wind-based installations. The energy and carbon intensity of wind energy applications in a life cycle perspective indicates that compared to other electricity production technologies wind energy offers a significant potential for achieving future GHG emissions reduction. In conclusion, all the above suggest that wind energy is definitely an effective power generating technology able to cover the increasing global energy demand and improve sustainability of future electricity markets.

References

Aden, N., A. Marty and M. Muller. 2010. Comparative Life-Cycle Assessment of Renewable Electricity Generation Technologies. http://gadgillab.berkeley.edu/wp-content/uploads/2013/02/Muller-M.2010.pdf.
DWIA, Danish Wind Industry Association. 2017. Industry Statistics–2015. http://www.windpower.org/en/knowledge/statistics/industry_statistics.html.
EIA, Energy Information Administration. 2017a. Wind Capacity. International Energy Statistics. http://www.eia.gov/.
EIA, Energy Information Administration. 2017b. Wind Electricity Net Generation. International Energy Statistics. http://www.eia.gov/.
EPI, Earth Policy Institute. 2014. World Cumulative Installed Wind Power Capacity and Net Annual Addition, 1980–2014. Data Center - Climate, Energy, and Transportation. http://www.earth-policy.org/data_center/C23.
Eurostat. 2017. Energy Statistics - Infrastructure - Electricity. http://ec.europa.eu/eurostat/web/energy/data/database.
EWEA. 2011. Wind in Our Sails - The Coming of Europe's Offshore Wind Energy Industry. http://www.ewea.org/fileadmin/files/library/publications/reports/Offshore_Report.pdf.
EWEA. 2017a. The European Offshore Wind Industry - Key Trends and Statistics 2011 to 2016. WindEurope. https://windeurope.org/about-wind/statistics/offshore/.
EWEA, European Wind Energy Association. 2017b. Wind in Power 2011–2016. WindEurope. https://windeurope.org/about-wind/statistics/european/.
GWEC, Global Wind Energy Council. 2017. Global Wind Reports 2006 to 2016. http://www.gwec.net/publications/global-wind-report-2/.
IEA, International Energy Agency. 2015. Projected Costs of Generating Electricity-2015 Edition. France. https://www.oecd-nea.org/ndd/pubs/2015/7057-proj-costs-electricity-2015.pdf.
IEA. 2016a. 2015 Wind Annual Report. https://www.ieawind.org/annual_reports_PDF/2015/2015%20IEA%20Wind%20AR_small.pdf.
IEA. 2016b. Key World Energy Statistics 2016. Paris: IEA. http://www.iea.org/publications/freepublications/publication/key-world-energy-statistics.html.
IPCC, Intergovernmental Panel on Climate Change, and Ottmar Edenhofer (eds.). 2014. Climate Change 2014: Mitigation of Climate Change: Working Group III Contribution to the Fifth Assessment Report of the Intergovernmental Panel on Climate Change. New York, NY: Cambridge University Press.

IRENA, International Renewable Energy Agency. 2015. Renewable Power Generation Costs in 2014. https://www.irena.org/DocumentDownloads/Publications/IRENA_RE_Power_Costs_2014_report.pdf.

IRENA. 2016a. Global Atlas for Renewable Energy. http://irena.masdar.ac.ae/#.

IRENA. 2016b. Renewable Energy Statistics 2016. UAE. http://www.irena.org/DocumentDownloads/Publications/IRENA_RE_Statistics_2016.pdf.

Jones, C.R. and J.R. Eiser. 2010. Understanding 'local' opposition to wind development in the UK: How big is a backyard? Energy Policy, The Role of Trust in Managing Uncertainties in the Transition to a Sustainable Energy Economy, Special Section with Regular Papers 38(6): 3106–17. doi:10.1016/j.enpol.2010.01.051.

Kaldellis, J.K. 2008. Integrated electrification solution for autonomous electrical networks on the basis of RES and energy storage configurations. Energy Conversion and Management 49(12): 3708–20. doi:10.1016/j.enconman.2008.06.034.

Kaldellis, J.K. and D. Zafirakis. 2011. The wind energy (R)evolution: A short review of a long history. Renewable Energy 36(7): 1887–1901. doi:10.1016/j.renene.2011.01.002.

Kaldellis, J.K., K. Garakis and M. Kapsali. 2012. Noise impact assessment on the basis of onsite acoustic noise immission measurements for a representative wind farm. Renewable Energy 41(May): 306–14. doi:10.1016/j.renene.2011.11.009.

Kaldellis, J.K. and M. Kapsali. 2013. Shifting towards offshore wind energy—recent activity and future development. Energy Policy 53(February): 136–48. doi:10.1016/j.enpol.2012.10.032.

Kaldellis, J.K. and D. Zafirakis. 2013. The influence of technical availability on the energy performance of wind farms: Overview of critical factors and development of a proxy prediction model. Journal of Wind Engineering and Industrial Aerodynamics 115(April): 65–81. doi:10.1016/j.jweia.2012.12.016.

Kaldellis, J.K. 2015. Overview of wind energy resource. pp. 1–30. In: Jinyue Yan (ed.). Handbook of Clean Energy Systems. Chichester, UK: John Wiley & Sons, Ltd. doi:10.1002/9781118991978.hces095.

Kaldellis, J.K., D. Apostolou, M. Kapsali and E. Kondili. 2016. Environmental and social footprint of offshore wind energy. Comparison with onshore counterpart. Renewable Energy 92(July): 543–56. doi:10.1016/j.renene.2016.02.018.

Kaldellis, J.K. and D. Apostolou. 2017. Life cycle energy and carbon footprint of offshore wind energy. Comparison with onshore counterpart. Renewable Energy 108(August): 72–84. doi:10.1016/j.renene.2017.02.039.

Kondili, E. and J.K. Kaldellis. 2012. Environmental-social benefits/impacts of wind power. pp. 503–39. In: Comprehensive Renewable Energy. Elsevier. http://linkinghub.elsevier.com/retrieve/pii/B9780080878720002195.

Kumar, Y., J. Ringenberg, S.S. Depuru, V.K. Devabhaktuni, J.W. Lee, E. Nikolaidis, B. Andersen and A. Afjeh. 2016. Wind energy: Trends and enabling technologies. Renewable and Sustainable Energy Reviews 53(January): 209–24. doi:10.1016/j.rser.2015.07.200.

REN21, Renewable Energy Policy Network. 2016. Renewables 2016 – Global Status Report. http://www.ren21.net/wp-content/uploads/2016/06/GSR_2016_Full_Report_REN21.pdf.

Serrano-González, J. and R. Lacal-Arántegui. 2016. Technological evolution of onshore wind turbines-a market-based analysis: technological evolution of onshore wind turbines. Wind Energy, n/a-n/a. doi:10.1002/we.1974.

Sovacool, B.K. 2009. Contextualizing avian mortality: A preliminary appraisal of bird and bat fatalities from wind, fossil-fuel, and nuclear electricity. Energy Policy, China Energy Efficiency 37(6): 2241–48. doi:10.1016/j.enpol.2009.02.011.

The Crown Estate. 2012. Offshore Wind Cost Reduction. Pathways Study. http://www.thecrownestate.co.uk/media/5493/ei-offshore-wind-cost-reduction-pathways-study.pdf.

Tricklebank, A.H., P.H. Halberstadt, B.J. Magee and A. Bromage. 2007. Concrete towers for onshore and offshore wind farms: conceptual design studies. Camberley, Surrey: Concrete Centre.

World Energy Council. 2004. Comparison of Energy Systems Using Life Cycle Assessment. London: World Energy Council.

World Energy Council. 2016. World Energy Resources 2016. UK: London.

Yue, C.-D., C.-M. Liu and E.M.L. Liou. 2001. A transition toward a sustainable energy future: feasibility assessment and development strategies of wind power in Taiwan. Energy Policy 29(12): 951–63. doi:10.1016/S0301-4215(01)00025-8.

CHAPTER 3

Failures and Defects in PV Systems

Review and Methods of Analysis

*Sonia Leva** and *Mohammadreza Aghaei*

1. Introduction

Photovoltaics (PV) solar energy has become more significant as compared to other kinds of renewable energy sources (RES). The unexpected growth of PV power plants and integrated facade of residential buildings with solar modules have resulted in an increase in demand for PV modules in the global market (Aghaei et al. 2015). Operation and maintenance (O&M) are defined as decisions and methods to control energy equipment and properties. Moreover, it represents a set of prediction and prevention actions for proactive and timely plant control solutions (Grimaccia et al. 2015, Trancossi 2011). The performance of PV systems should be monitored to keep electricity generation at an optimal level in PV plants. Failures and defects identification are the first steps for keeping the PV system in high-performance condition.

Monitoring of PV systems is a crucial task for assessment of PV power plants performance to manage its functioning and to continuously optimize output. Accurate monitoring of a PV system is complicated and expensive due to the capabilities of the available monitoring equipment. However, if the acquired data is used for analysis and modeling, it is very important to be as precise and comprehensive as possible. Unreliable monitoring equipment can affect results due to missing, noisy and inaccurate performance measurements. Moreover, it is very likely to come

[1] Politecnico di Milano - Department of Energy, Piazza L. Da Vinci, 32 – 20133 Milano – Italy.
* Corresponding author: sonia.leva@polimi.it

to the wrong conclusions about a PV system's status based on inaccurate data. Therefore, reliable and trustworthy PV monitoring equipment is worth the cost.

Analysis of failure mode in PV plants should be performed for each part separately. However, in the literature, most of the publications focused on the failures modes in PV modules and just a few of them considered an overall analysis of the PV system.

In order to reduce the O&M costs, early fault diagnosis plays a significant role by enabling the long effective life of PV arrays. Traditionally many different methods have been used for solar plant inspection and monitoring. Visual inspection is still used for cracks, corrosion, snail trails, discoloration and similar phenomena (El-Ghetany et al. 2002). The IR method is able to extend the detectable range of defects to resistive soldering, shunting paths, bypass diodes, disconnections, etc. (King et al. 2000). These methods are often time consuming, not practical for PV plants integrated in roof of buildings and sometimes it is difficult to evaluate the impact of the defects on the performance of the power plant. Further, most of the time it is not possible to recognize or distinguish the origin of the defect from others defects.

On the other hand, laboratory tests are not feasible in terms of economic impact on the O&M budget, and often not cost effective and reasonable for large PV array fields, unless particular purposes are necessary. Therefore, for instance, accelerated tests and I-V curves are not practical during normal operation in plants, even if a portable analyzer is available in the field for the maintenance team. The same is valid for other methods like electroluminescence and ultrasonic inspection.

The choice of a suitable inspection method is critical to detect any fault in the PV plant. Moreover, early failure detection plays a significant role to optimize PV plants performance during operation condition. Hence, the rapid recognition of defects on modules can keep PV system performances in the highest level. In the future, it is expected to observe a growing attention to research on reliable and cost effective methods for PV devices monitoring. Investigation about novel monitoring methods is a crucial issue in the solar energy market growth (Quater et al. 2014, Buerhop et al. 2012).

The goal of this chapter is to present the possible methods for monitoring and analysis of the overall PV system and in particular for the single PV module with the aim of detecting possible defects or failure. Furthermore, the main causes in PV plant failures are listed and reported based on the experiences of some operation & maintenance operators and also experimental tests made in some laboratories. In current chapter, a lot of example will be reported from various results of industries and laboratories side.

This chapter is subdivided into four sections. Section 1 expresses some background about challenges and opportunities for operation and maintenance of PV power plants. In addition, it illustrates that there is a big gap in the literature about failure and defects analysis and also inspection methods for PV systems.

Section 2 reports typical methods for defects and failures detection on PV Plants. Furthermore, again Section 2 introduces an innovative method using Unmanned Aerial Vehicle (UAV) to make automatic the entire steps of inspection and prognostic procedure and provide accurate and reliable information on operating conditions of PV plants. Section 3 explains about common failures and faults in the main components of PV plants (PV modules, inverters and Bias of system (BoS)). In Section 3, some statistical distributions are presented based on the previous experimental tests in the literature. Section 4 focuses on PV plants comparison based on different indications (e.g., Array Yield, Reference Yield, Final Yield and Performance Ratio) as a way to evaluate the performance of PV plant. Finally, Section 4 concludes the results and discussion of the present chapter.

2. Typical Methods for Defects and Failures Detection on PV Plants

Generally, any effect on a PV module or device which decreases the performance of the plant, or even influences the module characteristics or plant working, is considered as a failure. Whereas, a defect can be defined as an unexpected or unusual thing that has not been observed before on the PV plant. However, defects often are not the cause of power losses in the PV plants: they affect—for example, the PV modules in terms of appearances (Quater et al. 2014).

There are many different diagnosis tools and methods to explore the defects and failures on the PV devices (Golnas 2013, Ndiaye et al. 2013) as detailed in the following.

2.1 Visual Assessment

The first step of the PV plant monitoring is the inspection by sight. The visual assessment is a very easy method to detect some failures or defects in particular on PV modules. In fact, visual monitoring allows to observe the most external stresses on the PV devices. In addition, this method can present an overview of PV systems condition. The most visible defects and failures in PV modules are bubbles, delamination, yellowing, browning, bended, breakage, burned, oxidized, scratched, broken or cracked cells, corrosion, discoloring, anti-reflection, misaligned, loose, brittle, detachment (see Fig. 1). Visual assessment should be carried out before and after module's installation to evaluate effect percentage of electrical, mechanical and environmental stresses on the PV module.

In converter and BOS devices, visual inspection allows to recognize disconnections, burning parts, defect in supporting structure (see Fig. 2) and also problems link to dirty and shading.

Fig. 1: Example of visual assessment for PV modules (corrosion, delamination in front and back sides, browning) (Köntges et al. 2014).

Fig. 2: Some defects in PV modules installations and ground installation detected by visual inspection.

2.2 Real Time Analysis and Data Acquisition Systems (DAS)

There are many monitoring systems now used in medium-large size PV power plant (nominal power higher than 100 kW). Such systems can give us useful information about general performance of the PV plant, detailed information about the working status of inverters, transformers, PV arrays and switches with direct measurements performed in the plant by using ad hoc instruments, or collected data by the on-site monitoring system if available. On the other hand, these systems are not able to detect problems related to single module faults and sometimes neither to a series of modules.

In various applications, data acquisition systems (DAS) are applied to store data in high precision for evaluation of system performance. Recently, many different DAS were developed to evaluate the performance of PV systems in operation condition. Performance data presents problems, failures or malfunction of PV systems in details.

However, the main purposes of a monitoring system using DAS are to measure energy yield, to assess PV system performance and to quickly identify design flaws or malfunctions. Therefore, it is necessary for a large-scale PV system to use accurate monitoring systems to prevent economic losses due to operational problems. Therefore, the performance of PV systems should be monitored precisely to ensure long-term operation of PV power plant. Generally, electrical measurement signals in PV array include power, voltage and current in DC and AC sides, which contain rapid fluctuations. These fluctuations effect on the accuracy of the data acquisition and they are not visible by typical monitoring systems. However, the metrological parameters must be measured as well like for example solar irradiance on array surface, array planes and ambient temperatures.

Figure 3 displays the schematic view of the procedure of PV power plants monitoring by a commercial data acquisition system. The test controller is high speed and precision instrument, which can be programmed. In addition, it can store the data in internal memory. Data can be transferred over the Ethernet interface. It can also be used as a web client or E-mail.

2.3 Thermovision Assessment

Thermography inspection is a popular method that can provide enrichment data about PV device status. Typically, it is carried out by infrared radiation (IR) imaging sensor. Thermal vision assessment is a harmless and contactless monitoring technique and it can diagnose some of the defects and failures on the PV modules, connectors, AC or DC converter and panels. Furthermore, this method can be applied during

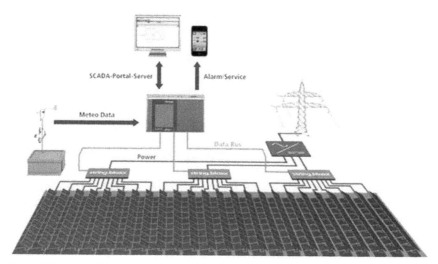

Fig. 3: Schematic of procedure of PV plants analysis and monitoring (Gantner Instrument 2012).

Fig. 4: (a) Multiple, (b) single by-pass failures and (c) Complete disconnected string with an empty module observed with IR sensor in some PV plants of about 1 MW.

normal operation of PV systems and it does not need to shut down the systems. The main task of thermography measurement is to find the defects or failures under temperature distribution of device.

Commonly, the thermovision assessment is carried out to identify open circuited modules, by pass diode problems (see Fig. 4), internal short circuits, potential induced degradation, delamination, complete or partial shadowing, cracks or micro-cracks, broken cell and hot spot. In addition, thermovision assessment can detect the local over heating in connectors, converters, transformers and some other devices.

2.4 PV Module or String I-V Curve Measurement

The main measurement parameters of PV modules are comprised of open-circuit, short-circuit current, fill factor and maximum power point.

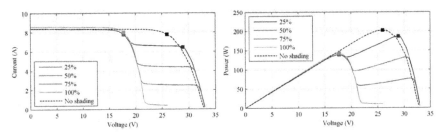

Fig. 5: I-V and P-V characteristics of mono-crystalline modules (referred to 1000 W/m²) as a function of the shading on cell.

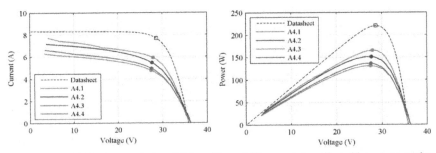

Fig. 6: I-V and P-V characteristics of PV multi-crystalline modules affected by micro-cracks.

I-V measurement curve gives sufficient information about PV module condition. Normally, the I-V curves is measured under Standard Test Condition (Cell temperature = 25°C, Irradiance = 1000 W/m², spectral distribution of irradiance AM = 1.5 G). Nonetheless, the PV modules do not meet the standard test conditions during the test measurement and usually the radiation is less than 1000 W/m², temperature is higher than 25°C and also usually, wind is blown with normal speed in outdoor. However, STC can be as an ideal reference for comparison of the PV modules in different conditions (Herman et al. 2012).

Figure 5 shows I-V and P-V measured characteristics reported to STC when some solar cells are subjected to shading. It can be noted the presence of bypass diode in the PV-module to avoid overheating of the PV-module itself, and also to reduce the loss in power generation due to shading (Dolara et al. 2013).

Figure 6 shows a comparison among the measured I-V curves and P-V curves of the 4 PV modules affected by micro-cracks defect and expected ones (Dolara et al. 2014, Dolara et al. 2016).

Currently, some commercial instruments allow measurement of I-V characteristic in real condition for string of 20 modules.

2.5 *Photoluminescence, Electroluminescence and UV Fluorescence Technique for PV Modules Analysis*

Photoluminescence (PL) and Electroluminescence (EL) are recent measurement methods which evaluates PV modules by luminescence images. In fact, the PL and EL measure the irradiative recombination of photons since carriers are excited into the solar cells. If this excitation is obtained by external injected current, the technique is called Electroluminescence. Otherwise, as the excitation is created by radiation incident, the technique is called Photoluminescence. Both measurement tests are non-destructive methods for PV module monitoring. Furthermore, Potential induced degradation, hot spot, white spot, cell finger metallization, humidity corrosion, cracks, micro-cracks, soldering, discoloration, snail trails and other defects and failure can be detected by means of these assessment techniques (Ebner et al. 2013, Potthoff et al. 2010).

Figure 7 shows the EL results for an area of a multi-crystalline module affected by microcracks. Also its thermal and visual images are reported. Black areas in EL images represent electrically separated sections. The positions of cell are indicated in terms of coordinate (row, column) within the PV module, e.g., position (1,1) is on the left, top. However, it is possible to identify a link among visual defects, hot parts, and electrically separated areas by comparison of these three images (Dolara et al. 2016).

Initially, UV fluorescence (FL) imaging technique was performed to detect EVA (Ethylene Vinyl Acetate) degradation. The most of materials degradation are found by using UV fluorescence imaging. FL imaging techniques is useful to detect cell micro-cracks but not along edge of the cell (see Fig. 8). However, EL technique is more appropriate for cell crack monitoring on the PV modules (Khatri et al. 2011).

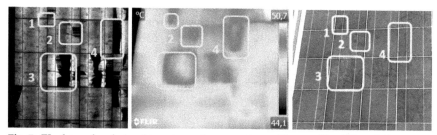

Fig. 7: EL, thermal and visual imagines of an area of a multi-crystalline PV-module (Dolara et al. 2016).

Fig. 8: PV cell monitoring using FL technique (No failure, cell cracks, insolated cell part and disconnected cells) (Köntges et al. 2014).

2.6 Novel Monitoring Methods

Figure 9 depicts a novel method including IR and visual images at the same time captured by the aerial perspective with direct measurements performed in the plant by using ad hoc instruments, and the collected data by the on-site monitoring system if available (Quater et al. 2014, Grimaccia et al. 2015). With a semi-automatic system (Aghaei et al. 2015), pictures of the georeferenced panel are reported in a failure map (Aghaei et al. 2016b, Aghaei et al. 2016a) which fully describes the plant status and it is possible to correlate performance indexes with parts of the overall plant in the same cases (Grimaccia et al. 2017).

This new monitoring method is used to provide quality assurance service for PV power plant in a short time and explore the defects or failures on PV system. The system can propose an appropriate solution for PV systems. In addition, the system can communicate with the ground monitoring cabinet and data acquisition system for analysis and evaluation of the plant electrical performance.

In this concept, the control system is integrated with multiple phases as inspection, recognition of the problem, processing, and decision

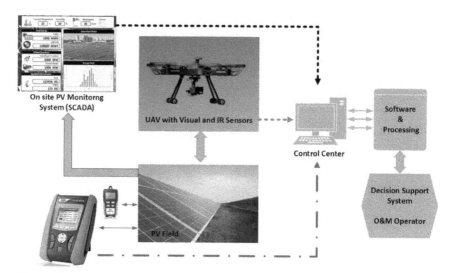

Fig. 9: Schematic view of the automated control monitoring system for PV power plants based on UAV.

making. Therefore, all the requirements for operation and maintenance are associated with such a system. As it is shown in Fig. 9, the visual and thermography assessments are carried out by mounted infrared sensor and visual camera on the UAV. Then, automatically the images captured are transferred to the Ground Control Station (GCS) by RF channel. Later on, the classified data are sent by the GCS to the database for the following processing. Subsequently, signal characteristics data (e.g., current, voltage, power, temperature, irradiation, energy production, etc.) are acquired by SCADA (Supervisory control and data acquisition). After performing the software analysis over the received data, all the processed information is transferred to the decision support system and O&M operator for future actions. In this phase, the system evaluates the information in order to detect particular defect or failure on the PV systems.

The capability of this new method is higher than the human inspector and other traditional methods. The time duration of monitoring procedure and decision making for the proper solution is faster than previous inspection methods, and the system can provide reliable information for the owner of PV plant in short time. The accuracy of the system is high, and it can identify not only defects or failures but also the location of the specifically degraded module in the PV plant since images captured by the UAV have GPS associated metadata.

The system can record all the monitoring information related to the PV system failures on the database for future actions based on the history of the plant, its track-record, and current performance. Moreover, the method is compatible with other standard methods for defects and failure detection (e.g., visual assessment, thermography analysis, etc.) and also with all of data acquisition systems.

Power forecasting methods can also be used to evaluate the correct working condition of PV plants (Leva et al. 2017, Faranda et al. 2014, Ogliari et al. 2016, Dolara et al. 2015).

3. PV Plants Failures

The failures in PV systems can be classified in two categories: those related to the overall PV system and those concerning single PV modules. Some of these failures occur because of transportation, installation, clamping, connector failures (fuse boxes, extension cables, inverters or combiner boxes) and lightning.

Figure 10 shows the main reasons for PV plant failures: they are mainly due to installations errors and planning & documentation errors. Among design errors, a very important cause of failure in PV plants is related to lighting and overvoltage systems. According to the literature, 30% of PV plants are subjected to this kind of problems in the first three years of operation. Figure 11 shows an example of damaging in PV module and inverter due to overvoltage and lightning.

The serious faults occurring during the PV plants installation (Leader et al. 2011) are due mainly to lack of heat dissipation in inverter and solar generator cabling not mechanically fastened. Some other faults can be due to the junction boxes or incorrect terminal connection in cables (see Fig. 12).

Figure 13 shows statistical distribution of reported faults for some general failure areas in PV systems during their life. It is important to underline that the results were acquired based on 3500 tickets which were issued for 350 PV systems by SunEdison in a period of 27 months. Figure 13

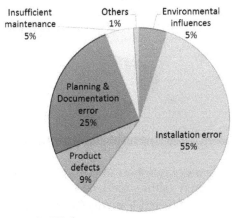

Fig. 10: Main failure causes in PV plants.

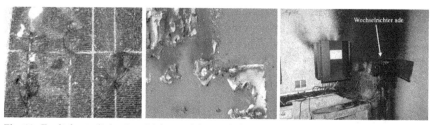

Fig. 11: Fault due to lightning: (a) front of the module, (b) back of the module and (c) inverter.

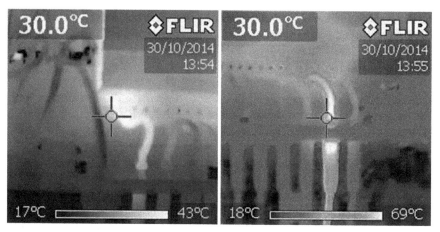

Fig. 12: IR analysis of a DC panel in 1 MW PV plant: it is evident an incorrect terminal connection in a DC cables.

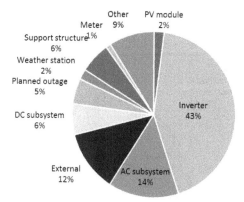

Fig. 13: Statistical distribution of general failure areas in PV systems.

illustrates that—in terms of numbers of occurrence not in terms of power or energy loss—the main failures occurred in PV inverter components rather than other subsystems in PV plants (King et al. 2000).

Although the number of PV modules faults is very low, PV modules failure represents a very important cause of power reduction. In fact, usually a great number of defects in modules is diffused in the PV plant and, when a problem occurs, the operation & maintenance operator once for all communicate it: this means one ticket for hundreds of PV module with defect. For examples in a PV (see plant B in (Grimaccia et al. 2017)) more than 75% of the considered 4010 PV modules were affected by snail trails or microcracks. Some of these modules have been analyzed and tested in (Dolara et al. 2016).

3.1 PV Modules

The solar module consists of PV cells, encapsulant, bypass diodes, connectors, frame, junction box, cable, glass on the front side of the module for protection, and glass or polymer film on the rear sheet of the module. These components can protect the cells against the climatic stress and various contacts (Ndiaye et al. 2013).

During the PV power plant operation, PV modules may be subjected to many different failures and defects (e.g., snail trails, hot spot, micro cracks, cell breakage, delamination, bubbles, yellowing, discoloration, oxidation, corrosion, etc.) due to the weather condition like wind, sand, humidity, high UV radiation and other internal and external stresses. However, most of these stresses cause power losses in the PV systems, hence investigating about inspection methods of PV module is a significant issue to identify the failures in the solar energy field. Thus, the lifetime of PV modules hugely depends on monitoring and maintenance and early defects detection can hamper to deploy of the degradation part on PV modules (Quater et al. 2014, Petrone et al. 2008).

Typically, any effect on PV module which decreases the performance of the module, or even influences on the module characteristics, is considered as a failure whereas, a defect can be defined as an unexpected or unusual thing which has not been observed before on the modules. However, defects often are not the cause of power losses in the PV fields (Golnas 2013, Grimaccia et al. 2017, Petrone et al. 2008).

The best performance of PV modules can be obtained at a standard temperature (25°C). Hence, any excess in PV module temperature can affect the standard performance. Temperature stress can accelerate the chemical degradation of the panels (Ferrara and Philipp 2012). Therefore, it can lead to creating defects in the PV modules, and the performance of PV systems may decline in a short time (Meyer and Van Dyk 2004).

According to the investigations, most visible failures appear in the PV modules due to the polymers' defects such as delamination, bubbles, cracking, or yellowing (Abdelhamid et al. 2014). Discoloration, oxidation, and corrosion of connectors' defects can also affect the PV modules' electrical characteristics, but their inspection is quite easy by visual observation. Other phenomena such as snail trails, shading, hot spots, micro-cracks, and cell breakage defects can have the highest influences on the performance of the PV modules. These kinds of failures can, in fact, be better detected also using thermal and infrared cameras (Grimaccia et al. 2017).

Adherence loss among PV modules' layers usually causes delamination. Typically, it happens between cells and front glass or between polymeric encapsulant and cells. This defect can increase reflection, and the water can then penetrate into the module itself. Nevertheless, a delamination defect in the borders of the PV module causes both electrical and installation risks and likely transmittance losses. On the other hand, bubble defect is more similar to delamination, while adherence losses occur only in some areas of PV modules due to chemical reactions. Bubble defects arise in the backside and not on the front side of the PV module. In fact, a bulk in the back cover or polymeric encapsulant prevents the dissipation of heat from the solar cells.

Yellowing and browning can appear in PV modules due to dry heat (e.g., due to desert climate), high UV radiations, and humidity. Moreover, it can occur because of insufficient adhesion between cells and glass material. However, this creates an obstacle between solar cells and sunlight, which leads to reduction in PV modules' voltage output.

The corrosion will occur in the PV modules' glass and metal because of the combination of gasses and humidity. Snow and wind can produce a higher static load; hence, they can break PV modules' glass due to the mechanical load for dynamic and static reasons. In the desert climate, for example, sand, wind, and dust significantly decrease the performance of PV modules (Ferrara and Philipp 2012, Meyer and Van Dyk 2004, Sánchez-Friera et al. 2011). Furthermore, glass breakage can be caused by bumps (see Fig. 14).

Micro-crack defects can appear as some different color lines on both sides of the PV module, and they can be detected only using special devices such as thermal and infrared cameras or by the optical method. Cracks and micro-cracks are formed in PV modules due to mechanical loads or during the process of lamination and soldering (Gabor et al. 2006, Reil et al. 2010). Cracks in solar cells can influence the performance of PV modules, thus investigation about the formation of the cracks is required (Dallas et al. 2007, Köntges et al. 2011, Grunow et al. 2005).

The PV modules can be subjected to a defect known as the snail trail phenomenon. The snail trail impact emerges on the PV modules' edge because of both environmental conditions and manufacturing process. They appear as dark and small lines or solar cell discoloration on the PV modules. Furthermore, snail trails can occur if the PV cell is produced as a

Fig. 14: Picture captured by a visual camera of a particular shock-defect.

thin thickness, and in this case, it cannot compromise efficiency too much (Meyer et al. 2013).

One of most significant defects is the hot spot phenomenon, which is defined as an area on the PV module with higher temperature. Typically, the reasons of the hot spot defect include mismatch of solar cells, partial shadowing, or any failure in the interconnection between the solar cells. Hot spots can easily be detected by thermal cameras. Figure 15 shows some hot spot detected by IR inspection and also the corresponding visual image.

Some failures occur within the first two years of the PV modules installation which impacts on the costs of PV modules installers and manufacturers because they should be responsible for these failures and defects. Figure 16 displays the statistic of common failure types in the first two years and is based on 2 million PV modules delivered by a German company between 2006 and 2010 (Köntges et al. 2014).

Optical failure, junction-boxes and cables defects are the main failure modes of PV modules within the first two-years after delivery.

Figure 17 illustrates the particular areas of defects in the PV modules based on the investigation of about 3500 tickets which were issued for 350 PV systems by SunEdison during the period of 27 months (Golnas 2013).

Furthermore, Table 1 reports the failure rates of most common observed defects, over the detected failures in PV modules, in the five heterogeneous plants inspected by the UAVs (Grimaccia et al. 2017).

Finally, according to the literature, PV modules with various technologies have different degradation rates annually. Table 2 summarizes

Fig. 15: Hot spot due to (a) corrosion, (b) dirty, (c) shading.

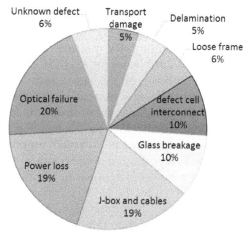

Fig. 16: Failure rates of PV modules in the first two year after delivery in 2006–2010.

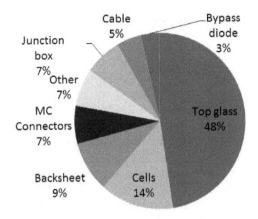

Fig. 17: Distribution of specific failure area in PV module.

Table 1: Failure rates of observed damaged modules by UAV inspection method.

Defect - Failure	% over only defected (24,254 total inspected modules)
Discoloration/browning	22.8%
Cracks in the cells	< 1%
Oxidation/corrosion	< 1%
Hot spot	12.05%
Shading	2.65%
Snail trails	20.73%
Dirty	3.24%
By-pass/disconnect.	10.92%
Other/minor	26.72%

Table 2: Degradation rate for different PV modules technologies.

Technology of PV module	Degradation rate (%/Year)
a-Si	1.34%
CdTe	1.70%
CIGS	1.86%
mono-c-Si	0.89%
Multi-c-Si	0.81%
Thin-Film	2.24%

the degradation rate in PV modules based on different technologies (Phinikarides et al. 2014).

3.2 PV Inverter and BOS Failures

The inverters are the most complicated and expensive component in PV systems, and they are considered as the brain of PV system (Petrone et al. 2008). The studies show that PV inverter is the weakest component of PV system as it is subjected to more failures and these faults of PV inverters are the main outage causes in PV plants (Kaplar et al. 2011, Navigant 2006). Therefore, the inverter failures are the most critical and frequent problems in PV plants. Typically, the failures in PV inverters are classified as electrical component failures, quality control problems, manufacturing and insufficient design. This inability of the PV inverter functionality can influence the entire PV modules connected to the PV system. Therefore, the failures in inverters are the greatest reason of lost productivity in PV plants (Golnas 2013). However, the inverter is one of the main component of PV systems so if the inverter gets off line, the error should be identified and corrected as soon as possible. Table 3 summarizes common inverter troubleshooting (Grunow et al. 2005).

Sandia National Laboratories reported that the IGBT should be considered as one of the main components of PV inverter which often are subjected to failure. In addition, Sandia National Laboratories stated that of the to 65% outage of 103 PV systems, 213 events were related to PV inverter failure (Kaplar et al. 2011). However, the failures of PV inverters are also dependent on the external stresses such as tropical situation and lightning effects (Ma and Thomas 2011). Furthermore, the mechanism of heat extraction should be considered in the design of inverters for cooling of switching components and capacitors due to the inverters are exposed to thermal and electrical stresses during the operation. Control software problem is one of other crucial factors of PV inverter failures which may occur at AC and DC side between the inverter and the grid or the PV strings (Kaplar et al. 2011, Ma and Thomas 2011). Figure 18 depicts the specific failure area in PV inverters according to the investigation of SunEdison (Golnas 2013).

Some problems in PV fields can affect transformers, switches in electrical panels, fuses in string boxes, wires, weather sensors, and so on. A statistical distribution of general failure is shown in Fig. 18. Figure 19 shows an example of a DC switches fault: this fault is not very common in terms of number but it is very important in terms of power generation losses. In this case, more than 12% of the PV plant was out of work. Faults frequently affected the fuses presents in string boxes: about one fuse per week have to be changed in a power plant of 1 MW of nominal power.

Table 3: Common inverter errors and failures.

General failures area	Solution
DC under voltage	Steps to diagnosing underperforming systems
DC over voltage	V_{oc} string testing
DC ground fault	Ground fault detection procedure
Gating fault	Check connections. Contact manufacturer
AC under voltage	Confirm all breakers are on Check ac voltage with voltmeter. If within range, perform a manual restart If outside of range, contact utility
AC overvoltage	Check ac voltage with voltmeter If within range, perform a manual restart If outside of range, contact utility
Low power	System is likely just shutting down because of lack of sun; if it is sunny, perform steps to diagnose underperforming systems
Over temperature fan not operating	Check power supply to fan if good, replace fan; if bad, replace power supply
Over temperature fan is operating	Check to confirm sensor readings—if bad, replace sensor; if good, investigate further
Over temperature fan is operating, sensors are accurate	Check intake and exhaust filters for excessive buildup, and clean or replace if necessary
Software fault	Contact manufacturer

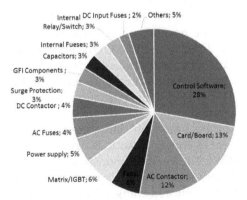

Fig. 18: Distribution of specific failure area in PV inverter.

According to research and analysis by Politecnico di Milano (SolarTechLab 2011) in over 20 PV plants with capacity of 1 MW, the results demonstrate that an error can occur on a pyranometer during the global radiation measurement up to 10% in comparison to a pyranometer in first class.

Fig. 19: DC switch fault.

4. Normalized Indices for PV Plants Comparison

PV systems with different configurations and placed in different locations can easily be compared by evaluating their normalized indices of system performance such as yields, losses and efficiencies. The amount of energy yields are normalized to the nominal power of the modules. Yields of the system are normalized to the area of the arrays and losses are calculated as the differences between the energy yields (CEI EN 61724 1999, IEA PVPS n.d.). The performance indices of systems connected to the distribution network of autonomous systems and hybrid systems may differ significantly because of their adaptation to loading and other special operating characteristics.

All the indexes are based on daily/monthly/annual mean yields. They are the quotient of energy quantities over the installed array's rated output power P_0 (kW). The yields demonstrate the amount of time during which the array of modules would be required to operate at P_0 to provide a particular amount of monitored energy. The energy yields indicate actual array operation relative to its rated capacity. The average daily energy performance is reported as the follows.

Array Yield

The Array Yield (Y_A) is calculated as the net daily/monthly/annual energy from the PV array $(E_{A,d})$ divided by the installed array's rated output power P_0.

$$Y_A = \frac{E_{A,d}}{P_0} = \tau_r \cdot \left(\sum_{day} \frac{P_A}{P_0} \right)$$

This yield indicates the daily/monthly/annual energy output per kW of the installed PV array. Y_A indicates the amount of time during which the array would be required to operate at P_0 to provide the monitored daily/monthly/annual energy.

Reference Yield

The Reference Yield (Y_R) can be calculated by dividing the total daily/monthly/annual in-plane irradiation $H_{I,d}$ (the monthly or yearly average value) by the module's in-plane reference irradiance G_{STC} (kW·m^{-2}):

$$Y_r = \tau_r \cdot \left(\sum_{day} \frac{G_I}{G_{STC}} \right)$$

This yield represents the number of hours per day/month/year during which the solar radiation would need to be at reference irradiance levels in order to produce the same incident daily/monthly/annual energy as it was monitored. Thus Y_R would be the number of peak sun-hours per day.

Final Yield

Final Yield (Y_F) is calculated dividing the total system output energy (E_{USE}) by the installed array's rated output power P_0.

The final PV system yield Y_F is the portion of the daily/monthly/annual net energy output of the entire PV plant which was supplied by the array per kW of the installed PV array:

$$Y_f = \frac{E_{use,r}}{P_0}$$

This yield indicates the number of hours per day/month/year that the array would need to operate at its rated output power P_0 to equal its monitored contribution to the total system output energy E_{USE}.

Performance Ratio

The Performance Ratio (PR) is calculated by dividing the Final Yield by the Reference Yield. This index indicates the overall effect of losses on the rated output power of the array:

$$PR = \frac{Y_f}{Y_r}$$

The *Performance Ratio* is a dimensionless quantity that indicates the amount of the net output energy of the PV system, compared with the theoretical one in input for a certain period. It doesn't represent the produced energy—in fact systems with low *PR* in a place with high radiation can produce more energy than a system with a high *PR* located in a place with little radiation, but the overall components efficiency by the actual operating conditions with the exception of the PV array.

The main factors of *PR* reduction are high temperature on the PV array, incomplete utilization of the irradiation and inefficiencies or failures of the system components.

4.1 Some Examples of Performance Indexes Evaluation

With respect to the aforementioned indexes, it is possible to assess the quality of PV plants, evaluate if they are in efficient condition or not, and also find some defects and/or the impact of a failure on the performance of the PV plant. In the following, the results of the analysis performed for different PV plants are reported with the aim of demonstrate how performance indexes can emphasize inefficiencies or failures of the system components.

4.1.1 Comparison of different technologies in PV module

The performance indexes were used to analyze a PV plant located in south of Italy made by six different fields. The total rated installed power is equal to 3.2 MWp, it has been working since May 2009 and it was characterized to use different kind of PV module (multi-crystalline and thin film cell technology) and inverter configurations (Faranda et al. 2010).

The main features of the field #1 to #5 are: CdTe thin film PV modules with a maximum 2 inverters/field. Polycrystalline silicon PV module and 7 inverters were used in field #6.

Table 4 shows the trends of the monthly *PR* for the thin film fields (#1 to #5, CdTe) and for the polycrystalline one (no. 6, SiPoli). Field #2 demonstrates the highest average of annual *PR*, the lowest deviation of the minimum monthly *PR* and the lowest deviation of the maximum *PR* from this value in comparison with other fields. The major differences are recorded for the polycrystalline in Field #6.

The low *PR* recorded for the field #6 in summer due to polycrystalline technology undergoes a greater influence of high temperature than the CdTe thin film. The first technology loses, on average, about 4–5% of *PR* against the second one. Indeed, the temperature coefficient for polycrystalline panels technology amounted to values of $-0.45\%°C^{-1}$ whereas, this value is equal to $-0.25\%°C^{-1}$ for CdTe modules. With regards to similar analysis for the winter months, especially for December, it is noticed that all the fields have recorded a lower *PR*. During this month, Field #6 has seen the

Table 4: Monthly PR, from April 2009 to March 2010, of the 6 PV fields. The minimum value for each month is highlighted in gray.

Month	Field 1 CdTe	Field 2 CdTe	Field 3 CdTe	Field 4 CdTe	Field 5 CdTe	Field 6 SiPoli
apr/09	91.75	93.83	90.95	87.87	91.82	91.24
may/09	89.79	93.37	91.43	91.33	91.20	89.27
june/09	93.66	92.00	91.34	90.48	90.28	86.90
july/09	91.94	92.38	91.81	83.70	90.39	86.70
aug/09	89.33	92.02	91.27	89.21	89.60	85.91
sept/09	91.46	91.66	90.95	87.25	93.99	88.17
oct/09	90.30	89.87	86.94	88.39	87.61	88.26
nov/09	88.42	91.25	89.14	87.63	87.48	85.20
dec/09	83.47	86.31	84.46	83.86	83.09	78.34
jan/10	87.44	89.85	86.25	87.29	85.95	81.54
feb/10	87.80	89.25	87.27	86.53	86.56	89.09
mar/10	91.10	91.93	87.28	88.70	89.23	91.65

lowest *PR* which was also lower than in comparison with other fields. In this case, the poor performance of Field #6 is due primarily to mutual shading between rows of photovoltaic modules. The mutual shadings are present in all the fields, but in the case of polycrystalline technology they are actually more burdensome. In fact, CdTe technology has a single diode for each cell, while the polycrystalline silicon modules used in Field #6 provide a total of three blocking diodes: in case of shading, the second type of modules cannot convert irradiance for a greater proportion than by using CdTe modules.

Figure 20 shows the output power measured from all inverters (WR E1 to WR E8) in Field #6 on the 13th of December, 2009. From Fig. 20 we can observe that early in the morning and at dusk, the conditions of more shading, the inverters WR E5 and E7 are the first to operate, while WR E6 is the last to get going nearly two hours later. The repetition of this phenomenon in most of other winter days effects on the PR in Field #6, rather than the other fields with CdTe modules.

4.1.2 Analysis of a PV plant during the first year of work

The performance indexes were also used to analyze a PV plant located in center of Italy made by multi-crystalline PV modules with a central inverter. The total rated installed power is equal to 662.7 kWp, it has been working since September 2012. Figure 21 shows the monthly *PR* during the first year of operation of the polycrystalline PV plant. Similar to the

FIELD 6 - 13/12/2009

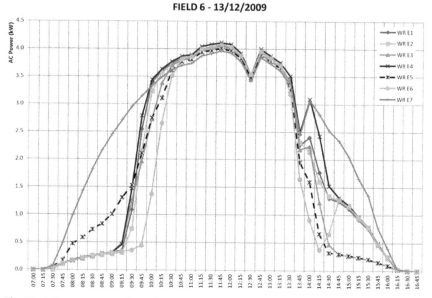

Fig. 20: AC power output from the inverters of Field #6 on 13/12/2009.

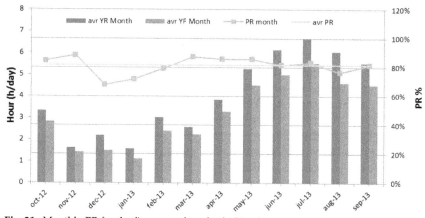

Fig. 21: Monthly PR for the first year of work of a PV plant in center of Italy.

previous case, for the winter months (especially in December) the PV filed has recorded the lower *PR*. This is primarily due to mutual shading between rows of photovoltaic modules. Furthermore, the *PR* is not so high in summer because of the influence of high temperature.

Moreover, the analysis of the daily operation shows a reduction of *PR* related to the operation of the inverter in the so called de-rating condition. In this case, Fig. 22 shows that the inverter converts a maximum power of about 300 kW corresponding to an irradiance of about 550 W/m².

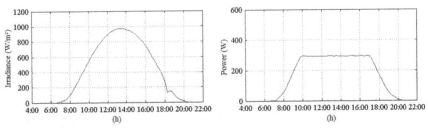

Fig. 22: Irradiance and output power in a day.

Nevertheless, higher irradiance is not converted into electrical power. Such behavior is evident in all the days characterized by high irradiation values and gives a lost in power conversion of the irradiance higher than 550 W/m². This behavior can be traced back to an error in the control software of the inverter made during the installation of the inverter itself.

4.1.3 Comparison of some PV plants after three years of work

The performance indexes were used to analyze six PV plants (SF01, 02, 03, 04, 05 and 08) located in center of Italy made by the same multi-crystalline PV module and inverters configuration (two inverters per each plant). Each PV plant has a capacity of 1 MWp roughly and it has been working since 2011. The only difference between the six plants is related to structures: in the plants SF01, 02 and 03 the modules were positioned at the minimum height of one meter from the ground; in SF04, 05 and 08 the minimum height is about 50 cm.

After three years of operation, the performance of the six PV plants has been evaluated. Figure 23 shows the monthly *PR* during 8 months

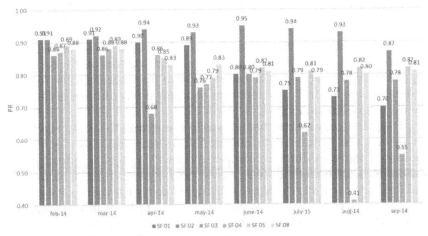

Fig. 23: Monthly PR for six PV plant in center of Italy.

Fig. 24: Fault in structure design: grass easily affects the performance in PV plant.

in 2014 for the six PV plants. According to the analysis of the PR trends follow this result.

- A low PR in SF03 in April and May. In these months some problems occurred in one of the two inverters: one inverter was out of order for about 1 week in April and two days in May.
- A very low PR in SF04 in July, August and September. One inverter was completely out of order in this period.
- A low PR in May in SF04 and SF05 in May because of the shadow due to high grass (see Fig. 24). As it was mentioned before, this could be considered as an error in structure design.
- A low PR since July in SF01 due to increasing number of strings out of work.

5. Conclusions

A growing demand of PV plants in the energy market shows a high potential of the photovoltaic technology to meet the future energy quality requirements. Effective operation and maintenance can guarantee the PV system, long term lifetime and payback for PV plants owner investment. Analysis of failure mode in PV plants should be performed for each part separately. However, in the literature, most of the researches just focused on the failures modes in PV modules and only a few of them considered the overall system analysis.

The main scope of this chapter was to summarize the typical methods for defects and failures detection in PV systems and also identify the significant faults in different components of PV plants. In addition, the various indication methods (e.g., Array Yield, Reference Yield, Final Yield and Performance Ratio) have been discussed above to evaluate the quality and performance of PV plants with some real examples in different seasons. Moreover, a detail statistical analysis has been presented for possible failures and defects of each component in PV systems individually.

The statistical distributions have been summarized according to many different analysis and experiences at small and large scale PV plants by various research institutions, universities, manufacturers and monitoring companies.

References

Abdelhamid, M., R. Singh and M. Omar. 2014. Review of microcrack detection techniques for silicon solar cells. IEEE J. Photovoltaics 4: 514–524. doi:10.1109/JPHOTOV.2013.2285622.

Aghaei, M., F. Grimaccia, C.A. Gonano and S. Leva. 2015. Innovative automated control system for PV fields inspection and remote control. IEEE Trans. Ind. Electron. 62: 7287–7296. doi:10.1109/TIE.2015.2475235.

Aghaei, M., A. Dolara, F. Grimaccia and S. Leva. 2016a. Image resolution and defects detection in PV inspection by unmanned technologies. 2016 IEEE Power Energy Soc. Gen. Meet. 0–4. doi:10.1109/PESGM.2016.7741605.

Aghaei, M., F. Grimaccia and S. Leva. 2016b. PV power plant inspection by image mosaicing techniques for IR real-time images. 2016 IEEE 43rd Photovolt. Spec. Conf. 3100–3105.

Buerhop, C., D. Schlegel, M. Niess, C. Vodermayer, R. Weißmann and C.J. Brabec. 2012. Reliability of IR-imaging of PV-plants under operating conditions. Sol. Energy Mater. Sol. Cells 107: 154–164. doi:10.1016/j.solmat.2012.07.011.

CEI EN 61724. 1999. Photovoltaic system performance monitoring. Guidelines for measurement, data exchange and analysis.

Dallas, W., O. Polupan and S. Ostapenko. 2007. Resonance ultrasonic vibrations for crack detection in photovoltaic silicon wafers. Meas. Sci. Technol. 18: 852–858. doi:10.1088/0957-0233/18/3/038.

Dolara, A., G.C. Lazaroiu, S. Leva and G. Manzolini. 2013. Experimental investigation of partial shading scenarios on PV (photovoltaic) modules. Energy 55: 466–475. doi:10.1016/j.energy.2013.04.009.

Dolara, A., S. Leva, G. Manzolini and E. Ogliari. 2014. Investigation on performance decay on photovoltaic modules: Snail trails and cell microcracks. IEEE J. Photovoltaics 4: 1204–1211. doi:10.1109/JPHOTOV.2014.2330495.

Dolara, A., F. Grimaccia, S. Leva, M. Mussetta and E. Ogliari. 2015. A physical hybrid artificial neural network for short term forecasting of PV plant power output. Energies 8. doi:10.3390/en8021138.

Dolara, A., G.C. Lazaroiu, S. Leva, G. Manzolini and L. Votta. 2016. Snail trails and cell microcrack impact on PV module maximum power and energy production. IEEE J. Photovoltaics 6. doi:10.1109/JPHOTOV.2016.2576682.

Ebner, R., B. Kubicek and G. Ujvari. 2013. Non-destructive techniques for quality control of PV modules: Infrared thermography, electro- and photoluminescence imaging. IECON Proc. Industrial Electron. Conf. 8104–8109. doi:10.1109/IECON.2013.6700488.

El-Ghetany, H.H., G.E. Ahmad, H.M.S. Hussein and M.A. Mohamad. 2002. Long-term performance of photovoltaic modules at different tilt angles and orientations. Energy Convers. Eng. Conf. 2002. IECEC '02. 2002 37th Intersoc. doi:10.1109/IECEC.2002.1392135.

Faranda, R., S. Leva, E. Ogliari and V. La Masa. 2010. Performance ratio of a PV power plant: Different panel technologies comparison. Proc. Sol. Energy Tech. 2010 12.

Faranda, R., H. Hafezi, S. Leva, M. Mussetta and E. Ogliari. 2014. Energy production estimation for suitable PV Planning. pp. 248–252. *In*: 2014 International Symposium on Power Electronics, Electrical Drives, Automation and Motion, SPEEDAM 2014. doi:10.1109/SPEEDAM.2014.6872104.

Ferrara, C. and D. Philipp. 2012. Why do PV modules fail? Energy Procedia 15: 379–387. doi:10.1016/j.egypro.2012.02.046.

Gabor, A., M. Ralli, S. Montminy, L. Alegria, C. Bordonaro, J. Woods and L. Felton. 2006. Soldering induced damage to thin Si solar cells and detection of cracked cells in modules. 21st Eur. Photovolt. Sol. Energy Conf.

Gantner Instrument. 2012. No Title [WWW Document]. PV Monit. Solut. URL http://www. gantnersolar.com/index.html.

Golnas, A. 2013. PV system reliability: An operator's perspective. IEEE J. Photovoltaics 3: 416–421. doi:10.1109/JPHOTOV.2012.2215015.

Grimaccia, F., M. Aghaei, M. Mussetta, S. Leva and P.B. Quater. 2015. Planning for PV plant performance monitoring by means of unmanned aerial systems (UAS). Int. J. Energy Environ. Eng. 6: 47–54. doi:10.1007/s40095-014-0149-6.

Grimaccia, F., S. Leva, A. Dolara and M. Aghaei. 2017. A survey on PV modules' common faults after an O&M flight extensive campaign over different plants in Italy. IEEE J. Photovoltaics.

Grunow, P., P. Clemens, V. Hoffmann, B. Litzenburger and L. Podlowski. 2005. Influence of micro cracks in multi-crystalline silicon solar cells on the reliability of PV modules. Proc. 20th EUPVSEC, WIP, Barcelona, Spain 2042–2047.

Herman, M., M. Jankovec and M. Topič. 2012. Optimal I-V curve scan time of solar cells and modules in light of irradiance level. Int. J. Photoenergy 2012. doi:10.1155/2012/151452.

IEA PVPS, n.d. IEA PVPS [WWW Document]. URL http://www.iea-pvps.org/.

Kaplar, R., R. Brock, S. DasGupta, M. Marinella, A. Starbuck, A. Fresquez, S. Gonzalez, J. Granata, M. Quintana, M. Smith and S. Atcitty. 2011. PV inverter performance and reliability: What is the role of the IGBT? 2011 37th IEEE Photovolt. Spec. Conf. doi:10.1109/PVSC.2011.6186311.

Khatri, R., S. Agarwal, I. Saha, S.K. Singh and B. Kumar. 2011. Study on long term reliability of photo-voltaic modules and analysis of power degradation using accelerated aging tests and electroluminescence technique. Energy Procedia 8: 396–401. doi:10.1016/j. egypro.2011.06.156.

King, D.L., J.A. Kratochvil, M.A. Quintana and T.J. McMahon. 2000. Applications for infrared imaging equipment in photovoltaic cell, module, and system testing. Conf. Rec. Twenty-Eighth IEEE Photovolt. Spec. Conf. - 2000 (Cat. No.00CH37036). doi:10.1109/ PVSC.2000.916175.

Köntges, M., S. Kajari-Schröder, I. Kunze and U. Jahn. 2011. Crack statistic of crystalline silicon photovoltaic modules. Eupvsec 26: 3290–3294. doi:10.4229/26thEUPVSEC2011-4EO.3.6.

Köntges, M., S. Kurtz, C.E. Packard, U. Jahn, K. Berger, K. Kato, T. Friesen, H. Liu and M. Van Iseghem. 2014. Review of failures of photovoltaic modules. IEA-Photovoltaic Power Systems Programme. doi:978-3-906042-16-9.

Leader, W., C. Assoc Theocharis Tsoutsos, Z. Gkouskos and S. Tournaki. 2011. Catalogue of common failures and improper practices on PV installations and maintenance. Eur. Photovolt. Ind. Assoc.

Leva, S., A. Dolara, F. Grimaccia, M. Mussetta and E. Ogliari. 2017. Analysis and validation of 24 hours ahead neural network forecasting of photovoltaic output power. Math. Comput. Simul. 131: 88–100. doi:10.1016/j.matcom.2015.05.010.

Ma, Z.J. and S. Thomas. 2011. Reliability and maintainability in photovoltaic inverter design. Proc. - Annu. Reliab. Maintainab. Symp. 0–4. doi:10.1109/RAMS.2011.5754523.

Meyer, E.L. and E.E. Van Dyk. 2004. Assessing the reliability and degradation of photovoltaic module performance parameters. IEEE Trans. Reliab. 53: 83–92. doi:10.1109/ TR.2004.824831.

Meyer, S., S. Richter, S. Timmel, M. Gläser, M. Werner, S. Swatek and C. Hagendorf. 2013. Snail trails: Root cause analysis and test procedures. Energy Procedia 38: 498–505. doi:10.1016/j.egypro.2013.07.309.

Navigant, N.R.E. 2006. A Review of PV Inverter Technology Cost and Performance Projections 1–100.

Ndiaye, A., A. Charki, A. Kobi, C.M.F. Kébé, P.A. Ndiaye and V. Sambou. 2013. Degradations of silicon photovoltaic modules: A literature review. Sol. Energy 96: 140–151. doi:10.1016/j.solener.2013.07.005.

Ogliari, E., A. Bolzoni, S. Leva and M. Mussetta. 2016. Day-ahead PV power forecast by hybrid ANN compared to the five parameters model estimated by particle filter algorithm, Lecture Notes in Computer Science (including subseries Lecture Notes in Artificial Intelligence and Lecture Notes in Bioinformatics). doi:10.1007/978-3-319-44781-0_35.

Petrone, G., G. Spagnuolo, R. Teodorescu, M. Veerachary and M. Vitelli. 2008. Reliability issues in photovoltaic power processing systems. IEEE Trans. Ind. Electron. 55: 2569–2580. doi:10.1109/TIE.2008.924016.

Phinikarides, A., N. Kindyni, G. Makrides and G.E. Georghiou. 2014. Review of photovoltaic degradation rate methodologies. Renew. Sustain. Energy Rev. 40: 143–152. doi:10.1016/j.rser.2014.07.155.

Potthoff, T., K. Bothe, U. Eitner, D. Hinken and M. Königes. 2010. Detection of the voltage distribution in photovoltaic modules by electroluminescence imaging. Prog. Photovoltaics Res. Appl. 18: 100–106. doi:10.1002/pip.941.

Quater, P.B., F. Grimaccia, S. Leva, M. Mussetta and M. Aghaei. 2014. Light Unmanned Aerial Vehicles (UAVs) for cooperative inspection of PV plants. IEEE J. Photovoltaics 4: 1107–1113. doi:10.1109/JPHOTOV.2014.2323714.

Reil, F., J. Althaus, W. Vaassen, W. Herrmann and K. Strohkend. 2010. The effect of transportation impacts and dynamic load tests on the mechanical and electrical behaviour of crystalline PV modules. pp. 3989–3992. *In*: EU PVSEC. Valencia, Spain.

Sánchez-Friera, P., M. Piliougine, J. Peláez, J. Carretero and M. Sidrach de Cardona. 2011. Analysis of degradation mechanisms of crystalline silicon PV modules after 12 years of operation in Southern Europe. Prog. Photovoltaics Res. Appl. 19: 658–666. doi:10.1002/pip.1083.

SolarTechLab. 2011. SolarTechLab [WWW Document].

Trancossi, M. 2011. Testing Performance, Weathering and Aging of Photovoltaic Modules.

CHAPTER 4

Tidal and Wave Power Systems

Luca Castellini,* Michele Martini and
Giacomo Alessandri

1. Introduction

Oceans, seas and rivers cover about 70% of the Earth's surface. By
definition, ocean energy includes resource associated to kinetic, potential,
thermal and chemical energy of the seawater. Most often, such resources
are used to generate electricity but there are other applications such as
production of desalinated water, cooling systems and production of
hydrogen through electrolysis. Other forms of energy, such as offshore
wind and solar energy from marine structures, are not considered part of
ocean energy as they do not belong intrinsically to seawater.

Two forms of ocean energy are especially promising in terms of energy
potential:

- Wave energy
- Tidal energy

Waves are a renewable energy source, as they derive from solar energy.
Indeed, sea waves are generated by a momentum exchange between the
ocean surface and the winds, which in turn result from a non-uniform
heating of air masses by the Sun. Storms generate local short-crested
wind waves that can travel long distances before reaching the shore as

Umbragroup spa, Via Valter Baldaccini 1, 06034 Foligno (PG), Italy.
Emails: lcastellini@umbragroup.com; mmartini@umbragroup.com;
galessandri@umbragroup.com

long-crested *swell waves*. Waves move offshore almost without energy losses, and dissipate energy at low water depths due to bottom-friction effects and wave breaking phenomena. Wave energy consists of both kinetic and potential energy.

Tides are the result of gravitational attraction of Earth, Sun and Moon that acts on the Earth's oceans. This causes the slow motion of large masses of water. Tidal energy consists of both potential energy, related to vertical variations of the sea level, and kinetic energy, related to the horizontal motion of the water column.

At first glance, power generation with wave and tidal energy has important advantages over other common renewable energy sources. First of all, they could allow powering coastal regions where 44% of the global population lives and where land availability for other renewable sources (wind, solar) is often scarce (United Nations 2016). Secondly, they are available throughout the day (unlike solar energy). Thirdly, they are highly predictable (2–3 days in advance for waves and weeks or months for tides, unlike wind energy). These points facilitate the delivery of electrical power to the grid, the integration with other sources of energy and the reduction of power cuts. In addition, the production of electricity from ocean energy is CO_2 free, unlike biomass and geothermal energy, and has limited visual impact on the shoreline. These factors make ocean energy very attractive, especially considering the worldwide energy potential.

2. The Energy Potential

For both wave and tidal resources, the energy potential is huge. Recently, the global wave power (theoretical) potential has been estimated as high as 2,100 GW (Gunn and Stock-Williams 2012), at 30 nautical miles from the coastline. The countries with the highest potential are Australia, the United States of America and Chile (see Table 1 and Fig. 1), as all of them have a very long coastline exposed to oceans. The highest resource is found in latitudes between 30 and 60 degrees, at a water depth higher than 40 meters.

Table 1: Global wave power potential, adapted from (Gunn and Stock-Williams 2012).

Continent	P (GW)	Country	P (GW)
North America	427 ± 18	Australia	280 ± 13
Oceania	400 ± 15	United States	223 ± 12
South America	374 ± 16	Chile	194 ± 11
Africa	324 ± 12	New Zealand	89 ± 6
Asia	318 ± 14	Canada	83 ± 7
Europe	270 ± 20	South Africa	69 ± 4

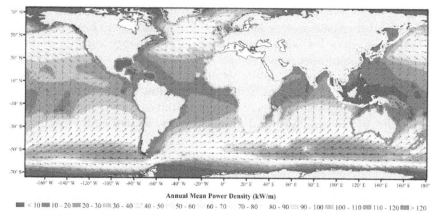

Fig. 1: Global wave energy potential, expressed as kW per unit length of wave crest (Gunn and Stock-Williams 2012).

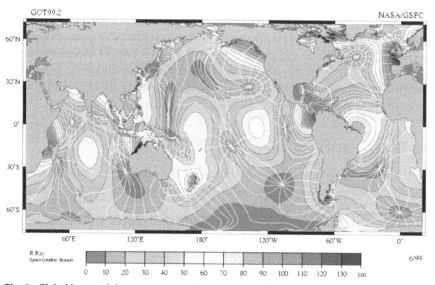

Fig. 2: Global lunar tidal range expressed in centimeters (Starobin 2007, Ray 1999).

The energy that can be theoretically harnessed, amounts to 29,500 TWh/year, whereof 1,000 TWh/year in Europe (International Renewable Energy Agency 2014a). The technical potential is lower, 2,000–4,000 TWh globally, according to other studies (Cruz 2008, Falcao 2010) and depends on many environmental and technological factors.

The tidal energy potential is largely unmapped and site-specific. However, the potential global resource is estimated to be approximately 3,700 GW (Munk and Wunsch 1998) (also see Fig. 2), which reduces to about

1,000 GW when considering its technically harvestable component (Lewis 2011). The theoretical energy potential was estimated as 7,800 TWh/year (Ocean Energy Systems 2011, International Renewable Energy Agency 2014b), which reduces to 1,200 TWh/year of technical potential (European Commission SETIS 2013). The potential for both wave and tidal energy is considerable, when compared to the global electricity consumption in 2013 that was 23,322 TWh (International Renewable Energy Agency 2016).

3. Methods for Energy Extraction

Wave and tidal energy can be transformed into usable electricity through a wide number of conversion principles. Nevertheless, for all of them the energy conversion chain can be generalized as illustrated in Fig. 3 (Penalba and Ringwood 2016):

Wave or tidal energy is first transformed into mechanical energy at the "absorption stage", where the kinetically and potential energy of the water particles is transferred to either a moving body or a physical medium (e.g., air).

At a second step, the "transmission stage", such mechanical energy is transferred to a mechanical transmission system or to a motor. The third step, "generation stage", converts mechanical energy into electrical energy using an electrical generator. The "transmission" and "generation" stages are typically performed by the so-called Power Take-Off (PTO) system.

Finally, when it is required, a fourth step ("conditioning stage") converts the electrical energy generated on the device into electrical energy with appropriate characteristics for feed-in into the electrical grid.

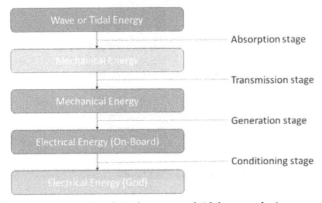

Fig. 3: The energy conversion chain for wave and tidal energy devices.

3.1 Wave Energy: Fundamentals

3.1.1 Conversion Principles

For what concerns electricity generation from sea waves, most of the devices developed in the last 20–30 years can be associated to eight extraction principles (see Fig. 4): attenuators, point absorbers, oscillating wave surge converters (OWSCs), oscillating water columns (OWCs), overtopping devices, pressure differential devices, bulge wave devices and other devices.

As illustrated in Fig. 5, the technology that gathered most of the R&D efforts (39%) is the point absorber, followed by oscillating water columns and attenuators.

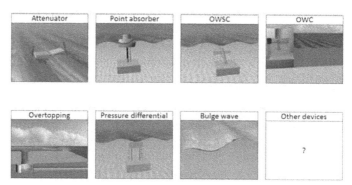

Fig. 4: Main wave energy converter concepts (adapted from Aquaret 2017).

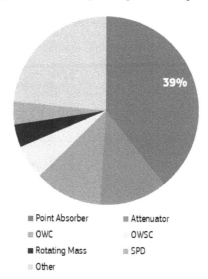

Fig. 5: R&D effort in wave energy (European Commission Joint Research Center 2016).

The mentioned extraction principles can be briefly introduced as follows.

- *Attenuators.* These converters use the energy of the incoming waves to induce a relative rotational motion between two or more adjacent structural components. Attenuators can be both surface floating and fully submerged and need to be moored to the seabed. Furthermore, they should be parallel to the main wave direction for working properly. As these systems are not restrained on the sea surface, they typically yaw automatically with the waves depending on the overall metocean conditions. Example: Pelamis device by Pelamis Wave Power;

- *Point absorbers.* In such devices, vertical hydrodynamic forces induce a relative heaving motion between a floating body and a secondary body. The latter can be either fixed, and thus connected to the sea bed through a foundation, or floating, and thus moored to the sea bed. Typically, the two bodies are axisymmetric and thus can receive energy independently of the wave direction. Example: PowerBuoy device by Ocean Power Technologies;

- *Oscillating wave surge converters (OWSCs).* These systems use the horizontal component of the wave particle motion to induce the rotation of one or more bodies relative to a secondary fixed or floating body. Oscillating wave surge converters are suitable for shallow-water locations, where the motion of the wave particles tends to have a large horizontal component. Example: WaveRoller device by AW-Energy Oy;

- *Oscillating water columns (OWCs).* These converters consist of a chamber, communicating with the sea water that is filled by water and air. Variations in wave height compress the air inside the chamber, which can be used to drive an air turbine. The water column acts therefore as a piston inside the chamber and the compressed air represents the medium between the waves and the PTO system (an air turbine, typically). Oscillating water columns may be both fixed on-shore and floating and moored offshore. Example: Marmok-A-5 device by Oceantec Energias Marinas;

- *Overtopping devices.* Sea water is collected into a reservoir slightly above the ocean surface and then flows through a low-head hydraulic turbine. These devices basically convert all the kinetic and potential energy of sea waves into potential energy. Overtopping devices may be either on-shore or floating offshore. Example: Wave Dragon device by Wave Dragon;

- *Pressure differential devices.* Such converters rely on the variations of hydrodynamic pressure caused by passing waves. They are typically submerged and may have two working principles: (1) the pressure differential is used to compress air that is sent to a turbine and (2) as for point absorbers Example: CETO device by Carnegie Clean Energy;

- *Bulge wave.* In this case, the converter consists of a flexible tube filled with a fluid and moored to the seabed. Pressure variations imposed by passing waves deform the tube and thus compress the fluid inside, forming a bulge wave that travels along the device and can drive a turbine placed at its end. Example: Anaconda device by Checkmate Energy;

- *Other devices.* Other concepts, such as floaters with an inner rotating mass (example: Penguin device by Wello Oy), floaters with a gyroscope (example: ISWEC by Wave for Energy) and point-pivoted buoys (example: Wavestar device by Wavestar A/S), have also been proposed (more than 1000 devices have been patented) (Clément 2002).

3.1.2 Power Take-Off systems

Any of the presented extraction principles aims at converting the energy of ocean waves into electrical energy. As mentioned, a relevant role in this process is given to the PTO system, which takes care of both the "transmission stage" and the "generation stage", see Fig. 3. At the moment, a large number of PTO systems have been proposed as a consequence of the very large number of extraction principles that can be found worldwide. As outlined in Fig. 6, typical PTO systems are:

- *Mechanical systems,* in which mechanical power is transmitted to a rotary electrical generator through a transmission system (e.g., a

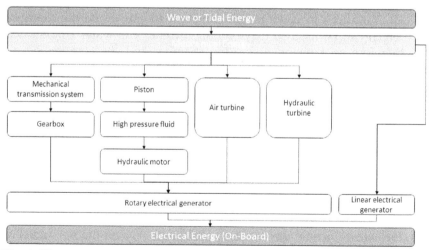

Fig. 6: Wave energy conversion methods and PTO systems, adapted from (European Commission Joint Research Center 2014).

lever mechanism, a ballscrew, etc.). These systems can be used for attenuators, point absorbers, submerged pressure differential devices, and OWSCs;

- *Hydraulic systems,* in which mechanical power is used to actuate a piston that compresses a hydraulic fluid. The fluid then expands on a hydraulic turbine connected to a rotary electrical generator. These systems can be used for attenuators, point absorbers, submerged pressure differential devices, and OWSCs;

- *Pneumatic systems,* in which a compressible fluid expands on an air turbine. In this case, during the "absorption stage", the energy of the water particles is transferred to a physical medium (e.g., air). These systems are typically used for OWCs and pressure differential devices;

- *Hydraulic turbines,* in which the flowing water is collected into a reservoir, and then the water flux actuates a hydraulic turbine. These systems are typically used for overtopping devices;

- *Linear electrical generators,* in which the mechanical energy is directly converted into electricity through linear induction systems. These systems can be used for attenuators, point absorbers, submerged pressure differential devices, and OWSCs;

All these PTO systems present a number of advantages and disadvantages. Some systems are inherited from other sectors (technology transfer) while other are being entirely designed for this specific application. It is important to keep in mind that a good power conversion train for wave energy should respond to the following characteristics:

- *Have a short energy conversion chain.* This is needed mainly for two reasons: first, any conversion step has intrinsic power losses; second, long-conversion chains typically require a large number of components which makes PTOs bulky and difficult to maintain in the long-term;

- *Be able to cope with high loads and low speeds.* Indeed, the dynamics of sea waves has such characteristics, which is far from ideal power generation conditions;

- *Be efficient in a wide range of working conditions.* Sea waves have variable seasonal patterns that give forces with very different amplitude and period throughout the year. This means that the working conditions of PTOs are extremely variable, and therefore energy conversion should also be efficient outside design conditions to maximize the amount of energy harnessed through the four seasons;

- *Be able to survive to high peak-to-mean load ratios.* This is indeed another typical characteristic of wave-induced loads;

- *Be able to survive the marine environment.* As these systems are often installed offshore to harness a larger energy resource, they should be designed for a lifetime of at least 20 years, with low-maintenance frequency as inspection, overhaul and repair activities may be dramatically costly and risky when taking place at sea. Corrosion effects typical of the marine environment increase the level of technical complexity as well.

3.2 Wave Energy: Relevant Devices and Pilot Systems

A number of pioneering companies deployed and operated pilot prototypes at sea over the last decade. Among them, the Pelamis Wave Power (European Marine Energy Centre 2017a) and Aquamarine Wave Power (European Marine Energy Centre 2017b) are worth mentioning. Unfortunately, these devices did not reach commercial maturity and the respective companies do not exist anymore. One of the main reasons behind these failures can be found in a "too big too soon" philosophy, where large capital-intensive projects have been developed to accelerate technology development without meeting performance requirements and often experiencing early failures at sea.

After the setbacks of Pelamis Wave Power and Aquamarine Power (to mention only two, but other companies also ceased trading in the same period), the wave energy sector faced a severe loss of confidence from private investors. This made other players in the sector humbler and set more achievable and realistic objectives when presenting any technology to the public or to investors. At present, public bodies and national governments have jumped-in so as to not lose the collected momentum, granting a long-term commitment to the sector and reviewing the targets for long-term installed capacity. This was for example the case of Wave Energy Scotland that was formed in 2014 at the request of the Scottish Government and financed 54 projects between 2014 and 2016, having awarded more than 22 million pounds (Wave Energy Scotland 2017).

Innovative and ambitious new players are developing several relevant projects nowadays. Successful pilot systems are proving the reliability and performance of new technologies, increasing the momentum of the wave energy sector. Among the most technologically-advanced projects are, Carnegie Clean Energy (which developed the CETO technology, a pressure differential device), Wello Oy (which developed the Penguin, an original inner rotating mass device) and AW-Energy (which developed the WaveRoller, an OWSC system). Other companies that deserve to be mentioned for their technical contribution include: Albatern, AWS Ocean Energy, Bio Power Systems, Eco Wave Power, Fred Olsen, Marine Power Systems, Ocean Power Technology, Seabased, Wave for Energy, and Wavepiston.

3.2.1 CETO (Carnegie Clean Energy)

Carnegie Clean Energy is an Australian company working in the field of renewable energy and micro-grids. It is the responsible for developing the CETO technology, a pressure differential device. The most important demonstration project carried out by Carnegie was the CETO5 Perth Wave Energy Project, where three 240 kW CETO devices were installed off Garden Island, Western Australia throughout 2015 (Fig. 7 and Fig. 8). The CETO5 device consists in a submerged buoy (11 m diameter) connected to hydraulic cylinders by a flexible tether, thereby creating a pumping system under the action of the waves. This was used to circulate a hydraulic fluid between the onshore and offshore component of the plant. Onshore, the pressurized fluid expanded on a hydraulic motor, which on turn rotated an electrical generator.

Fig. 7: Aerial view of the CETO5 device (Carnegie Wave Energy 2017).

Fig. 8: Aerial view of the CETO5 array (Carnegie Wave Energy 2017).

The project achieved over 14,000 hours (Carnegie Wave Energy 2017) of cumulative operation across the 4 seasons, making it the longest continuous period of operation of any wave energy project. Currently, the company is developing the CETO6 device, a 1 MW system that aims at producing electricity on board through a buoy with a diameter of 20 m. Carnegie aims at installing a prototype in the Wave Hub test-site in Cornwall, England and off Garden Island, Western Australia (Carnegie Wave Energy 2017).

3.2.2 Penguin (Wello Oy)

Founded in 2008, Wello Oy is a Finnish company that developed the unique Penguin device. This system captures the rotational energy generated by the motion of an asymmetrical hull, which rolls, heaves and pitches under the action of the waves. This motion induces the rotation of a spinning flywheel located inside the hull, which turns a direct drive electrical generator to produce electricity exported through a subsea cable. The Penguin device has two advantages compared to other systems: first, the flywheel tends to turn always in the same direction that improves the efficiency of the electrical generator; second, the PTO system is entirely located inside the hull that avoids problems related to the marine environment (corrosion, bio fouling). A 1 MW grid-connected device was installed in Spring 2017 at the EMEC test-site in Orkney, Scotland (European Marine Energy Centre 2017c). The hull is around 30 m long and 9 m high, and weighs approximately 1,600 tonnes (see Fig. 9 and Fig. 10).

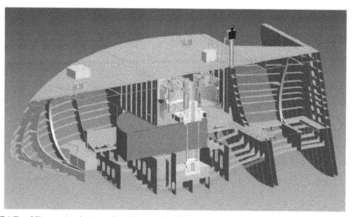

Fig. 9: CAD of Penguin device developed by Wello Oy (Wello 2017).

Fig. 10: View of Penguin device developed by Wello Oy (Wello 2017).

3.2.3 WaveRoller (AW-Energy Oy)

The company AW-Energy Oy was founded in Finland in 2002 and designed an OWSC device named WaveRoller. This consists of a number of plates anchored at the sea bottom, where the back and forth motion of tidal surge moves the plates, transferring kinetic energy to a piston pump. A system with three–100 kW flaps operated in Peniche, Portugal between 2011 and 2013 (see Fig. 12). The three flaps had a size of 4.5 m by 3.5 m, and the foundation was installed on the seabed filling the lateral tanks with water. The system survived waves with a significant wave height of 5 m, and was able to produce 500 kWh in 24 hours with a wave height of approximately 2.5 m. After these tests, the WaveRoller obtained an external performance verification by DNV-GL, the first of its kind in wave energy (Maki et al. 2014).

3.3 Tidal Energy: Fundamentals

Tidal energy can be harness essentially through two principles: tidal stream devices, that harness the kinetic energy of sea currents, and tidal range devices, which rely on the potential energy resulting in different water levels between high and low tides.

Tidal range devices were demonstrated to be a reliable technology, and large power plants of this kind already exist today. Among them, it

Fig. 11: View of Waveroller device developed by AW-Energy Oy before installation (AW-Energy Oy 2017).

Fig. 12: View of Waveroller device developed by AW-Energy Oy during installation (AW-Energy Oy 2017).

is worth remembering the Rance (France, 240 MW, operating since 1966) and the Sihwa Lake plant (South Korea, 254 MW, operating since 2011). As tidal range devices can be considered an established technology, these systems are not discussed in this chapter.

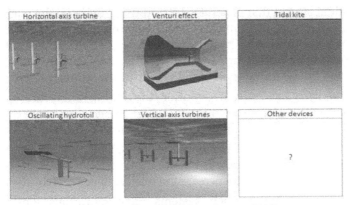

Fig. 13: Tidal stream devices concepts (adapted from Aquaret 2018).

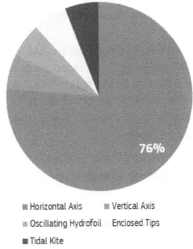

Fig. 14: R&D effort for tidal technologies tested at full scale (European Commission Joint Research Center 2016).

As opposite, tidal stream devices are still under development. Research and industrial players are investigating six different extraction principles: horizontal axis turbines, Venturi effect systems, tidal kites, oscillating hydrofoil, vertical axis turbines and other devices (see Fig. 13).

As illustrated in Fig. 14, the technology that gathered most of the R&D efforts (76%) is the horizontal axis turbine.

Each of the mentioned extraction principles can be briefly introduced as follows.

- *Horizontal axis turbines*. This system works under the same principle of wind turbines, with the difference that energy is contained in seawater, a 1000 times denser medium compared to air. A turbine is placed in the tidal stream that generates lift and drag forces on the blades. Two main effects result from this water-blade interaction: a torque, which makes the turbine spin and drive a rotary electrical generator, and a trust, which gives important forces on the structure that must be reacted by the foundation of the system at the seabed;

- *Venturi effect devices.* In this solution, the tidal stream is directed inside a duct which concentrates the flow and creates a pressure difference that can be exploited either by a water or an air turbine.

- *Tidal kites.* A tidal kite is moored to the seabed and carries a turbine below the wing. The kite is able to move into the tidal stream, on a figure-of-eight-shape to increase the relative speed seen by the turbine;

- *Oscillating hydrofoil.* Such systems have hydrofoils which move upwards and downwards normal to the tidal stream. The resulting oscillating motion is used to produce power that can be harnessed by a hydraulic cylinder or any other system able to convert linear motion into electricity;

- *Vertical axis turbines.* These turbines use the same principle as horizontal axis turbines, but the rotor spins around an axis that is normal to the seabed. A turbine is placed in the tidal stream and rotates to generate power through a rotary electrical generator;

- *Other devices.* These include systems not listed above, such as those based on undulating carpets.

While wave energy devices have not reached convergence towards a single technological solution, tidal devices are converging towards the use of horizontal axis turbines, through a path similar to that of wind technology. Therefore, there is a strong convergence towards the use of direct drive rotary electrical generators for tidal devices. Nevertheless, it is interesting to see how emerging innovative concepts (such as those based on undulating carpets) are involving linear or direct drive electrical generators.

3.4 Tidal Energy: Relevant Stream Devices and Pilot Systems

Tidal energy has been rapidly taking off during the past few years. The first tidal farms have been commissioned and constructed. Among the companies that have made giant steps, it is important to include Atlantis Resources Ltd., Nova Innovation Ltd., Sabella, Schottel, OpenHydro, Scotrenewables and EEL-Energy. It is worth detailing the activities of Atlantis Resources Ltd., which is developing one of the most ambitious projects using horizontal-axis devices, and EEL-Energy, which developed a promising innovative concept that falls beyond the classical categories of tidal energy.

3.4.1 MeyGen Project (Atlantis Resources Ltd.)

Atlantis Resources Ltd. is a Scottish company that developed one of the most advanced horizontal axis tidal turbines. In 2010, the Crown Estate awarded an agreement for lease to the project MeyGen, to develop a tidal stream project of up to 398 MW between the northern coast of Scotland and the Island of Stroma. The project was split into several phases: 1A,

Fig. 15: One of the 1.5 MW turbine of the MeyGen project during installation - lifting (Atlantis Resources Ltd. 2017).

Fig. 16: One of the 1.5 MW turbine of the MeyGen project during installation – deployment of foundation (Atlantis Resources Ltd. 2017).

1B, 1C, 2 and 3. Phase 1A of the project involved the deployment of four 1.5 MW turbines (three developed by Andritz Hydro and one by Lockheed Martin), as a precursor to the development of other 86 MW. Each turbine is located on an individual foundation that weighs between 250 and 300 tonnes, coupled with six ballast blocks (1,200 tonnes) that provide horizontal stability to the system. Each turbine has a subsea array cable laid on the seabed, brought to the shore through a directionally drilled borehole. The turbines are connected to the onshore grid, where the low voltage supply is converted to a 33 kV level for the network.

Phase 1B will involve the deployment of four additional 1.5 MW turbines, to further demonstrate the technology. Phase 1C will build additional 49 turbines, and the farm will finally reach the planned 389 MW capacity through Phases 2 and 3 (Atlantis Resources Ltd. 2017).

3.4.2 EEL (EEL Energy)

A very peculiar and innovative device is being developed by EEL Energy, a small French start-up that was recently granted important financial support. This concept consist in a carpet that undulates under the action of the tidal stream. A cable, shorter than the carpet, connects the carpet ends and ensures that the undulating motion takes place. Energy is extracted through a number of ballscrew generators placed on top of the carpet, which harness the relative motions given by the undulation. This innovative concept is at an early stage of development but has already shown important experimental results (EEL Energy 2017). A prototype is now being built for installation at the SEENEOH test site in Bordeaux (France) in 2017 and off the coast of France in 2018.

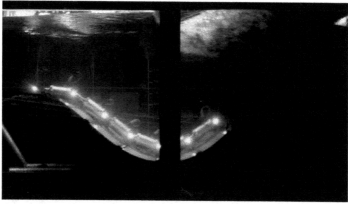

Fig. 17: The tidal undulating membrane developed by EEL during flume tests (Ocean Power Matrix 2017).

Criticalities

At the moment, wave and tidal energy devices are not economically competitive with respect to other renewable energy sources. The Levelized Cost of Energy (LCOE) is in the range 340–650 €/MWh for wave devices (International Renewable Energy Agency 2014) and 250–470 €/MWh for tidal devices (International Renewable Energy Agency 2014b). These are still high, especially if compared with offshore wind which has already reached a price of 106–176 €/MWh and various GW of global installed power (International Renewable Energy Agency 2015). As the installed capacity increases, for both technologies, costs are expected to reduce as illustrated in Fig. 18 and Fig. 19.

The setback of some of the major technology developers has also decreased the initial, large interest of private investors to the wave and tidal energy sector. Currently, large public investments are taking place to facilitate the commercialization of such technologies and bring back the interest of private investors. The development of wave and tidal energy systems has been slowed down by a number of criticalities related to the marine environment, to technological requirements, to survivability issues, to control and modelling difficulties. Some of these criticalities are shared between wave and tidal systems, while others belong to each specific technology. It is however important to mention that tidal systems are

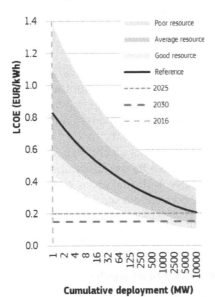

Fig. 18: Trend for LCOE of wave farms with cumulative deployment (European Commission Joint Research Center 2016).

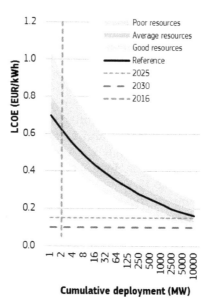

Fig. 19: Trend for LCOE of tidal farms with cumulative deployment (European Commission Joint Research Center 2016).

being developed at a faster pace and are likely to reach commercialization earlier than wave energy systems.

3.5 The Marine Environment

Both wave and tidal devices typically have structural steel sub-systems that are installed in the marine environment. This means that such components must be designed to prevent marine corrosion. Corrosion is an electro-chemical process that converts a refined metal into a more chemically stable form such as its oxide, sulfide or hydroxide. This process can have a rapid impact on the structural strength of the system and thus dramatically reduce its reliability and survivability. Appropriate protection measures have to be taken in order to ensure a long-lasting service to any metallic component working in water. These include:

- Protection by painting, which isolates the metallic surface from seawater;
- Cathodic protection, which enable changing the system potential through the use of sacrificial anodes;
- Inhibition, which reduces the corrosion rate using a chemical compound;
- Use of corrosion resistant alloys.

Fig. 20: Example of biofouling - Current measurement instrument encrusted with zebra mussels (Wikipedia 2017).

Though these solutions are already well-known in other marine-related fields (e.g., naval or oil and gas industry), their application for renewable energy applications is not trivial as they often concern moving parts and they might increase capital costs which are already higher than many other renewable sources.

A second important challenge brought by the marine environment is the possible onset of biofouling. This process consists in the accumulation of biologic material on a wetted surface. The major problem arising from the buildup of biofouling on marine systems is represented by the consequent increase of hydrodynamic volume and friction. As an example, for the shipping industry, this can significantly increase the use of fuels. For marine energy devices, this may affect the conversion efficiency and the loads acting on the moving bodies. Research activities are currently going on to understand this topic and quantify its impact. Anti-fouling solutions include the use of biocides, non-toxic coatings or energy methods. For what concerns both corrosion and biofouling, finding a tradeoff between reliability and affordability is of uppermost importance for the marine energy sector.

A third relevant challenge for offshore devices is given by the installation of structures out at sea. This involves complex and risky operations, including towing and load-out of heavy structures that must take place respecting rigorous health and safety standards. Such activities typically take place using large, expensive vessels borrowed from other sectors (offshore wind, nautical, oil and gas). To drive down installation costs per unit installed power, device developers tend to increase the rated power of the prototypes for sea trials. However, this enhances the technological challenges of such systems and pushes the limits of engineering design.

3.6 Technology Requirements

Wave and tidal systems share also some technology requirements. In most cases, moving parts in water are needed to convert hydraulic energy into mechanical energy. Therefore, appropriate sealing systems are required to ensure no external agents enter the conversion system and pressurized fluids (if present) do not leak out, while not affecting the overall performance of the system. Besides seals, bellows and telescopic cylinders may also be used depending on the application.

The technology requirements for wave energy systems deserve some more time, as they are difficult to overcome and this sector has been slowing down for many years. This is mainly due to the intrinsic properties of sea waves. Firstly, a great challenge is set by the slow-motion and high-force nature of waves, opposite to what is typically required for power generation (high speed and low force). Appropriate energy conversion systems (gearboxes, lever mechanisms, etc.) are thus necessary, which decrease the overall system efficiency and reliability and increase the system complexity. Secondly, waves are periodic and thus typically induce oscillating forces with frequent zero-crossings, for which energy conversion is not efficient. In addition, waves induce large peak-to-mean forces in the short-term (and thus large peak-to-mean power ratios), which impose to design large mechanical and electrical systems for small average power flows. To overcome these challenges, great efforts are being put into the development of PTOs with appropriate control systems, able to convert the mechanical energy of waves into electricity.

Finally, it is important to remark that often both wave and tidal technologies consist of floating structures, which must be moored with appropriate mooring systems and anchors to secure the devices in place especially in case of stormy weather.

3.7 Reliability and Survivability

As the typical design life of wave and tidal devices is twenty years, such systems must withstand with high reliability the corresponding long-term loading conditions. All the components should be designed to survive both extreme events and fatigue loading. While this is fairly a standard engineering process for tidal devices, as tide variations can be predicted in a deterministic way, the same does not happen for wave devices. Indeed, waves are not predictable in the long term and extreme loads are typically evaluated using statistical models and considering waves with a 50-years or 100-years return period. Besides possessing a certain level of uncertainty, extreme loads can be several times higher than operational loads and thus represent a great challenge from the design point of view. Furthermore, such systems should be designed for a long fatigue life— considering an average wave period of 10 seconds, this results in more than 60 million cycles during the device lifetime.

Reliability is another key-factor for wave and tidal energy devices. It can be defined as the level of confidence in the device to perform to design specifications without interruption to operation. Increasing the time between failures and reducing the down time of components and sub-systems is essential to decrease the LCOE of the technology. In first place, when a fault occurs, energy production is interrupted and thus the same happens for the associated incomes. Having good maintenance procedures in place would allow restoring energy production sooner. In second place, repairing components installed at sea is costly as it relies on the use of expensive vessels. Weather conditions can also increase the accessibility and downtime, thus delaying the restoration of energy production. Increasing reliability levels of wave and tidal devices are therefore fundamental to drive down its associated costs.

3.8 Control and Power Conditioning

The purpose of a control system is to make the wave or tidal energy converter behave so that its components meet optimal performance specifications.

Tidal devices have the advantage of inheriting the control technology already used for wind turbines. For variable-speed generators, this consist in using a torque-control below rated flow speed, to optimize power production, and torque-control and blade-pitch control above rated flow-speed, to regulate the electrical power output. For fixed-speed generators, power regulation typically occurs through passive stall at the rotor blades. In both cases, such strategies result in the well-known power curve typical of wind turbines.

The control problem in wave energy is much more complex. The large fluctuations in the wave elevation, and the non-causal relationship between wave elevation and wave force, make it difficult to identify an optimal control strategy—also considering the wide variety of wave energy concepts introduced in Section 3.1.1. In addition, strongly non-linear phenomena, such as hydrodynamic friction or excitation forces and PTO saturation characteristics, increase the complexity in the definition of suitable control laws. Powerful algorithms such as Model Predictive Control (MPC) and pseudo-spectral methods are gaining confidence in application to many systems (Bacelli and Ringwood 2015, Hals et al. 2011).

3.9 Modelling and Design

The modelling and design of marine energy systems presents several difficulties. In first place, at present, there are no existing standards dedicated to the design of a marine energy converter. A small number of documents were redacted to preliminary define the procedures for the certification of wave and tidal energy converters; these include:

- Det Norske Veritas, Certification of Tidal and Wave Energy Converters, October 2008;
- Bureau Veritas, Certification Scheme for Marine Renewable Energy Technologies, November 2016.

Secondly, another difficulty is represented by the uncertainty of site-specific wave and tidal current extreme conditions. Extreme loads are typically extrapolated through statistical techniques based on a certain return period of wave or current conditions. This brings a certain level of uncertainty in the evaluation due to extreme environmental conditions at sea. The combination of physical and numerical models may represent high-value yet low-cost means to overcome such difficulties.

3.10 Technology Development

An important criticality for both wave and tidal systems is related to the technology development path chosen by the developers. A very critical gap is represented by passing directly from small-scale tank tests to sea trials. This step has often been a challenge for most of the players from the engineering, economic and licensing points of view. Technically, sea trials involve testing devices under real conditions that are often far from those that can be tested in a wave tank. Costs can also be very high and difficult to predict, especially for what concerns the very first deployment of a new technology. In addition, one should consider that granting licenses and permits for experimental tests at sea can be very difficult, especially in those countries that do not have a legislator framework flexible enough. To partially overcome some of these issues, it is very common today to consider using "test sites" for early sea trials. These locations often dispose of pre-consented offshore areas, providing grid connection, mooring system and marine services for sea trials. This can dramatically reduce the risks, timing and costs for advancing the readiness level of marine devices. Among the most important test sites worldwide, it is important to mention the European Marine Energy Centre (EMEC) in Scotland, the BIMEP test site in the Basque Country, the SEM-REV test site in France, the PLOCAN test site in Spain and the WaveHub in England.

4. Future Scenario

There is a strong ambition worldwide to bring wave and tidal energy to commercialization. Both national and international public bodies are investing large amounts of money to learn from past errors and increase, through small steps, the technology readiness of the most promising technologies. In addition, technology developers are being more cautious and trying to build confidence within private investors. Technology development is taking place at a decreased pace, often through the help of third-party certification bodies and based on a risk-reduction philosophy (rather than a first-to-market approach). Efforts are being put also in reducing the gap between laboratory tests and sea trials, which is a recognized bottleneck in technology development. At the same time, companies are eying markets with large energy costs, such as isolated communities or islands, to test and bring to market their technologies. National and international agencies are developing ocean energy roadmaps to accurately shape the growing marine energy sector (Ocean Energy Forum 2016). The successful development of a healthy marine energy sector is envisaged to be dependent upon the following factors:

- *Long-lasting commitment of public bodies.* Financial commitment is needed to help developers increasing the readiness level of their devices while reducing the economic risks. Possible policy measures are feed-in tariffs, investment tax credits, power purchase agreements. Also, specific planning and licensing procedures should be adopted at regional level to speed-up the time-to-sea of different technologies;
- *Technology transfer from academia to the industry.* The research activities at academic level should find its continuation in the development of products at industrial level. The collaboration between universities and industry can help in building an integrated value chain which may speed-up the technological development of such complex systems;
- *Development and test of disruptive sub-systems to favor a step-change.* The need for affordable yet highly reliable systems implies that innovative technological solutions should be pursued, aiming to go beyond

state-of-the-art technology and simple technology transfer; These for example include the development of alternative WEC concepts, such as those based on the use of electroactive elastomers, which convert large deformations to electricity (see Fig. 21).

- *Evaluation of environmental and social benefits together with economic benefits.* Very preliminary studies stated that marine energy has limited environmental impact and induces benign landscape implications. In fact, as the harvesting devices absorb part of the wave energy, the waves hitting the shoreline are reduced, thus decreasing erosion phenomena. On the other hand, foundations and buoys have unknown impacts on the marine life. Similarly, the electromagnetic fields produced from sea cables may influence a range of marine organisms. Further studies are needed, together with data gathered from the operation of full-scale devices.

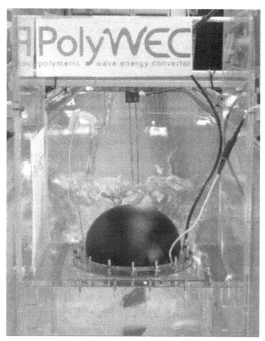

Fig. 21: Use of electroactive elastomers for electricity generation – application to an oscillating water column (Polywec 2017).

References

Aquaret. 2018, April. Retrieved from www.aquaret.com.

Atlantis Resources Ltd. 2017, April. Retrieved from: www.atlantisresourcesltd.com.

AW-Energy. 2017, April. Retrieved from www.aw-energy.com.

Bacelli, G. and J. Ringwood. 2015. Numerical optimal control of wave energy converters. IEEE Trans. Sustainable Energy.

Carnegie Wave Energy. 2017. Annual Report 2016.

Clément, A.e. 2002. Wave energy in Europe: current status and perspectives. Renewable and Sustainable Energy Reviews, 405–431.

Cruz, J. 2008. Ocean Wave Energy - Current Status and Future Perspectives. Springer.

EEL Energy. 2017, April. Retrieved from www.eel-energy.fr.

European Commission Joint Research Center. 2014. 2014 JRC Ocean Energy Status Report.

European Commission Joint Research Center. 2016. 2016 JRC Ocean Energy Status Report.

European Commission SETIS. 2013. Ocean Energy: Technology Information Sheet.

European Marine Energy Centre. 2017a, April. Retrieved from http://www.emec.org.uk/about-us/wave-clients/pelamis-wave-power/.

European Marine Energy Centre. 2017b, April. Retrieved from www.emec.org.uk/about-us/wave-clients/aquamarine-power/.

European Marine Energy Centre. 2017c, April. Wello Oy. Retrieved from http://www.emec.org.uk/about-us/wave-clients/wello-oy/.

Falcao, A. 2010. Wave energy utilization: a review of the technologies. Renewable and Sustainable Energy Reviews 14: 899–918.

Gunn, K. and C. Stock-Williams. 2012. Quantifying the global wave power resource. Renewable Energy 44: 296–304.

Hals, J., J. Falnes and T. Moan. 2011. Constrained optimal control of a heaving buoy wave-energy converter. Journal of Offshore, Mechanical and Arctic Engineering.

International Renewable Energy Agency. 2012. Renewable Energy Technology: Cost Analysis Series - Wind Power.

International Renewable Energy Agency. 2014a. Wave Energy - Technology Brief.

International Renewable Energy Agency. 2014b. Tidal Energy - Technology Brief.

International Renewable Energy Agency. 2015. Renewable Power Generation Costs in 2014.

International Renewable Energy Agency. 2016. Key world energy statistics 2015.

Lewis, T. 2011. Ocean energy. *In*: IPCC Special Report on Renewable Energy Sources and Climate Change Mitigation. Cambridge and New York: Cambridge University Press.

Maki, T., M. Vuorinen and T. Mucha. 2014. WaveRoller - One of the leading technologies for wave energy conversion. International Conference on Ocean Energy. Halifax, Canada.

Munk, W. and C. Wunsch. 1998. Abyssal recipes II: energetics of tidal and wind mixing. Deep Sea Research Part I: Oceanographic Research Papers 45: 1977–2010.

Ocean Energy Forum. 2016. Ocean energy strategic roadmap 2016, building ocean energy for Europe.

Ocean Energy Systems. 2011. An International Vision for Ocean Energy.

Ocean Power Matrix. 2017, April. Retrieved from www.oceanpowermatrix.com.

Penalba, M. and J. Ringwood. 2016. A review of wave-to-wire models for wave energy converters. Energies.

PolyWEC. 2017, April. Retrieved from www.polywec.org.

Ray, R. 1999. A global ocean tide model from TOPEX/POSEIDON altimetry: GOT99.2. National Aeronautics and Space Administration, Greenbelt, Maryland.

Starobin, M. 2007. TOPEX/POSEIDON: revealing hidden tidal energy. (National Aeronautics and Space Administration). Retrieved from http://svs.gsfc.nasa.gov/stories/topex/tides.html.

United Nations. 2017, April. Human settlements on the coast. Retrieved from http://www.oceansatlas.org/servlet/CDSServlet?status=ND0xODc3JjY9ZW4mMzM9KiYzNz1rb3M~.

Wave Energy Scotland. 2017, April. Retrieved from http://www.waveenergyscotland.co.uk.

Wello Oy. 2018, April. Retrieved from www.wello.eu.

Wikipedia. 2017, April. Biofouling. Retrieved from https://en.wikipedia.org/wiki/Biofouling.

Yemm, R., D. Pizer, C. Retzler and R. Henderson. 2012. Pelamis: experience from concept to connection. Philosophical Transactions of the Royal Society.

CHAPTER 5

Electrical and Electrochemical Energy Storage Applications

*Dominic Bresser** and *Stefano Passerini**

1. Introduction

There is a great push for satisfying the steadily rising energy demand of modern society for renewable sources due to the depletion of fossil fuels and the increasing awareness of the overall CO_2 emission effect on the climate. Nonetheless, beside the technological challenges associated to the energy conversion earlier discussed in this book, one of the greatest challenges is the intermittency of the resulting energy supply, especially the renewable ones, requiring the employment of efficient energy storage technologies. In this regard, there are to date only a few technologies that have gained practical relevance. The most important—also in terms of energy stored worldwide—is certainly pumped hydropower (PHP) with a worldwide contribution of around 99% (Dunn et al. 2011) because of its relatively high energy storage efficiency (ranging from about 70 to 85%) and the highest potential power ratings, typically from 100 MW up to 3000 MW (Chen et al. 2009). Apart from these highly advantageous characteristics, however, this technology heavily relies on suitable geographic conditions, providing

Helmholtz Institute Ulm (HIU), Karlsruhe Institute of Technology (KIT), Helmholtzstrasse 11,89081 Ulm, Germany.
* Corresponding authors: dominic.bresser@kit.edu; stefano.passerini@kit.edu

the opportunity to use two large water reservoirs at different heights (for instance, lifting 1 m³ of water by 1 m is equivalent to an energy storage of 3 Wh (Larcher and Tarascon 2015)) and the inclusion of a suitable dam system accompanied by relatively high investments (Chen et al. 2009). Similarly, the utilization of compressed air energy storage (CAES) devices, also capable of providing power ratings exceeding 100 MW, is dependent on certain geographic conditions, i.e., the presence of salt caverns, rock mines, or depleted gas fields to store the large amounts of compressed air. More importantly even, this technology can be combined basically only with gas turbine power plants, rendering it unsuitable for any other type of power plants, including those based on renewables (Chen et al. 2009). For these reasons, electrical and electrochemical energy storage—even though limited with respect to the achievable power ratings to several MW—has recently gained increasing interest, offering geographical independence, relatively lower installation cost, enhanced response times and energy storage efficiencies up to 95%, as well as the independence of the type of power plant. As a matter of fact, this rising interest is not least the result of the great success of electrochemical energy storage devices for comparably smaller applications such as portable electronics and (hybrid) electric vehicles, which would not have been possible or even conceivable without batteries. In fact, the targeted switch from combustion engine powered transportation to purely electric vehicles will ideally also contribute to the large-scale energy storage, if forward-looking approaches like the vehicle-to-grid concept may turn into common reality one day ((Yang et al. 2011).

Accordingly, this chapter presents either the most employed or most promising electrochemical energy storage technologies in the order of their potential response time and power ratings. Starting from capacitive energy storage in supercapacitors, faradaic storage in lithium-ion batteries, a mature technology for portable electronic devices and electric vehicles, sodium-ion batteries as a potential cost-efficient replacement for stationary applications, and high-temperature sodium batteries will be discussed. Eventually, redox-flow batteries, the battery technology that may provide the relatively highest power ratings due to the decoupling of power and energy density by storing the electrochemically active compounds in large separated tanks, will be presented. Not discussed herein will be—even though still of industrial importance (Doughty et al. 2010, Ferreira et al. 2013)—the nickel-cadmium (Ni-Cd) and the lead-acid battery technologies, since the accompanying environmental and health issues related to the presence of the heavy metals, cadmium and lead (Järup 2003), has already or presumably will eventually lead to a ban from several important consumer markets as, for instance, the European one.

2. Energy Storage Technologies as a Function of Scale and Response Time

One may generally differentiate the energy storage technologies on the basis of their power ratings and potential response time, i.e., one may roughly divide them into two classes: Energy and Power applications. While the first class includes applications requiring charge-discharge times ranging from one to several hours and performing up to a few cycles per day, energy storage technologies capable of up-taking and delivering energy within a few seconds to a few minutes, i.e., running multiple cycles per hour, fall into the second class. Based on these characteristics, Energy-oriented storage technologies may be used, e.g., for extended peak shaving or load leveling, while the Power-oriented ones are more suitable for, e.g., frequency and voltage regulation or power quality improvement (Doughty et al. 2010). Even though this classification appears rather simple at a first glance, the discrimination between the two classes is rather blurry when electrochemical energy storage technologies are considered. In fact, their performance eventually depends on the final design of the devices, also including the employed chemistry. Thus, we will follow herein some rather general considerations with respect to the suitability of the different technologies to store certain amounts of energy, while at the same time considering the inherent response time, i.e., the general rate capability (or in other words, the relationship between rated power and energy as illustrated in Fig. 1), as these two characteristics are commonly contradictory.

In this regard, the fundamental basics of the different technologies are later described with a particular focus on the already existing and potential future applications of these technologies, including the impact of the employed materials and compounds, as well as their current and/ or intrinsic limitations.

2.1 Capacitive Energy Storage

Strictly speaking, capacitors are the sole purely electrical energy storage technology discussed in this chapter due to the absence of any faradaic process. All other technologies, referred to as electrochemical applications, include the conversion of electrical energy into chemical energy and *vice versa* upon charge and discharge, respectively. Nonetheless, the two terms are frequently used synonymously in literature and summarized among either one of these two terms. This is also related to the continuous progress in the field, as will be discussed later on.

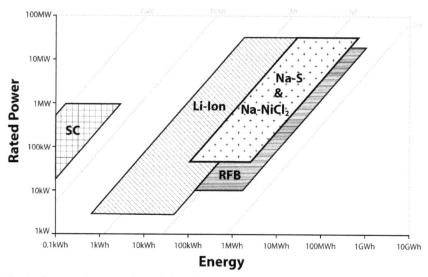

Fig. 1: Comparative overview of the electrochemical technologies for reversible energy storage as a function of rated power and energy content: SC = supercapacitors; Li-Ion = lithium-ion batteries; Na-S = sodium-sulfur batteries; Na-NiCl$_2$ = sodium-nickel chloride batteries; RFB = redox-flow batteries (double logarithmic scale; adapted from (IEC 2011)).

Generally, a capacitor is composed of two purely electronic conductors separated by a thin layer of a dielectric material. In these devices the energy is stored by the application of an electrostatic field, i.e., by polarizing the two electronic conductors, as such they are traditionally used in electronics. However, the potentially stored energy is rather small in such cases. In order to increase the stored energy content, electrical double-layer capacitors (EDLCs)—more often referred to as supercapacitors (SCs) or ultracapacitors—were developed, classically comprising two electrodes made of high specific surface area (SSA) activated carbons, separated by a liquid electrolyte (Béguin et al. 2014). In this case, the energy is stored by polarizing the two electrodes, which is accompanied by the electrostatic accumulation of reversely charged ions dissolved in the electrolyte solvent (Fig. 2).

Not considering some rather recent findings concerning the importance of tailoring the electrodes' pore structure to specific dimensions in order to match the volume of the desolvated ions (the interested reader is referred to the following two review articles: (Béguin et al. 2014, Simon and Gogotsi 2008)), the achievable capacitance (*C*) is, basically, directly related to the available electrode surface area. The overall energy (*E*) is then the product of the capacitance and the square of the cell voltage (*V*), according to the formula:

$$E = \tfrac{1}{2}\,C \cdot V^2$$

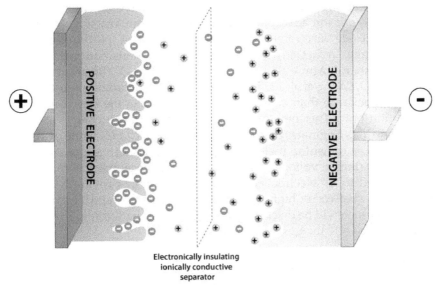

Fig. 2: Schematic illustration of a symmetric electrical double layer capacitor in the charged state, as indicated by the electrostatic adsorption of reversely charged ions at the negative and positive electrode surface (adapted from (Béguin et al. 2014)).

This dependence on the square of the cell voltage has led to the replacement of aqueous electrolytes by organic ones (mainly acetonitrile or propylene carbonate), since these latter reveal electrochemical stabilities up to three times higher than that of water-based systems (i.e., 2.5–2.7 V vs. ca. 0.9 V). Since the response time of these devices is mostly limited by the reorientation (in some case short-length migration) of the ions in the electrolyte, these systems are capable of being dis-/charged in the timescale of seconds, providing extremely high specific power (ca. 15 kW kg^{-1}), high reliability, very little maintenance efforts, high roundtrip efficiencies (> than 90%), and a lifetime of several hundred thousand up to a million dis-/charge cycles (Béguin et al. 2014, Simon and Gogotsi 2008). Their energy, however, remains rather limited (< 10 Wh kg^{-1} (Béguin et al. 2014)) and their suitability for long-term energy storage is restricted by the relatively fast self-discharge (ca. 5–40% per day) and high investment cost (Luo et al. 2015, Zakeri and Syri 2015). For these reasons, SCs manufactured by several companies in, e.g., Australia, Japan, USA, or Germany, are presently employed mainly to ensure the power quality and buffer short-term pulse power requirements (Luo et al. 2015).

Eventually, we may get back to the earlier mentioned blurring boundaries between solely electrical energy storage technologies (like supercapacitors) and electrochemical ones (such as batteries). A very good example for this is the approach to enhance the energy density of supercapacitors by switching from purely non-faradaic charge storage (i.e., electrostatic ion adsorption) to superficial redox-reactions occurring at the electrode/electrolyte interface, the so-called pseudo-capacitive charge storage. In fact, it was found that several transition metal oxides like, for instance, MnO_2 (Bélanger et al. 2008), Nb_2O_5 (Brezesinski et al. 2010), or TiO_2 (Wang et al. 2007, Zukalová et al. 2005), provide substantially higher capacitance than expected when considering the electrostatic ion adsorption only. This potential to reversibly exchange charge with the electrolyte ions originates *inter alia* from the ease of changing the oxidation state for these transition metals and the availability of various oxidation states in a suitable voltage range. As a consequence, pseudo-capacitors, also referred to as 'electrochemical capacitors', may provide a bridging technology between high-power supercapacitors and high-energy batteries (which will be discussed in the following paragraphs), including the associated advantages and challenges.

2.2 Lithium-Ion Batteries

Different from supercapacitors, lithium-ion batteries (LIBs), firstly commercialized in 1991 by SONY (Nagaura and Tozawa 1990), can be considered purely electrochemical energy storage devices, since electrical energy is converted into chemical energy upon charge and *vice versa* upon discharge. The general concept of state-of-the-art LIBs is illustrated in Fig. 3, employing commonly two intercalation-type lithium-ion host materials as electrodes and a liquid organic electrolyte—typically a solution of $LiPF_6$ as conducting salt and organic carbonates as solvents (Bresser et al. 2014a).

On the anode side, graphite is used in most cases due to its rather low de-/intercalation voltage (close to that of Li/Li^+) and high reversibility of the lithium uptake/release in combination with a relatively high theoretical specific capacity of 372 mAh g^{-1}, i.e., about twice the capacity of state-of-the-art cathodes. Regarding the latter, there is more variation in chemistry, depending on the origin and manufacturer of the cells. The most commonly employed active materials are presently $Li[Ni_{1/3}Mn_{1/3}Co_{1/3}]O_2$ (NMC), $LiCoO_2$ (LCO), $Li[Ni_{0.8}Co_{0.15}Al_{0.05}]O_2$ (NCA), or $LiFePO_4$ (LFP). More recently, companies are also using NMC with a higher nickel content to achieve higher energy densities compared to the 1:1:1 ratio as indicated above (e.g., 5:3:2). The working principle is rather simple at first glance, as basically lithium ions are shuttling from the cathode to the anode upon charge and *vice versa* upon discharge (the latter process is indicated in Fig. 3 by the arrows). Following this mechanism, lithium-ion batteries were initially referred to as "rocking chair batteries" (Scrosati 1992)—a term, which is not often used anymore, though. Nonetheless, this de-/intercalation mechanism and the accompanying rather small structural and volume changes are key to the exceptionally

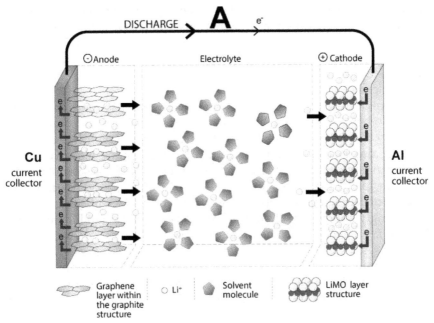

Fig. 3: Illustrative scheme of a lithium-ion battery with the anode being a graphitic carbon, providing the opportunity for lithium ions to be intercalated in-between the graphene layers upon charge (and deintercalated upon discharge), the cathode being a layered lithium transition metal oxide, equally providing the opportunity for the lithium ions to be hosted within the layered structure, and a liquid organic electrolyte comprising a lithium salt—the lithium cations being solvated by the organic solvent molecules (adapted from (Bresser et al. 2014a)).

high roundtrip efficiency, exceeding 95% in the best case, and the outstanding reversibility, i.e., long cycle life, of these batteries, providing (depending on the eventual cell chemistry) up to several thousand dis-/charge cycles. Generally, the stored energy is given by the following formula, indicating also the differences to the purely electrical energy storage in supercapacitors, for which I is the applied or delivered current, V_{avg} being averaged the electromotive force in volt, E the total energy, t the time of dis-/charge, and P_{avg} the provided power:

$$E = P_{avg} \cdot t = V_{avg} \cdot I \cdot t$$

In addition to these advantageous properties, lithium-ion batteries outperform all other existing battery technologies in terms of volumetric and gravimetric energy and power density (Bresser et al. 2014a, Tarascon and Armand 2001), benefitting from the ultimately low redox potential of lithium $E_0(Li/Li^+) = -3.04$ V vs. SHE, i.e., the standard hydrogen electrode, and consequently cell voltages of around 3.7 V in dependence of the eventually comprised cathode material, i.e., cell voltages being two- to three-times higher than those of other battery systems like nickel-cadmium or nickel-metal hydride. For these reasons, LIBs have revolutionized the market for portable electronic devices, (hybrid) electric vehicles, and are also about to play a dominant role in the field of stationary electrochemical energy storage ((Thielmann et al. 2016); see also Fig. 4 for the current applications of LIBs).

As a matter of fact, the development of today's smartphones and tablets, among others, would not have been feasible without this technology. Similarly, they have enabled the development of electric vehicles, currently entering also the mass market as a result of the continuous improvement of this technology. While the general requirements for all these applications are principally the same, there are some differences concerning the priority of certain characteristics. Cost and cycle life, for instance, become increasingly important with the eventual size and expected lifetime of the battery, i.e., increasing in the order portable electronics < hybrid electric vehicles < fully electric vehicles < stationary storage, while safety, weight, and volume are of particular importance for the automotive sector. Accordingly, the eventual cell chemistries may vary for the different applications. While there are extensive research efforts undertaken in

Fig. 4: Schematic presentation of the various applications for lithium-ion batteries, ranging from portable electronic devices over e-bikes and e-scooters to (hybrid) electric vehicles and stationary storage devices for renewables (adapted from (Bresser et al. 2016)).

the automotive sector to improve the energy density by switching, for example, from NMC (5:3:2) to NMC (8:1:1) cathodes and from graphite to silicon/carbon anodes (Bresser et al. 2014b), stationary applications may be served best at present with relatively lower energy density LFP cathodes and titanium-based anodes like TiO_2 or $Li_4Ti_5O_{12}$ (LTO), offering substantially enhanced power capability (dis-/charge times in the range of minutes) and cycling stability of up to more than 10,000 cycles (Yang et al. 2011). In fact, the share of LIBs towards stationary electrochemical energy storage, totaling 1,639 MW worldwide in 2016 according to the U.S. Department of Energy (out of 171,600 MW overall, including also the earlier mentioned PHP and CAES), has risen already to 1,134 MW, i.e., almost 70%, while other technologies like high-temperature sodium batteries (206 MW) and redox-flow batteries (74 MW) play currently only a minor role.[1] With a specific focus on the combination with renewable energy sources, these installations include, for example, a 32 MW facility with a storage capability of 8 MWh located at the Laurel Mountain Wind Farm in Moraine, Ohio, U.S.A. (Zakeri and Syri 2015) or the 80 MWh grid-scale battery farm in Southern California (U.S.A.) with a power rating of 20 MW for peak shaving purposes.[2] Further increased installations for up to 129 MWh are planned in Australia by the end of 2017.[3]

Besides extensive technical progress in recent years, this rather recent deployment of LIBs for large-scale applications too is largely driven by the substantial cost reduction. A very comprehensive study (Nykvist and Nilsson 2015), indeed, revealed that the cost for EV battery packs, for instance, was already about US$300 per kWh in 2014, i.e., far lower than commonly assumed, and projected a further annual cost reduction by about 7–10%, not least due to the installation of so-called Gigafactories in the U.S.A. or, as very recently announced, in Sweden[4] and Australia.[5]

This continuously increasing importance for a wide range of applications, however, also highlights the remaining issues of this technology, which are safety, cost, sustainability, and environmental friendliness (Kalhoff et al. 2015, Larcher and Tarascon 2015). While enhancing the safety is especially addressed by replacing the easily

---

[1] U.S. Department of Energy: Global Energy Storage Database, 2016 (http://www.energystorageexchange.org/projects/data_visualization).

[2] https://www.theguardian.com/sustainable-business/2017/jan/31/tesla-battery-farm-california-energy-elon-musk.

[3] https://www.theguardian.com/australia-news/2017/jul/07/tesla-to-build-worlds-biggest-lithium-ion-battery-in-south-australia.

[4] http://www.nordicgreen.net/startups/energy-storage/sgf-energy-ab-swedish-gigafactory; accessed 07/10/2017.

[5] http://reneweconomy.com.au/boston-energy-consortium-advances-plans-for-queensland-battery-gigafactory-63187/; accessed 07/10/2017.

flammable liquid organic electrolytes by, e.g., ionic liquids, polymer, or ceramics (Kalhoff et al. 2015), improvements towards cost, sustainability, and environmental friendliness can be addressed *inter alia* by replacing critical elements like cobalt, developing advanced active material preparation methods with a lower CO_2 footprint (Barnhart and Benson 2013, Larcher and Tarascon 2015) utilizing water rather than toxic and costly N-methyl-2-pyrrolidone (NMP) for the cathode processing (Loeffler et al. 2016, Moretti et al. 2013), and establishing efficient recycling strategies (Dewulf et al. 2010). Concerning the latter, the European Commission has recently published the Battery Directive 2006/66 EG, fixing a minimum recycling rate of 50% at the cell level.

Advantageous in this regard are also the anticipated next generation lithium battery technologies as lithium-sulfur or lithium-oxygen (Bresser et al. 2013, Bruce et al. 2012, Grande et al. 2015), comprising solely abundant, readily available, and cost-efficient elements as cathode materials. Lithium-sulfur batteries, in fact, have been commercialized already for unmanned aerial vehicles like drones, for which long cycle life is not a major requirement, thus providing well-grounded hope that this technology may also find its way to larger scale commercial applications, while lithium-oxygen batteries (the "holy grail" of battery research) still need to overcome a plethora of challenges (Thackeray et al. 2012).

Last but not least, fully organic batteries are recently attracting rising interest by academics and companies, theoretically paving the way for almost completely sustainable battery fabrication processes, though the storable energy will presumably be rather limited (Armand and Tarascon 2008, Iordache et al. 2016, 2017, Poizot and Dolhem 2011) small-scale applications, as, for example, wearable or portable devices.

2.3 *Room-Temperature Sodium-Ion Batteries*

Based on the great success of lithium-ion batteries and the potential cost increase for lithium resources (Tarascon 2010), room-temperature sodium-ion batteries (SIBs) have recently experienced a renascent scientific interest and great progress has been achieved in the past few years (Slater et al. 2013, Yabuuchi et al. 2014)—not least because it is considered a drop-in technology, following the same concept as LIBs. However, there are also some fundamental differences including the suitability of graphite as anode, since the reversible intercalation of sodium ions has been shown to be at least limited (Stevens and Dahn 2001) without expanding the interlayer distance, e.g., by initially oxidizing the graphite particles (Wen et al. 2014). Instead, hard carbons revealed to be, at present, the most suitable anode, providing reversible capacities in the range of 250 to 300 mAh g^{-1} and good cycling stability (Dahbi et al. 2014). Nonetheless, it is anticipated that SIBs will deliver at their best about 70% of the energy density of LIBs due to the larger mass and volume of the sodium cation, thus intrinsically limiting the gravimetric and volumetric energy density. The great promise, however, results from the economic perspective, as sodium is widely available, well distributed worldwide (compared to the rather concentrated lithium resources (Tarascon 2010)), and may enable the replacement of rather costly materials like the copper current collector by an aluminum one, since sodium does not form an alloy with the latter (Peters et al. 2016). For this reason, SIBs appear particularly promising for large-scale stationary storage applications, for which cost is of major importance, and recent results on well performing sodium-ion prototype cells underline this promise (Mu et al. 2015) and,[6] even though the technology is not commercial, yet.

An approach that targets the utilization of even more abundant electrode materials, i.e., sodium-comprising seawater as the cathode, combined, for instance, with a hard carbon or tin/carbon composite (Kim et al. 2014) anode, may be mentioned here briefly as well. Even though not having reached the commercial level either, this new concept nicely illustrates the great chances of eventually realizing electrochemical energy storage devices, which are solely based on abundant active materials with an almost negligible CO_2 footprint.

2.4 High-Temperature Sodium Batteries

Different from room-temperature sodium-ion batteries, high-temperature sodium batteries have already reached the commercial stage, specifically sodium-sulfur batteries. These batteries are based on inexpensive, non-toxic, and easy to recycle molten sodium and sulfur as anode and cathode, respectively. The core element, however, is the solid sodium β-alumina ($NaAl_{11}O_{17}$) ceramic electrolyte (see Fig. 5a for the general cell design). This ceramic single-ion conductor (i.e., only sodium cations are mobile) provides exceptional ionic conductivities at elevated temperatures of ca. 300°C, similar to aqueous H_2SO_4 electrolytes (Yao and Kummer 1967), rendering it one of the best solid ceramic electrolytes for alkali metal ions. Moreover, it allows for the utilization of molten active materials, thus, providing high ionic transport within the electrodes. The electrochemical reaction upon discharge is based on the formation of sodium polysulfides (Na_2S_x), providing an output cell voltage of ca. 2.0 V. In addition, these batteries can typically be dis-/charged for about 4,000 to 5,000 cycles, applying dis-/charging times between 5 and 10 hours, with a roundtrip efficiency of around 75% and only little self-discharge when kept in rest conditions (Dunn et al. 2011, Hueso et al. 2013, Sudworth et al. 2000).

---

[6] https://news.cnrs.fr/articles/a-battery-revolution-in-motion.

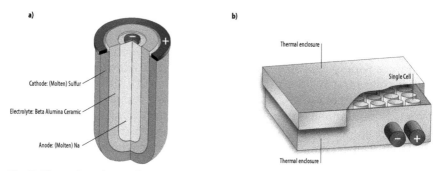

Fig. 5: Illustration of a single Na-S cell (a) and a common battery module design (b), i.e., multiple cells cased in a thermal enclosure (typically 50 kW and 300–360 kWh) to take advantage of the heat released upon operation to maintain the required high temperature of about 300–350°C (adapted from (IEC 2011)).

Nevertheless, beside the fundamental requirement of producing high-quality ceramics on large-scale—also with respect to safety issues and the corrosive nature of the molten active compounds, especially the polysulfides formed upon discharge, the particular challenge of this technology is the thermal management of these cells in rest conditions (Doughty et al. 2010, Sudworth et al. 2000). While the proper sealing of a series of cells in a thermal enclosure (Fig. 5b) provides the opportunity to benefit the exothermic and endothermic nature of the discharge and charge reactions, respectively, in order to maintain a constant temperature of about 300 to 350°C (Bit 2005) (i.e., keeping the electrodes in the liquid state), leaving the cell in rest mode results in the need of extensive temperature control (Dunn et al. 2011). For these reasons, the practical utilization of these batteries is especially suitable, if the cells are dis-/charged on a daily basis as, e.g., in case of stationary storage for load leveling and peak-shaving purposes, for which they are presently employed in particular in Japan (ca. 200 sites), but also in the U.S.A., France, and Germany, with an average power rating of 2 MW per installation (Doughty et al. 2010, IEC 2011).

Another high-temperature sodium battery type, likewise based on the use of sodium β-alumina as solid electrolyte, is the so-called ZEBRA (Zero-Emission Battery Research Activities) battery, in which the sulfur cathode is replaced by a transition metal chloride, i.e., $NiCl_2$. This alternative technology, developed firstly in the mid-1980s (Coetzer 1986) and presently commercialized by FIAMM (Italy),[7] provides several advantages over sodium-sulfur batteries. In particular, they operate at a slightly lower temperature of around 270°C, the electrochemically active components are less corrosive, and the cell voltage is slightly higher (ca. 2.6 V). Furthermore,

---

[7] http://www.off-grid-europe.com/fiamm-sonick-48tl200.

the cells can be assembled in the discharged state (i.e., metallic nickel vs. sodium chloride, NaCl), which, as a consequence, substantially facilitates the fabrication (Doughty et al. 2010, Sudworth 2001). However, the overall energy and power density remains lower than for sodium-sulfur batteries, which has so far hindered their initially envisioned application in electric vehicles (Dustmann 2004). As there is presently no dramatic improvement foreseen for this technology, the reader, interested in further details concerning the cell chemistry (including *inter alia* the addition of aluminum and iron sulfide to the cathode), is referred to comprehensive review articles (Hueso et al. 2013, Lu et al. 2010, Sudworth 2001).

2.5 Redox-Flow Batteries

Redox-flow batteries, even though also being a purely electrochemical energy storage technology, provide one great difference to the previous kinds of batteries, as they allow for the decoupling of energy and power (Skyllas-Kazacos et al. 2011). This decoupling is realized by storing the energy, i.e., the electrochemically active species, for instance, a transition metal cation in solution, in large reservoirs while the electrochemical reaction itself occurs in a separate cell. Accordingly, the power rating is mostly determined by the hydraulic pumps ensuring the transport of the dissolved electrochemically active species (i.e., the anolyte and catholyte) and the surface area of the ion-selective membrane acting as separator in this cell, while the energy is determined by the amount of anolyte/catholyte, i.e., the size of the storage tanks.

The most common redox-flow battery (RFB) is the all-vanadium based one (V-RFB), pioneered at the University of New South Wales in Australia in the mid-1980s (Skyllas-Kazacos et al. 1986, Skyllas-Kazacos and Grossmith 1987), which is schematically illustrated in Fig. 6.

For these V-RFBs, the active species are solely vanadium cations. The anodic and cathodic reaction are based on the V^{2+}/V^{3+} and V^{4+}/V^{5+} redox couple, respectively; both being dissolved in mild sulfuric acid. The two electrolytes are separated by the aforementioned ion-selective membrane, which is permeable basically only for single-charged ions like protons, acting as charge carrying media to ensure charge neutrality. Such cell provides an overall cell voltage of around 1.3 V and a roundtrip efficiency of about 80% for more than 10,000 cycles (i.e., an estimated lifetime of ca. 20 years) (Liu et al. 2013, Stenzel et al. 2015, Yang et al. 2011). The great advantage of this RFB type is the prevention of ionic cross-contamination in case any ions different from protons pass the separator membrane. In that case, the only detrimental effect would be a decrease in energy, indicated by self-discharge, rather than severe cell degradation and loss of cycle life, as observed for RFBs, which are based on two different elemental species (Baxter et al. 2009). The overall specific energy,

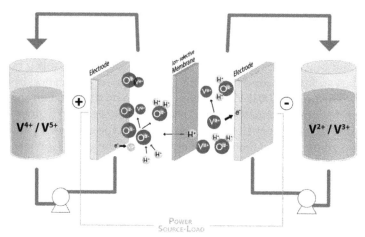

Fig. 6: Schematic of an all-vanadium redox-flow battery, for which the active materials consist of dissolved vanadium cations having different oxidation states, stored in large external tanks (adapted from (IEC 2011)).

however, remains rather low, about 6–10 Wh kg^{-1}, as a result of the limited solubility of the vanadium salts in the aqueous electrolyte. This limits the application of these batteries to stationary storage only, for which the weight and volume of the storage unit are less important. Another challenge is the careful control of the state of charge in order to avoid side reactions like the formation and evolution of hydrogen (Stenzel et al. 2015) and the toxicity of the vanadium species (Domingo 1996).

In an attempt to overcome these challenges, especially the rather low specific energy, researchers have developed a so-called hybrid RFB, for which one of the electrochemically active species reversibly turns solid upon operation. The most investigated technology of this class is the Zn-Br RFB, offering a gravimetric energy of around 32–45 Wh kg^{-1}, i.e., a value three- to four-times that of the vanadium RFB. This substantial improvement in specific energy, however, comes at the expense of a significantly lower roundtrip efficiency of only ca. 70% and a limited cycle life of only around 5 years (Stenzel et al. 2015) due to the detrimental long-term effect of the continuous zinc plating and dissolution upon charge and discharge, respectively, leading to dendrite formation (Luo et al. 2015). Similarly, the formation of elemental bromine results in serious corrosion issues, further affecting the lifetime of the battery (Luo et al. 2015). Nonetheless, the singular characteristic of decoupling energy and power has recently triggered a renascent interest in these batteries and has led to their commercialization, for example by Sumitomo Electric in Japan,[8] as well as the installation of several experimental large-scale

---

[8] http://global-sei.com/products/redox/.

facilities, which are presently tested; *inter alia* a 3 MW installation in Japan (Poullikkas 2013) and a 2 MW one in Germany.[9]

3. Conclusive Remarks

The steadily increasing demand for clean energy, including CO_2-free mobility and transportation, is accompanied by the simultaneously rising demand for energy storage and not least due to its flexibility, high efficiency, and continuous technological improvement electrical/electrochemical energy storage is one of the most evolving ones. In this regard, further stimulus is expected from approaches like the vehicle-to-grid concept as significant part of future smart grids particularly with the increasing availability of relatively low-cost electric vehicles, offering suitable driving ranges.[10] At present, the electrochemical energy storage technology with the greatest impact in this regard is certainly the lithium-ion one and it is anticipated that this technology will continue to dominate the electrochemical energy storage at least for small- to medium-scale applications—for many years to come. As a matter of fact, even though not immediately intuitive, a recent comprehensive study, comparing LIBs, Na-S batteries, as well as vanadium and Zn-Br RFBs revealed that LIBs provide the highest ESOI (energy stored on invested) index (Barnhart and Benson 2013), meaning that they are the most sustainable energy storage technology in this regard. Moreover, the earlier mentioned increasing awareness of researchers to further improve the CO_2 footprint of lithium-ion batteries, increase the energy density, prolong the cycle life, and develop efficient recycling strategies will presumably further strengthen their outstanding position.

Especially for large-scale applications like stationary storage, however, other technologies like room-temperature sodium-ion or redox-flow batteries may provide significant economic advantages, when properly improved and eventually being manufactured at large scale, thus, benefitting also of potentially reduced cost. In fact, a certain variety in energy storage technologies would be advantageous—not only with respect to the spectrum of potentially employed chemical elements and resources, but moreover with regard to the intrinsic storage characteristics in terms of energy density, power capability, and storage duration suitability as eventually summarized in Table 1.

---

[9] https://www.swr.de/odysso/die-megabatterien-der-zukunft/-/id=1046894/did=18982566/nid=1046894/1gd746x/index.html;   https://www.ict.fraunhofer.de/de/komp/ae/RFBWind.html.

[10] https://www.theguardian.com/technology/2017/jul/29/elon-musk-hands-over-first-tesla-model-3-electric-cars-to-buyers.

Table 1: Comparison of the herein discussed electrical and electrochemical energy storage systems with respect to their suitability concerning specific energy storage applications as a function of the storage duration (adapted from (Durand et al. 2017)).

Storage segment	Storage type	Storage duration	SC	Li-Ion	Na-Ion	Na-S	Na-NiCl$_2$	RFB
Fast Acting Storage	Power Quality	≤ 1 min	++	0	0	–	–	–
	Power System Quality	1–15 min	+	+	+	0	0	–
Power Storage		15–60 min	0	+	+	+	+	0
Energy Storage	Daily	ca. 6 h	–	+	+	+	+	+
	Weekly	ca. 30–40 h	–	+	+	+	+	+
	Monthly	168–720 h	–	0	0	0	0	0
	Yearly	≥ 720 h	–	–	–	–	–	–

[++ = very suitable; + = suitable; 0 = less suitable; – = not suitable]

4. Acknowledgments

The authors would like to acknowledge the Vector Foundation as well as the Helmholtz Association for financial support.

References

Armand, M. and J.-M. Tarascon. 2008. Building better batteries. Nature 451: 652–657.

Barnhart, C.J. and S.M. Benson. 2013. On the importance of reducing the energetic and material demands of electrical energy storage. Energy Environ. Sci. 6: 1083–1092.

Baxter, J., Z. Bian, G. Chen, D. Danielson, M.S. Dresselhaus, A.G. Fedorov, T.S. Fisher, C.W. Jones, E. Maginn, U. Kortshagen, A. Manthiram, A. Nozik, D.R. Rolison, T. Sands, L. Shi, D. Sholl and Y. Wu. 2009. Nanoscale design to enable the revolution in renewable energy. Energy Environ. Sci. 2: 559–588.

Béguin, F., V. Presser, A. Balducci and E. Frackowiak. 2014. Carbons and electrolytes for advanced supercapacitors. Adv. Mater. 26: 2219–2251.

Bélanger, D., T. Brousse and J.W. Long. 2008. Maganese oxides: Battery materials make the leap to electrochemical capacitors. Electrochem. Soc. Interface 17: 49–52.

Bito, A. 2005. Overview of the sodium-sulfur battery for the IEEE stationary battery committee.

Bresser, D., S. Passerini and B. Scrosati. 2013. Recent progress and remaining challenges in sulfur-based lithium secondary batteries—a review. Chem. Commun. 49: 10545–10562.

Bresser, D., E. Paillard and S. Passerini. 2014a. Chapter 6: Lithium-ion batteries for medium- and large-scale energy storage: current cell materials and components. pp. 634. *In*: Menictas, C., M. Skyllas-Kazacos, T.M. Lim and N. Ann (eds.). Advances in Batteries for Large- and Medium-Scale Energy Storage: Applications in Power Systems and Electric Vehicles. Woodhead Publishing, Cambridge, UK.

Bresser, D., E. Paillard and S. Passerini. 2014b. Chapter 7: Lithium-ion batteries for medium- and large-scale energy storage: emerging cell materials and components. pp. 634. *In*:

Menictas, C., M. Skyllas-Kazacos, T.M. Lim and N. Ann (eds.). Advances in Batteries for Large- and Medium-Scale Energy Storage: Applications in Power Systems and Electric Vehicles. Woodhead Publishing, Cambridge, UK.

Bresser, D., S. Passerini and B. Scrosati. 2016. Leveraging valuable synergies by combining alloying and conversion for lithium-ion anodes. Energy Environ. Sci. 9: 3348–3367.

Brezesinski, K., J. Wang, J. Haetge, C. Reitz, S.O. Steinmueller, S.H. Tolbert, B.M. Smarsly, B. Dunn and T. Brezesinski. 2010. Pseudocapacitive contributions to charge storage in highly ordered mesoporous group V transition metal oxides with iso-oriented layered nanocrystalline domains. J. Am. Chem. Soc. 132: 6982–6990.

Bruce, P.G., S.A. Freunberger, L.J. Hardwick and J.-M.Tarascon. 2012. Li-O_2 and Li-S batteries with high energy storage. Nat. Mater. 11: 19–29.

Chen, H., T.N. Cong, W. Yang, C. Tan, Y. Li and Y. Ding. 2009. Progress in electrical energy storage system: A critical review. Prog. Nat. Sci. 19: 291–312.

Coetzer, J. 1986. A new high energy density battery system. J. Power Sources 18: 377–380.

Dahbi, M., N. Yabuuchi, K. Kubota, K. Tokiwa and S. Komaba. 2014. Negative electrodes for Na-ion batteries. Phys. Chem. Chem. Phys. 16: 15007–15028.

Dewulf, J., G. Van der Vorst, K. Denturck, H. Van Langenhove, W. Ghyoot, J. Tytgat and K. Vandeputte. 2010. Recycling rechargeable lithium ion batteries: Critical analysis of natural resource savings. Resour. Conserv. Recycl. 54: 229–234.

Domingo, J.L. 1996. Vanadium: A review of the reproductive and developmental toxicity. Reprod. Toxicol. 10: 175–182.

Doughty, D.H., P.C. Butler, A.A. Akhil, N.H. Clark and J.D. Boyes. 2010. Batteries for large-scale stationary electrical energy storage. Electrochem. Soc. Interface 19: 49–53.

Dunn, B., H. Kamath and J.-M. Tarascon. 2011. Electrical energy storage for the grid: A battery of choices. Science 334: 928–935.

Durand, J.-M., M.J. Duarte and P. Clerens. 2017. European Energy Storage Technology Development Roadmap Towards 2030. EASE/EERA.

Dustmann, C.-H. 2004. Advances in ZEBRA batteries. J. Power Sources 127: 85–92.

Ferreira, H.L., R. Garde, G. Fulli, W. Kling and J.P. Lopes. 2013. Characterisation of electrical energy storage technologies. Energy 53: 288–298.

Grande, L., E. Paillard, J. Hassoun, J.-B. Park, Y.-J. Lee, Y.-K. Sun, S. Passerini and B. Scrosati. 2015. The lithium/Air battery: Still an emerging system or a practical reality? Adv. Mater. 27: 784–800.

Hueso, K.B., M. Armand and T. Rojo. 2013. High temperature sodium batteries: status, challenges and future trends. Energy Environ. Sci. 6: 734–749.

IEC. 2011. Electrical Energy Storage—White Paper.

Iordache, A., V. Delhorbe, M. Bardet, L. Dubois, T. Gutel and L. Picard. 2016. Perylene-based all-organic redox battery with excellent cycling stability. ACS Appl. Mater. Interfaces 8: 22762–22767.

Iordache, A., D. Bresser, S. Solan, M. Retegan, M. Bardet, J. Skrzypski, L. Picard, L. Dubois and T. Gutel. 2017. From an enhanced understanding to commercially viable electrodes: The case of PTCLi4 as sustainable organic lithium-ion anode material. Adv. Sustain. Syst. 1600032.

Järup, L. 2003. Hazards of heavy metal contamination. Br. Med. Bull. 68: 167–182.

Kalhoff, J., G.G. Eshetu, D. Bresser and S. Passerini. 2015. Safer electrolytes for lithium-ion batteries: State of the art and perspectives. ChemSusChem. 8: 2154–2175.

Kim, J.-K., F. Mueller, H. Kim, D. Bresser, J.-S. Park, D.-H. Lim, G.-T. Kim, S. Passerini and Y. Kim. 2014. Rechargeable hybrid seawater fuel cell. NPG Asia Mater. 6: e144.

Larcher, D. and J.-M. Tarascon. 2015. Towards greener and more sustainable batteries for electrical energy storage. Nat. Chem. 7: 19–29.

Liu, J., J.-G. Zhang, Z. Yang, J.P. Lemmon, C. Imhoff, G.L. Graff, L. Li, J. Hu, C. Wang, J. Xiao, G. Xia, V.V. Viswanathan, S. Baskaran, V. Sprenkle, X. Li, Y. Shao and B. Schwenzer. 2013. Materials science and materials chemistry for large scale electrochemical energy storage: from transportation to electrical grid. Adv. Funct. Mater. 23: 929–946.

Loeffler, N., G.-T. Kim, F. Mueller, T. Diemant, J.-K. Kim, R.J. Behm and S. Passerini. 2016. *In Situ* Coating of Li[Ni0.33Mn0.33Co0.33]O2 particles to enable aqueous electrode processing. ChemSusChem 9: 1112–1117.

Lu, X., G. Xia, J.P. Lemmon and Z. Yang. 2010. Advanced materials for sodium-beta alumina batteries: Status, challenges and perspectives. J. Power Sources 195: 2431–2442.

Luo, X., J. Wang, M. Dooner and J. Clarke. 2015. Overview of current development in electrical energy storage technologies and the application potential in power system operation. Appl. Energy 137: 511–536.

Moretti, A., G.-T. Kim, D. Bresser, K. Renger, E. Paillard, R. Marassi, M. Winter and S. Passerini. 2013. Investigation of different binding agents for nanocrystalline anatase TiO$_2$ anodes and its application in a novel, green lithium-ion battery. J. Power Sources 221: 419–426.

Mu, L., S. Xu, Y. Li, Y.-S. Hu, H. Li, L. Chen and X. Huang. 2015. Prototype sodium-ion batteries using an air-stable and Co/Ni-Free O$_3$-layered metal oxide cathode. Adv. Mater. 27: 6928–6933.

Nagaura, T. and K. Tozawa. 1990. Lithium ion rechargeable battery. Prog. Batter. Sol. Cells 9: 209–217.

Nykvist, B. and M. Nilsson. 2015. Rapidly falling costs of battery packs for electric vehicles. Nat. Clim. Change 5: 329–332.

Peters, J., D. Buchholz, S. Passerini and M. Weil. 2016. Life cycle assessment of sodium-ion batteries. Energy Environ. Sci. 9: 1744–1751.

Poizot, P. and F. Dolhem. 2011. Clean energy new deal for a sustainable world: from non-CO$_2$ generating energy sources to greener electrochemical storage devices. Energy Env. Sci. 4: 2003–2019.

Poullikkas, A. 2013. A comparative overview of large-scale battery systems for electricity storage. Renew. Sustain. Energy Rev. 27: 778–788.

Scrosati, B. 1992. Lithium rocking chair batteries: An old concept? J. Electrochem. Soc. 139: 2776–2781.

Simon, P. and Y. Gogotsi. 2008. Materials for electrochemical capacitors. Nat. Mater. 7: 845–854.

Skyllas-Kazacos, M., M. Rychcik, R.G. Robins, A.G. Fane and M.A. Green. 1986. New all-vanadium redox flow cell. J. Electrochem. Soc. 133: 1057–1058.

Skyllas-Kazacos, M. and F. Grossmith. 1987. Efficient vanadium redox flow cell. J. Electrochem. Soc. 134: 2950–2953.

Skyllas-Kazacos, M., M.H. Chakrabarti, S.A. Hajimolana, F.S. Mjalli and M. Saleem. 2011. Progress in flow battery research and development. J. Electrochem. Soc. 158: R55–R79.

Slater, M.D., D. Kim, E. Lee and C.S. Johnson. 2013. Sodium-ion batteries. Adv. Funct. Mater. 23: 947–958.

Stenzel, P., J. Fleer and J. Linssen. 2015. Kapitel 10. Elektrochemische speicher. pp. 157–214. *In*: Wietschel, M., S. Ullrich, P. Markewitz, F. Schulte and F. Genoese (eds.). Energietechnologien Der Zukunft—Erzeugung, Speicherung, Effizienz Und Netze. Springer Vieweg, Wiesbaden, Germany.

Stevens, D.A. and J.R. Dahn. 2001. The mechanisms of lithium and sodium insertion in carbon materials. J. Electrochem. Soc. 148: A803–A811.

Sudworth, J. 2001. The sodium/nickel chloride (ZEBRA) battery. J. Power Sources 100: 149–163.

Sudworth, J.L., P. Barrow, W. Dong, B. Dunn, G.C. Farrington and J.O. Thomas. 2000. Toward commercialization of the beta-alumina family of ionic conductors. MRS Bull. 25: 22–26.

Tarascon, J.-M. and M. Armand. 2001. Issues and challenges facing rechargeable lithium batteries. Nature 414: 359–367.

Tarascon, J.-M. 2010. Is lithium the new gold? Nat. Chem. 2: 510.

Thackeray, M.M., C. Wolverton and E.D. Isaacs. 2012. Electrical energy storage for transportation-approaching the limits of, and going beyond, lithium-ion batteries. Energy Environ. Sci. 5: 7854–7863.

Thielmann, A., N. Friedrichsen, T. Hettesheimer, T. Hummen, A. Sauer, C. Schneider and M. Wietschel. 2016. Energiespeicher-Monitoring 2016. Deutschland auf dem Weg zum Leitmarkt und Leitanbieter? Fraunhofer ISI, Karlsruhe, Germany.

Wang, J., J. Polleux, J. Lim and B. Dunn. 2007. Pseudocapacitive contributions to electrochemical energy storage in TiO_2 (Anatase) nanoparticles. J. Phys. Chem. C 111: 14925–14931.

Wen, Y., K. He, Y. Zhu, F. Han, Y. Xu, I. Matsuda, Y. Ishii, J. Cumings and C. Wang. 2014. Expanded graphite as superior anode for sodium-ion batteries. Nat. Commun. 5: 4033.

Yabuuchi, N., K. Kubota, M. Dahbi and S. Komaba. 2014. Research development on sodium-ion batteries. Chem. Rev. 114: 11636–11682.

Yang, Z., J. Zhang, M.C.W. Kintner-Meyer, X. Lu, D. Choi, J.P. Lemmon and J. Liu. 2011. Electrochemical energy storage for green grid. Chem. Rev. 111: 3577–3613.

Yao, Y.-F.Y. and J.T. Kummer. 1967. Ion exchange properties of and rates of ionic diffusion in beta-alumina. J. Inorg. Nucl. Chem. 29: 2453–2475.

Zakeri, B. and S. Syri. 2015. Electrical energy storage systems: A comparative life cycle cost analysis. Renew. Sustain. Energy Rev. 42: 569–596.

Zukalová, M., M. Kalbáč, L. Kavan, I. Exnar and M. Graetzel. 2005. Pseudocapacitive lithium storage in $TiO_2(B)$. Chem. Mater. 17: 1248–1255.

CHAPTER 6

Modern Substation Technologies

Alexandra Khalyasmaa, Stanislav Eroshenko
and Rustam Valiev*

1. Introduction

Today intelligent power systems with active and adaptive networks is a new paradigm, which presupposes the association of electric networks at the technological level with electric power consumers and producers with unified automated power systems. The basic purpose of such systems is to monitor the operating modes of all parties involved in electric power generation, transmission and consumption in real time. By receiving on-line signals through an extensive sensor system, the intelligent system has to respond automatically to all changes taking place in the power network, making optimal solutions for emergency avoidance and implementation of new power supply system elements with maximum reliability and economic efficiency. Thus the intelligent network is defined as set of firmware, connected to the generation sources, substations elements and electrical installations of the consumers, including information-analytical and control systems, providing reliable and efficient transmission of electric energy from a power generation source to the final customer in due time and in necessary quantity.

Therefore, the development of an intelligent system of electric power transmission and distribution should be carried out by means of remote

Ural Federal University Named after the First President of Russia B.N. Yeltsin, Ekaterinburg, Russia 620002, Russian Federation, Ekaterinburg, Mira Str. 19.
Email: stas_ersh@mail.ru; rust-1202@yandex.ru
* Corresponding author: lkhalyasmaa@mail.ru

monitoring, development of automatic control and protection systems, and the improvement of power transmission system efficiency (IDGC of Urals 2016). The key technological components of intelligent electric power systems are:

- **Development and automation of electric power transmission systems.** Digital substations, monitoring systems of the transient operation modes (WAMS), technologies of flexible alternating current transmission (FACTS).
- **Development and automation of electric power distribution systems.** Digital substations, systems of emergency automation, intelligent systems of diagnostics and control of electric energy quality parameters.
- **Intelligent metering.** Intelligent metering systems, data collection from metering devices, data processing, integration into the retail and wholesale markets of electric energy.
- **Information systems and analytics.** Systems of data acquisition (SCADA), geographic information systems (GIS), energy management systems (EMS), distribution management systems (DMS), operational work management systems (OMS), billing systems (BMS), assets management systems (AMS).
- **Cyber security.** The software and means of information protection of power generation, transmission and distribution facilities.

The main objective while developing the intelligent system is to implement information technologies and distributed systems of automatic control, providing accurate computational algorithms and instructions for the power network's active participants in real time on the basis of power network's technological process analysis.

The important role of the developers of intelligent systems and the corresponding software presupposes the development of predictive analytical tools, providing preliminary analysis of the electric network operation mode and power equipment technical state, decreasing the risk of the technological disturbances, the violation of stated parameters of electric energy quality, etc. The implementation of such systems results in substantial increase of power system objects observability, providing remote control without direct human participation.

New production technologies of modern control systems have already shifted scientific research and experiments to practical application. Modern communication standards of information exchange have been developed and are currently being introduced. Digital devices of protection and automation equipment are widely used and essential development of hardware and software of control systems is observed.

Introduction of new international standards and development of modern information technologies opens innovative possibilities for the solution of problems of power system facilities automation and control, allowing to create a new type of substation—digital substation. The digital substation is a substation with high-level control automation of technological processes, equipped with developed IT and control systems (relay protection and automation, ECA, information acquisition and processing systems, metering systems, emergency registration, and fault localization), in which all processes of informational exchange between the elements of the substation, external systems and dispatch control are carried out in a digital form on the basis of the IEC protocols.

Digital substations are characterized by intelligent microprocessor devices, which are built into the primary equipment, application of local area networks for communications, a digital way of information processing and transmission, automation of substation operation and management processes. The application of uniform protocols of information exchange provides compatibility of application of various devices regardless of the hardware and firmware producer and standardizes the design of IT systems. This technology application provides improvement in observability of substation technological processes, monitoring of technical state of power equipment, remote technical diagnostics, providing at the same time essential reduction of operational expenses. One of the key technologies in construction of digital substations is the reliable monitoring and diagnostics that is integrated into the structure of power equipment and supporting digital data.

Taking into account the degree of technological processes automation and the amount of the processed data, the current trend of development is shifted from automated systems with fixed logic towards the intelligent systems, which are adaptive to external influences, giving the possibility to make decisions on further object operation in automatic mode on the basis of data and knowledge, gained as a result of intelligent processing of primary information flow.

1.1 Monitoring Systems of the Power Equipment Technical State

Currently the questions of electrical equipment technical state assessment at power plants and substations as an integral component of intelligent systems implementation are of crucial importance. It is connected with the fact that the major part of the main electrical network equipment has exceeded the service lifetime, determined by technical documents, provided by the producer. Therefore, the major part of the power equipment fleet is operated at the top of the opportunities (Ekaterinburg Electric Grid Company 2012, IDGC of Urals 2012, Rosseti 2009).

Technical state assessment of the power network primary equipment, allocated at the switchgear of power stations and substations, is carried out by means of specific hardware and software packages. Such packages are intended for registration of various data, regarding technical parameters of power equipment units and their components and its operation conditions. The analysis of the power equipment state is carried out on the basis of the obtained data.

It should be noted, that the cost of the monitoring system of the primary equipment is about 5–8% of the total cost of the electrical network object. Considering the cost reduction connected with scheduled and emergency repairs of the power equipment, the payback period of such systems does not exceed five years (Galkin 2006). However even at such payback periods and rather moderate cost, the economic feasibility (and sometimes the technical one) of monitoring systems installation on each substation (each generation unit) is often not proved, especially in towns or small settlements. In such a case the expert system of technical diagnostics data and test results analysis and/or any available aggregated technical information on the power network equipment parameters, obtained during its operation, can be a good alternative for the monitoring the sub-system, aimed at power equipment technical state assessment. There are various definitions of electric equipment diagnostics and monitoring.

In this chapter we define diagnostics as an intelligent grouping of the linked data, including parameter statistics and trends (changes), which are afterwards processed by the expert system to provide the output regarding the power equipment unit technical state and recommended actions. It is possible to define monitoring as accumulation of basic data (Sparling 2005).

1.2 Power Equipment Diagnostics

The main objective of technical diagnostics is recognition of the object's technical state under the conditions of information shortage and, as a result, adequate assessment of residual resource of the system under consideration and reliability improvement (Davidenko 1997). Various technical systems have various structures and applications. Therefore, it is impossible to apply the same type of technical diagnostics to all systems.

In this context the monitoring system is referred to as technical state assessment tool of the main electric equipment without secondary technological sub-systems (relay protection and emergency control automation, disturbances recording, electric energy quality control, etc.).

Certainly, the monitoring system has a number of advantages in comparison with technical state assessment systems, providing analysis of technical diagnostics data:

(1) the possibility of alive power equipment diagnostics;
(2) power equipment current state real-time control;
(3) high reliability of the output data;
(4) the possibility to detect interrelations in the array of primary data.

In respect to complex systems' technical state assessment, for example, of the substation or a power plant, the monitoring system has a number of disadvantages:

- the monitoring system rarely includes a comprehensive set of data on all need parameters. It is aimed at monitoring of the main elements, mainly—the power transformers and the switching equipment;
- output information represents a data set, the analysis and processing of which makes a separate task with the corresponding mathematical apparatus and specific software.

Besides, certain technical and economic conditions for installation of monitoring systems at power plants and substations are needed:

- the equipment has to have rather high investment cost;
- the damage expectancy due to emergency failures has to be considerable;
- diagnostic test of alive power equipment cannot be quickly and reliably made by mobile means and demands significantly more expensive diagnostic equipment (Davidenko 1997).

The analysis of hydraulic units' condition, when it is necessary to provide simultaneous control of temperature, electric, mechanical and hydraulic parameters is an example of the latter condition.

Today the methods of technical diagnostics (on the basis of non-destructive control methods) are widely applied for power equipment technical state assessment at power plants and substations.

1.3 Equipment Diagnostics Methods

Practically for any kind and type of equipment there is at least one method of non-destructive control by means of which it is possible to obtain data on the equipment condition with frequency of monitoring procedure at least once a year. It gives the chance to gain retrospective information on parameters of the equipment condition that allows not only to watch time history of these parameters but also to predict possible damages and defects on the basis of its analysis.

The main methods of equipment diagnostics are presented in (ISO/IEC 17025:2017). On the basis of general classification, all diagnosing (control) methods of electric equipment can be divided into two groups:

- methods of non-destructive control which do not demand destruction of samples of the test material (product);
- methods of destructive control which demand destruction of samples of the test material (product).

The main methods of assessment of the equipment condition that are in practice are:

- **Magnetic control methods.** Based on registration of the stray magnetic fields arising over defects or while defining magnetic properties of controlled products.
- **Electric control methods.** Based on registration of parameters of the electric field interacting with controlled object or the field arising in a controlled object as a result of external influence.
- **Eddy current control method.** Based on the analysis of the external electromagnetic field interaction with the electromagnetic field of the eddy currents induced by the exciting coil in a conductive part of a controlled object.
- **Radiowave control method.** Based on the analysis of the wave band electromagnetic emission interaction with a controlled object. These methods can be applied to controlling of the objects made of materials that do not "muffle" a radio wave—dielectrics (ceramics), semiconductors, magnetodielectrics and thin-walled metal objects.
- **Visual control methods.** Based on optical radiation interaction with a controlled object. By means of optical control method vacancies, interstices, stratifications, cracks, foreign inclusions, geometrical deviations and internal stresses in controlled objects are found. Information parameters of methods are integrated and spectral photometric characteristics of radiation.
- **Radiation control methods.** Based on registration and analysis of the penetrating ionizing radiation after interaction with a controlled object. Typically the gamma rays and x-rays that can reveal nearly any defect (both internal and surface) are used for control. The radiant flux density increasing in places of defects acts as the primary informative parameter.
- **Acoustic control methods.** Based on application of the elastic vibrations excited in a controlled object. Depending on the nature

of interaction with a controlled object, the passive and active control methods are allocated. In the first case the waves arising in the object are registered. The active methods include ones based on measurement of intensity of the acoustic signal passed or reflected by an object.

- **Capillary control methods.** Based on capillary penetration of tracing fluids into cavities of surface and through discontinuity of a controlled object material and registration of the formed indications visually or by means of the converter.

- **Assessment methods based on the partial discharge measurement** (Ahmed et al. 2010). The partial discharge (PD) is a spark discharge of very small power that is formed inside the isolation or on its surface, in the equipment of medium and high voltage classes. In case of long influence of the partial discharge dielectric properties of the equipment isolation can considerably worsen and lead to its complete failure. PD can be found and measured using piezoelectric, fiber-optical sensors and microwave detectors. PD measurement is widely used for an assessment of isolation of transformers and high-voltage cables (Gulski et al. 2004).

- **Assessment methods with use of the dissolved gas analysis** (Arakelian 2002). Even during the normal operation of the oil-immersed equipment, various gases are formed in oil. At beginning of faults, concentration of gases increases, in particular in case of defects in transformers such gases as hydrogen (H_2), methane (CH_4), acetylene (C_2H_2), ethylene (C_2H_4) and ethane (C_2H_6), etc., are formed. A certain composition of gases corresponds to each type of defects (Ahmed et al. 2010). The analysis of gases structure and concentration allows defining the defects of the equipment but the main problem consists in the analysis of gases composition with several defects in the equipment.

- **Assessment of a condition with the use of the thermal analysis** (Ahmed et al. 2010). Most defects cause change of the equipment temperature. The thermal analysis of the equipment can provide useful information about its condition and allows to find defects at the initial stage of its beginning. Abnormal conditions of operation can be found by the measurement of temperature at the most heated point. The most widespread state defined by means of the thermal analysis is an equipment overload. Life cycle of the equipment is defined by its operating mode (including temperature condition) in many respects; for forecasting of abnormal operating modes (temperature rise above standard values in the most heated point) temperature models (maps) of the equipment are used.

- **Assessment of the equipment condition with the use of vibration monitoring** (Van Horenbeek 2017). In industrial systems, vibration signal is one of the key methods of an assessment of the equipment condition. On the basis of information on vibration the spectral analysis of data can be made and defect in the equipment can be revealed. The main problem of the received analysis assessment is in data interpretation which in many respects depends on qualification of the expert responsible for this stage of work. The vibration analysis can be used for various types of the equipment so for the rotating machines shifts of elements, speed and acceleration are measured, which then will be transformed into a frequency spectrum on the basis of which damages are defined. In case of the transformer equipment, vibrations correspond to the basic structural elements (Ahmed et al. 2010):

  - Vibrations connected with the core;
  - Vibrations connected with a winding;
  - Vibrations connected with on-load tap changing.

All these vibrations pass through transformer oil and then are transferred to transformer tank walls through which, by means of sensors, are read out. Analyzing characteristics of vibrations received in this way it is possible to obtain information on technical condition of the transformer.

- **Use of the frequency analysis for an assessment of the equipment condition** (Ahmed et al. 2010). In case of heavy currents flowing through the equipment separate structural elements can be damaged because of considerable mechanical influences, so in the transformer windings can be deformed. Deformation of windings leads to change of inductive and capacitive resistance of the transformer which in turn leads to changes of frequency characteristics. Comparison of frequency characteristics of the studied and standard transformer allows estimating a winding damage degree.

1.4 Key Factors for Technical State Assessment Application

The power industry is characterized by the high cost of the primary equipment and its relatively long operation life, which defines the necessity of integrated control of the equipment life cycle starting from the manufacturing stage (State Standards Committee of the USSR 1980). The assessment of the technical state at certain stages of the power equipment life cycle will serve not only for identification of the state itself, but also for possible extension of the power equipment operation life time. Certainly, control of the power equipment life cycle at early stages is problematic if the enterprise does not handle the functions of production and operation simultaneously. Therefore, the problem of power equipment technical state assessment is solved under the conditions of information shortage and the lack of basic data.

Modeling of power equipment life cycle is a hardly realized task, for at any stage there are a number of factors, whose influence cannot be considered or predicted: human factor, environment (climatic conditions), service conditions, etc. In practice, life cycle control of the power equipment is carried out from the moment of its commissioning.

The influence of technical diagnostics methods application on the power equipment life cycle is presented in the example of the power transformer of 110 kV (Khalyasmaa et al. 2014). The analysis results are presented in the form of the functional relation between the equipment state (from the moment of its commissioning) and its time of operation (Fig. 1).

Technical state interpretation is made using linguistic variables (Brom 2013) instead of the numerical one:

- Operable condition D_1, when the object completely meets all requirements of technical documentation. Such condition is considered to be efficient;
- Operable, but faulted condition D_2, when only those properties of the object, that characterize the possibility to perform the given functions, meet the requirements of the technical standard documentation;
- Faulted, but repairable condition D_3, when the object cannot perform the given functions, but shifting to operable condition is possible after repair actions, which are technically and economically feasible;
- Faulted and non-repairable condition D_4, when the object cannot perform the given functions and shifting to operable state is impossible, because service is technically impossible or economically infeasible.

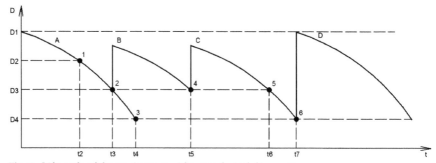

Fig. 1: Life cycle of the equipment without technical diagnostics.

In Fig. 1 life cycle of the power transformer begins from the D_1 condition. Further operation is described by the A curve up to a point 2, which corresponds to the first capital repair, which is performed according to the analysis of 12-year statistical data of the power transformer operation. Usually the transformer condition in service corresponds to D_3.

After the first capital repair the transformer condition is near the D_1 condition. Its further operation is described by the B curve. Time of the second capital repair (point 4) comes a little bit earlier than the first one and makes about 10 years, and the transformer condition at the time of the second capital repair also corresponds to the D_3 condition.

After the second capital repair the operation of the power transformer is described by the C curve. Each repair restores the equipment to operable, but not initial condition.

The example considers only capital repairs of the power transformer (without regard to current repairs, etc.), because this type of repair is the most labor-intensive from both technical and economic points of view.

Then the equipment reaches the D_3 condition again (point 5) and then it is already economically unfeasible to make capital repairs and in D_4 condition (point 6) the full replacement of the power transformer is performed.

In this example the averaged life cycle model of the power transformer of 110 kV is presented. It is worth noticing that there are cases when the number of capital repairs for the transformer of 110 kV is more than two and timing of repairs can differ from that presented in the example.

In Fig. 2 the graphic model of the transformer life cycle when using methods of technical diagnosing is represented. In this example the data obtained as a result of thermal imaging diagnostics have been considered for the analysis of the influence of technical diagnosing methods application on the equipment life cycle.

Diagnostics of the equipment is carried out at earlier stage, in between the conditions $D_1 - D_2$ (point 1), but not at the estimated time of defect detection D_3 (point 2). Therefore the operation process, described by curves D, unlike in the previous case, is reached, since a point 1 and the condition D_3 are not allowed.

It can be said, that the main objective of technical diagnostics is to support the equipment condition in an interval $D_1 - D_2$ by an assessment of its technical state and allocation of the defects. Technical diagnostics is made at average one time in 4–5 years; therefore, the average number of diagnostics equal to 5 is represented in figure (points 1, 3, 7, 8, 9).

Similar to the previous figure in a certain point the equipment reaches the D_2 condition again (point 10), when it is already economically unfeasible to carry out diagnostics of the power transformer, and after achievement of the D_3 condition (point 11) capital repair is performed.

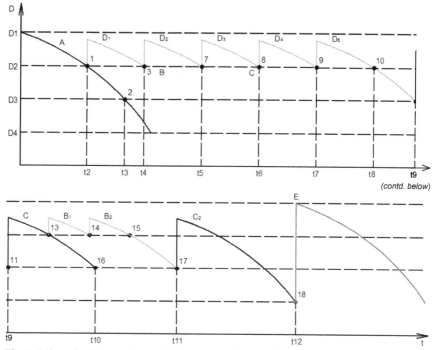

Fig. 2: Life cycle of the equipment when using technical diagnostics methods.

The operation after capital repair is described by the C curve, the diagnostics of the equipment, described by B_1 и B_2 curves is resumed. Operation time before the first capital repair and between the first and second capital repairs considerably increases.

Thus, the main advantage of using technical diagnostics methods is the essential increase in life cycle of the power network equipment. Technical diagnostics data allows to get a rather broad view of the technical state of the majority of the types of substation power equipment.

Use of technical diagnostics methods for the assessment of power equipment technical state not only makes it possible to correct the equipment's life cycle, but also prolongs it, thereby considerably increasing reliability of the equipment's operation.

Modern systems of power network elements technical state assessment are automated expert systems, directed generally on the solution of two types of problems—definition of the equipment state for the purpose of defects and failures detection and optimization of the management actions to increase reliability of the equipment operation and the life time of the system under consideration (Osotov 2000).

There are rather big differences in regulatory documentation on service, diagnostics, tests, and operation of the power equipment, which do not allow to develop integrated multi-purpose system of power equipment technical state assessment.

Similar expert systems used on real power facilities often do not suit the requirements completely as most of them only provide a solution of technical state assessment problem for separate types of electrical equipment, for example, only power transformers or only circuit switches.

In modern practice, depending on the volume and integrity of the basic dataset, different mathematical approaches are used, which are described in more detail in (Birger 1978):

- logical methods of curves recognition;
- separation methods in feature space;
- methods based on determination of features value;
- methods based on probability theory;
- methods based on information theory;
- statistical methods of recognition, etc.

Each method has both advantages and disadvantages. However, the application of most of them in real systems is difficult, and the integral assessment of technical state of all possible types of power equipment requires the solution of the multi-objective decision-making problem on further equipment operation on the basis of existing, often dissimilar and incomplete data.

The analysis of modern expert systems functionality and the principles of their construction for diagnostics of the high-voltage oil-filled equipment state are considered in (Davidenko 2000). One should notice that the specific number of transformer failures is about 2%, i.e., about 5–6% of the fleet can have critical defects, leading to abrupt failures (Shutenko 2010).

Thus, despite all unquestionable advantages of technical state assessment systems, they have a number of essential disadvantages:

- they are focused on the solution of specific problems of the equipment owner (specific schemes, unique equipment, etc.) and, as a rule, cannot be used for other objects without significant changes (Kokin 2013);
- they use multi-scale information of varying accuracy, which can result in uncertainty of the final decision (Dimitriyev 2007);
- they do not consider the history of the evaluation criteria, which are used for technical state assessment, i.e., such systems are not able to learn (Moshinsky 2011).

All this proves the insufficient universality of up-to-date systems of power equipment technical state assessment. The current situation forces the industry not to improve existing systems, but to look for new methodologies and approaches of such systems modeling. In other words, new tools of qualitative and quantitative assessment of power equipment technical state are required. The next generation systems are expected to have the following functionality:

- collecting statistics on equipment defects and failures, analysis and searching for data correlations and regularities;
- using various methods of diagnostics, including the application of information technologies for optimal solutions identification on operation and repair; making effective control actions.
- energy utilities asset management support functions: risk management, investment planning, repair planning, etc.;
- universality of the problem solution: the possibility to provide power equipment technical state assessment for various types of electrical network equipment without significant changes;
- accounting for the possible changes of evaluation criteria of power equipment technical state assessment, taking into account operational experience (learning ability).

2. Modern Systems of the Equipment Condition Assessment

The modern analytical systems of power equipment technical state assessment described below are implemented by the leading world companies, specializing in both production and diagnostics. In practice, systems of power equipment technical state assessment being used vary in functionality and provide not only data collection and analysis, but also offer possible scenarios of further equipment operation. Examples of modern systems of power equipment technical state assessment are:

2.1 ABB T-Monitor

Automated monitoring system of oil-filled power transformers in operation, produced by ABB and other companies. The key feature of the system is the possibility to represent the measurement results by means of mathematical models. At the same time the main distinctive feature of this system is the use of the threshold values, that vary depending on power transformer operation conditions. The special calculation algorithm, which provides the link between the power transformer condition and its real operating conditions, acts as a model. During the power transformer operation the model processes signals from sensors and provides the event forecast and possible decisions on further actions (Fantana and Pettersson 2000). The client can make decisions based on the results, presented by the system or in case there is no such opportunity, ABB can provide the contract for remote observation and trace results together with the client.

2.2 Siemens Transformer Monitoring and Diagnostic System (TMDS)™

Represents the system of technical state assessment, comprising technical models with methods of statistical learning, which allows using non-stationary threshold values, depending on operation conditions. The system can be built into an industrial control system of substations, providing recommendations for operating personnel (Siemens TMDS™ 2010). The application of specific algorithms and analysis blocks gives the possibility to make strategic decisions in the field of power equipment maintenance and repair. TMDS indicates deviations from a normal operation mode and carries out the trend analysis on the basis of the logical models, taking into account guidelines of IEEE/ANSI and Siemens experience of production, service and diagnostics of the power equipment. Similar approach allows revealing interrelation between the calculated and measured data. The messages given by the system during operation contain recommendations about power equipment maintenance and operation and also information about possible consequences of the revealed deviation. In addition, TMDS can provide control of the cooling system for power transformer temperature modes optimization. The TMDS system can be installed on new or old power transformers units from any producer with the use of the existing sensors.

2.3 General Electric Energy HYDRAN

The HYDRAN sensors developed by a division of General electric energy are used for power transformers condition monitoring in real time (de Pablo et al. 2011). They give the possibility to measure moisture and gases content in transformer oil. Sensors can be connected to the monitoring system of the power transformer, which are presented by two main products: Faraday tMEDIC and Faraday Transformer Nursing Unit. Faraday tMEDIC can be integrated into technological secondary systems of power substations, trace the transformer condition and perform its diagnostics in real time; similar realization allows to define the major part of emergencies (except for events of probabilistic character such as various natural disasters) before their occurrence, thereby avoiding expensive recovery from the accident. Faraday Transformer Nursing Unit unites the control and monitoring systems and allows to define and to illustrate the reasons of the emergencies.

2.4 *Serveron® TM8 Multi Gas On-line Dissolved Gas Monitor*

Is the system of power transformer condition monitoring based on the analysis of the composition of the gases, dissolved in oil (SERVERONTM8 2015). The system allows tracing eight key gases, nitrogen content, humidity, etc. For the analysis of the dissolved gases composition the neural networks are used which gives the possibility to increase analysis accuracy.

The functionality of the described systems shows that the existing decisions not just measure parameters of the oil-filled equipment, but also allow to reveal the type of the failure. Such decision support systems provide solution for optimal control of power supply assets from both economic and technological points of view.

It should be noted that most of the systems of power equipment technical state assessment are mainly developed for power transformer units and require highly qualitative information regarding the technical state of the system elements (substations), assuming either development of the integrated monitoring and technical state assessment system or the system of data collection, analysis and processing with subsequent decision-making tools design.

3. Conclusion

Successful implementation of the intelligent active and adaptive network concept as a new stage in power system development depends on the organization of collecting, analysis and exchange of information between the power system objects. Digital substations are the basic components of the specified development trend. Their implementation will give the possibility to increase efficiency, to reduce the occupied space, to increase functionality, to increase assets reliability and, which is especially important, to increase the safety level of the operating staff, providing at the same time collection and analysis of the data regarding power network equipment condition with subsequent operation mode adjustment using processed information. Digital substations use advantages of digital technologies of protection, management and communication, reflecting a trend of informational exchange unification and standardization in power system monitoring and control, which is observed in many other industrial branches.

The existing trends of intelligent systems development assume the development of all areas of the power industry. Necessity of automation of both monitoring and control systems requires the detailed accounting of the power equipment functional state. One of the most promising directions, demanding intelligent networks development, is the assessment of the power equipment technical state at power plants and substations.

The problem of technical state assessment of power network equipment belongs to the class of multi-objective problems of decision-making. The existing systems of technical state assessment as a rule allow rather accurate estimation of the equipment life cycle. Moreover, they form the list of necessary control actions based on data, received from sensors and monitoring systems; however, it is fair mainly for re-installed power equipment units. In case of technical state assessment of the power equipment unit that have operated for a long period of time, it is required to carry out diagnostics under the conditions of historical data incompleteness, which imposes certain restrictions at making decisions on power equipment unit maintenance. The specified problem requires development of diagnostic systems capable to give rather exact results in the conditions of historical data incompleteness, which assumes the application of artificial intelligence systems, thereby opening a new stage in power systems development.

References

Ahmed, E.B. Abu-Elanien and M.M.A. Salama. 2010. Asset management techniques for transformers. Electric Power Systems Research 80: 456–464.

Arakelian, V.G. 2002. Effective diagnostics for oil-filled equipment. IEEE Electrical Insulation Magazine. 18: 26–38.

Birger, I.A. 1978. Technical diagnostics. Birger, I.A. (ed.). Moscow: Mechanical Engineering, 1978. 240 p.

Brom, A.E. 2013. Basic model of the power equipment life cycle cost. Brom, A.E., O.V. Belova and A. Cicinho (eds.). Humanitarian Bulletin 10: 1–11.

Davidenko, I.V. 1997. Structure of an expert, diagnostic and information system of an assessment of the high-voltage equipment condition. Davidenko, I.V., V.P. Golubev, V.I. Komarov and V.N. Osotov (eds.). Power Plants: Monthly Technological Magazine 6: 25–27.

Davidenko, I.V. 2000. System of computer diagnostics of the oil-immersed equipment within a power supply system. Davidenko, I.V., V.P. Golubev, V.I. Komarov, V.N. Osotov and S.V. Turkevich (eds.). Power Engineer. 11: 52–56.

de Pablo, A.F., W. Ferguson, A. Mudryk and D. Golovan. 2011. On-line condition monitoring of power transformers. Electrical Insulation Conference. 285–288.

Dmitriyev, S.A. 2007. Monitoring of power supply system in the megalopolis on the basis of object-oriented graph model: Cand. Tech. Sc. Thesis: 05.14.02. Dmitriyev Stepan Aleksandrovich. Yekaterinburg 174 p.

Ekaterinburg Electric Grid Company. 2012. The annual report of JSC EESK of 2012 [Electronic resource]: official website. Access mode: http://www.eesk.ru/actioners/Otchetnie_dokumenti/Ezhegodnaja_otchetnost. The title from the screen (accessed date 05.02.2015).

Fantana, N.L. and L. Pettersson. 2000. Condition-based evaluation. A new platform for power equipment life management. ABB Review. 4.

Galkin, V.S. 2006. Questions of the automated monitoring systems design for electric equipment on substations of 500-220 kV taking into account ensuring electric networks reliability. Galkin, V.S., T.M. Langbort, V.A. Lipatkin and V.A. Smirnov (eds.). Power Plants 7: 66–67.

Gulski, E., F.J. Wester, Wester Ph., E.R.S. Groot and J.W. van Doelan. 2004. Condition assessment of high voltage power cables. Power System Technology 1–5.

IDGC of Urals. 2012. Annual report of 2012 [Electronic resource]: official website. Access mode: http://report2012.mrsk-ural.ru/reports/mrskural/annual/ 2012/gb/ Russian/9030.html. – The title from the screen (accessed date 05.01.2015).

IDGC of Urals. 2016. The program of innovative development of JSC IDGC of Urals for 2016–2020 with prospect till 2025. [Electronic resource]: official website. Access mode: http://www.rosseti.ru/investment/policy_innovation_development/doc/innovation__program.pdf (accessed date 10.04.2017).

ISO/IEC DIS 17025:2017 General requirements for the competence of testing and calibration laboratories.

Khalyasmaa, A.I. 2014. Electrical equipment life cycle monitoring. Khalyasmaa, A.I., S.A. Dmitriev, D.A. Glushkov, D.A. Baltin and N.A. Babushkina (eds.). Advanced Materials Research 1008-1009: 536–539.

Kokin, S.E. 2013. Energoinformational models of functioning and development of power supply systems in big cities: D. Tech. Sc. Thesis: 05.14.02.Kokin Sergey Evgenyevich. Yekaterinburg, 367 p.

Moshinsky, O.B. 2011. Development of model of the functional condition assessment of power supply system in megalopolises: Cand. Tech. Sc. Thesis: 05.14.02/Moshinsky Oleg Borisovich. Yekaterinburg, 199 p.

Osotov, V.N. 2000. Some aspects of optimization of power electric equipment diagnostics system on the example of Sverdlovenergo: Cand. Tech. Sc. Thesis: 05.14.02/Osotov Vadim Nikiforovich. Yekaterinburg, 31 p.

Rosseti. 2009. Annual report of 2009 [Electronic resource]: official website. Access mode: http://www.rosseti.ru/investors/info/year/. The title from the screen (accessed date 05.01.2015).

SERVERON TM8. 2015. On-line DGA Monitor. Technical brochure.

Shutenko, O.V. 2010. The functionality analysis of the expert systems used for diagnostics of the high-voltage oil-immersed equipment condition [Electronic resource]. Shutenko, O.V. and D.V. Baklay (eds.). Bulletin of National Technical University. Kharkov Polytechnic Institute, pp. 179–193.

Siemens, TMDS™. 2010. transformer monitoring and diagnostic system. 2010. Technical Brochure.

Sparling, B.D. 2005. Increase in monitoring and diagnostics level for optimization of the electric power transmission and distribution to improve financial performance. Sparling, B.D. Methods and means of the power equipment condition assessment. Tadzhibayev, A.I. and V.N. Osotov (eds.). St. Petersburg 28: 178–202.

State Standards Committee USSR. 1980. Nondestructive control. Capillary methods. General requirements: GOST 18442 - 80: confirmed by Resolution of the State Standards Committee of the USSR 15.05.80. Moscow, 16 p.

Van Horenbeek, A. 2017. Vibration spectral analysis to predict asset failures by integrating R in SAS® asset performance analytics. SAS Global forum 2017: 1–15.

CHAPTER 7

Power System Stability

Antonio Carlos Zambroni de Souza,*
Diogo Marujo and *Marcos Vinicus Santos*

1. Introduction

During the commencement of electric power systems, the generation-load balance was easily met due to the large active and reactive power availability by the generation units. The transmission system operated with low levels of stress and the major concern was related to the development of reliable equipment for transmission and distribution systems. Such systems were considerably small and easily operated.

The modern operation of the Power System faces a very different scenario. The loading demands center is usually far away from generation center, requiring long transmission systems. In addition, the constant loading growth in areas with a low level of generation capacity has promoted the interconnection between different utilities in order to ensure a reliable and uninterrupted supply of the loading demands. This scenario may change for different reasons, like the use of energy efficient equipment and the penetration of distributed generation, turning the current distribution systems into active systems. The interconnection of systems, in this case, may help the reliability of the system as a whole, but several implications deserve special attention.

There are several advantages to interconnecting systems. Minimizing the risks of interruption in energy supply and reducing operating costs are some of the main advantages. However, one of the major challenges of system interconnection is to maintain proper operation during load

Federal University of Itajubá, MG, Brazil, Av. Pinheirinho 1303, 37400-903 Itajubá, Brazil.
* Corresponding author: zambroni@unifei.edu.br

variations, disturbances such as faults, or transmission lines outages and loss of loads or generation. In this case, some studies must be carried out in order to evaluate the performance of the system under various operating conditions. Since stability is of fundamental importance in power system operation, some theoretical concepts and case studies are presented below.

1.1 Power System Stability Definition and Classification

Power System Stability is defined as the ability of an electrical power system, given an initial operation condition, to recover a state of equilibrium that is feasible after being subjected to a disturbance, such that practically the entire system remains intact (Kundur et al. 2004).

Due to the large size and complexity of electrical systems, studies on power system stability are commonly divided into some classes. This division allows considering some simplifications to analyze specific types of stability, by using an appropriate degree of detail in the system modeling and an adequate solution technique. According to Kundur (1994), this classification is based on the following considerations:

- The physical nature of the resulting instability;
- The size of the disturbance considered, which directly affects the most appropriate method for stability analysis;
- The devices, processes and time span that must be taken into account in order to evaluate the stability.

According to the aforementioned characteristics, several definitions and classifications have been proposed in the literature. However, a task force composed of the Institute of Electrical and Electronics Engineers (IEEE) and International Council for Large Electric Systems (CIGRE) committees defined a more comprehensive and well-accepted classification for the stability. The purpose of this division was to obtain a definition that fully reflects the current needs of the industry. The classification and its subdivisions are shown in Fig. 1 (Kundur et al. 2004).

In order to facilitate the understanding of the three main divisions, a brief description of each class is given below. The mathematical approach to determining and evaluating stability will be addressed in the following sections.

a) Rotor Angle Stability

The rotor angle stability is also known as angular stability. The objective of this chapter is to evaluate the ability of the system to keep its generating units operating in synchronism. This kind of stability is associated with the ability to maintain or restore the balance between the electrical and mechanical torque of each synchronous machine connected to the

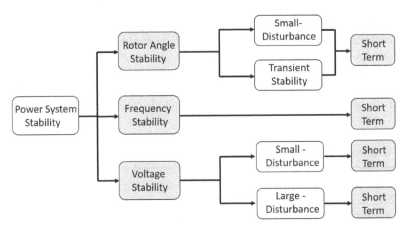

Fig. 1: Power system stability classification.

system. The instability occurs by increasing the angular difference of the generators as a function of the electrical power supplied. This mechanism can lead to loss of synchronism with other generators connected to the system (Kundur et al. 2004).

The rotor angle stability is subdivided into two classes: small-signal stability and transient stability. The first subdivision is associated with the occurrence of small disturbances, such that it is possible to use a linearized model of the system. Such linearization is discussed in Subsection 2.1. Normal load increases represent an example of this type of disturbance. For large perturbation, the linearized model of the system does not represent the appropriate approach. In this case, the transient stability is evaluated through the modeling of the system via algebraic-differential equations. Faults, transmission lines outages and loss of loads or generation are examples of large disturbances. In both subdivisions, the time frame is some seconds.

b) Frequency Stability

The Frequency Stability is the ability of a power system to keep the frequency constant after the system undergoes a disturbance that has resulted in imbalance between load and generation, with minimum unintentional loss of load. The instability can be observed as sustained frequency oscillations, which can lead to the disconnection of generating units and loads. In general, the frequency instability is associated with inadequate equipment responses, poor coordination of protection and control devices or insufficient generation reserve (Kundur et al. 2004).

c) Voltage Stability

The original definition of Voltage Stability refers to the ability of the system to maintain adequate voltage levels on all system buses under normal operating conditions and after having undergone a disturbance. However, regardless the voltage level, voltage stability is also associated with the correct effects of a control action. In this sense, a low voltage level may flag voltage stability by itself. Rather than that, attempts to recover the voltage level may indicate that condition. Section 4 deals with this problem that may lead power systems to voltage collapse. The main factor associated with this kind of instability is the inability of the system to meet the required reactive power demand, reducing the voltage level in the buses. Instability by voltage increasing is also possible, although it is an uncommon situation. Voltage instability is a local phenomenon. On the other hand, the voltage collapse represents a more complex phenomenon that can lead the system to a blackout or a low voltage profile in a significant part of the system. While angular stability focuses on the angular behavior of the synchronous machines, the voltage stability is more influenced by the behavior of the loads. After a disturbance, the power consumed by the loads tends to be restored by adjusting the motors slip, acting the voltage regulators and under-load tap changers. However, restored loads increase the power flow in the network. Consequently, the consumption of reactive power increases, leading to a low voltage profile (Kundur et al. 2004, Kundur 1994).

The voltage stability can be separated into two subclasses: small-disturbance and large-disturbance voltage stability. Again, the first is associated with the linearized model behind an operation point. The second subclass evaluates the transient or long-term performance of the system for a period sufficient to consider the effect of LTCs, thermostatic loads, field current limiters, and so on. In this case, simulation time should be chosen from a few seconds to several minutes.

1.2 Time Scale in Power System Stability

The simulation of large electrical power systems has been one of the major challenges through the years. When the first algorithms came up, the main difficulty was the limitations regarding data storage and processing by using the available computers.

A simulation with all the power system components and controls modeled in detail would represent an ideal scenario. This situation would allow operators to manage the system in an even more optimized way, with reduced costs and greater reliability. This complete representation is not possible in practice. The limitations mentioned above remain a drawback, even with the advancement of algorithms and computers. The availability of reliable data also currently precludes the approach with all the equipment represented with the highest level of detail.

In order to represent the power system with a greater level of detail and less computational effort, power system simulations are divided into time scales from milliseconds to hours. Three categories are considered (Kundur 1994):

- Short-Term or Transient: from 0 to 10 seconds;
- Mid-Term: from 10 seconds up to few minutes;
- Long Term: from a few minutes to hours.

The exact definition between the mid-term and the long-term is still imprecise. For periods of analysis greater than the transient, the choice of the model considered is based on the phenomenon under analysis rather than the equipment time response. For this reason, in many cases, the mid-term concept is neglected, such as the entire period superior to the transient be classified as the long term.

The time scales can be separated according to the equipment operation time in the power system. Synchronous generators, automatic voltage regulators, governors, and HVDC are short-term devices and controls. Under-load tap changers, overexcitation limiters, switched capacitors/inductors, automatic generation control, secondary voltage control and generation scheduling stands out for long term control systems and devices. From the load point of view, induction motors are classified into the short-term period and thermostatic loads belong to the long term scale. Static loads and the network are commonly considered as instantaneous (Cutsem and Vournas 1996, Cutsem and Vournas 2008).

The time scale division is motivated by the knowledge of the impact that the model to be considered results when a disturbance occurs, in accordance with the time span considered. When a specific model does not influence considerably, the latter can be considered in steady state (for fast time constants) or the device associated with the model simply has not acted yet (for slow time constants). This consideration reduces the size of the problem and the time required to perform the simulation, without major loss of generality.

2. Small-Signal Stability

The dynamic behavior of any system may be represented mathematically by a set of Differential-Algebraic Equations (DAE), which represents the several physical responses of the elements related to the system analyzed. Depending on the type of phenomenon under study, the set of algebraic-differential equations can be formed by linear or non-linear equations.

Given the nonlinear characteristic of the Power Systems, the mathematical representation may be written by a set of differential-algebraic equations of the following form:

$$\dot{x} = f(x,y)$$
$$0 = g(x,y)$$
(1.1)

where f is the set of differential equations, x is the state vector which includes the differential variables or state variables, related to the dynamic of machines and controls, g is the set of algebraic equations and y is a vector which includes the algebraic variables, such as network bus voltage and angle.

A system whose differential equations are simultaneously zero, is considered to be in an equilibrium or singular point. The ability of the system to remain in an equilibrium point or regain one after a disturbance is called Stability.

The stability of a nonlinear system depends on the type and magnitude of the disturbance, as well as the initial state of the system. When the disturbances at the power system are sufficiently small such that the system is able to maintain its stability, the DAE can be linearized in order to easily assess the power system responses. For such small contingencies, the stability problem is usually called Small-Signal Stability (Kundur 1994).

2.1 Linearization

In order to easily assess the Small-Signal Stability of a Power System, the Equation 1.1 may be linearized around the equilibrium point, as discussed previously. Considering a small disturbance in which the state and algebraic variables suffer small deviations from their equilibrium point, such as:

$$x = x_0 + \Delta x$$
$$y = y_0 + \Delta y$$
(1.2)

Hence, the set of DAE at Equation 1.1 can be rewritten as:

$$\dot{x} = x_0 + \Delta \dot{x} = f[(x_0 + \Delta x), (y_0 + \Delta y)]$$
$$0 = g[(x_0 + \Delta x), (y_0 + \Delta y)]$$
(1.3)

Using Taylor's series expansion, the nonlinear functions can be approximated around the equilibrium point. Therefore, considering the ith state variable and neglecting the second and higher order of Taylor's series, its differential equation may be written as:

$$\dot{x}_i = \dot{x}_{i0} + \Delta \dot{x}_i = f_i(x_0, y_0) + \frac{\partial f_i}{\partial x_1} \cdot \Delta x_1 + \cdots + \frac{\partial f_i}{\partial x_n} \cdot \Delta x_n + \frac{\partial f_i}{\partial y_i} \cdot \Delta y_i + \cdots + \frac{\partial f_i}{\partial y_m} \cdot \Delta y_m$$

(1.4)

Likewise, for the *i*th algebraic equation, it may be written as:

$$0 = g_i(x_0, y_0) + \frac{\partial g_i}{\partial x_1} \cdot \Delta x_1 + \cdots + \frac{\partial g_i}{\partial x_n} \cdot \Delta x_n + \frac{\partial g_i}{\partial y_i} \cdot \Delta y_i + \cdots + \frac{\partial g_i}{\partial y_m} \cdot \Delta y_m \quad (1.5)$$

Since $\dot{x}_{i0} = f_i(x_0, y_0) = 0$ and $g_i(x_0, y_0) = 0$, then:

$$\Delta \dot{x}_i = \frac{\partial f_i}{\partial x_1} \cdot \Delta x_1 + \cdots + \frac{\partial f_i}{\partial x_n} \cdot \Delta x_n + \frac{\partial f_i}{\partial y_i} \cdot \Delta y_i + \cdots + \frac{\partial f_i}{\partial y_m} \cdot \Delta y_m \quad (1.6)$$

$$0 = \frac{\partial g_i}{\partial x_1} \cdot \Delta x_1 + \cdots + \frac{\partial g_i}{\partial x_n} \cdot \Delta x_n + \frac{\partial g_i}{\partial y_i} \cdot \Delta y_i + \cdots + \frac{\partial g_i}{\partial y_m} \cdot \Delta y_m \quad (1.7)$$

This may be written using the state space representation:

$$\Delta \dot{x} = A \cdot \Delta x + B \cdot \Delta y$$
$$0 = C \cdot \Delta x + D \cdot \Delta y$$

(1.8)

$$\begin{bmatrix} \Delta \dot{x} \\ 0 \end{bmatrix} = \begin{bmatrix} A & B \\ C & D \end{bmatrix} \cdot \begin{bmatrix} \Delta x \\ \Delta y \end{bmatrix}$$

(1.9)

where *A* is a *n×n* matrix of the partial derivatives of the dynamic equations related to the dynamic variables, *B* is a *m×n* matrix of the partial derivatives of the dynamic equations related to the algebraic variables. *C* is a *n×m* matrix of the partial derivatives of algebraic equations related to the dynamic variables and *B* is a *m×m* matrix of the partial derivatives of the algebraic equations related to the algebraic variables.

2.2 *Eigenvectors and Eigenvalues*

As presented before, the linearized power system equations can be represented by a set of matrices. The manipulation of such matrices can provide valuable information about the response and behavior of the system under small perturbations. The analyses of the system's matrices are done by examining the eigenvalues and the eigenvectors of such matrices.

In general, a vector changes its direction when multiplied by a matrix. However, when the multiplication of the matrix and the vector maintains the same direction as the original vector, such vector is called eigenvector of such matrix.

Consider a $n \times 1$ vector, ϕ, which is multiplied by a scalar λ, and a $n \times n$ matrix, A, which is also multiplied by the same vector. If the equality shown by equation 1.10 is satisfied, then ϕ is called the eigenvector of matrix A and the scalar λ is the eigenvalue associated with the eigenvector.

$$A \cdot \phi = \lambda \cdot \phi \qquad (1.10)$$

For a $n \times n$ matrix A, the system has n eigenvectors and, consequently, n eigenvalues. To find such eigenvectors and eigenvalues, the equation 1.10 can be written as:

$$(A - \lambda. I) \cdot \phi = 0 \qquad (1.11)$$

Solving the equation 1.11 for a non-trivial solution:

$$\det(A - \lambda. I) = 0 \qquad (1.12)$$

As one can see, the equation 1.12 involves only λ. The solution of such equation gives the n eigenvalues of the system. Furthermore, solving Equation 1.11 for each λ obtained, the set of eigenvectors of the system may be obtained.

In matrix notation, the set of eigenvectors and eigenvalues can be written as:

$$A. \Phi = \Phi. \Lambda \qquad (1.13)$$

where Φ is the $n \times n$ matrix containing the set of eigenvectors of A and Λ is the $n \times n$ matrix containing the eigenvalues at the diagonal elements.

The ith column of matrix Φ is the ith eigenvector A, e.g., ϕ_i. The eigenvectors of such matrix are called right eigenvectors of A as they stand in the right side of the matrix in the equation. Equations 1.14 and 1.15 show the composition the matrices Φ and Λ.

$$\Lambda = \begin{bmatrix} \lambda_1 & 0 & \cdots & 0 \\ 0 & \lambda_2 & \cdots & 0 \\ \vdots & \vdots & \ddots & \vdots \\ 0 & 0 & \cdots & \lambda_n \end{bmatrix} \qquad (1.14)$$

$$\Phi = [\Phi_1 \ \Phi_2 \ \cdots \ \Phi_n] \qquad (1.15)$$

Now, consider Φ to be a normalized matrix and Ψ also being a normalized matrix, which satisfy the follow equation:

$$\Psi. \Phi = I \qquad (1.16)$$

Hence, rearranging equation 1.13 one can write as follows:

$$\Psi \cdot A = \Lambda \cdot \Psi \tag{1.17}$$

where Ψ is the $n \times n$ matrix also containing the set of eigenvectors of A.

Accordingly, the ith row of matrix Ψ is the ith eigenvector of A. In this case, as the eigenvectors matrix stands in the left side, these eigenvectors are called left eigenvectors of the matrix A.

Considering the associations created in equations 1.13, 1.16 and 1.17, one can write:

$$A = \Phi \cdot \Lambda \cdot \Psi \tag{1.18}$$

Hence, from equation 1.18 one may write any matrix in terms of its left and right eigenvectors and its eigenvalues. The use of such characteristics allows a fast and effective option for analyzing the system behavior under small disturbances. Hence, considering a set of the linearized equation:

$$\Delta \dot{x} = A. \Delta x \tag{1.19}$$

$$\Delta \dot{x} = \Phi \cdot \Lambda \cdot \Psi \cdot \Delta x \tag{1.20}$$

2.3 Participation Factors

The analysis of the right and left eigenvectors can provide valuable information about the correlation between the system's state variables and the eigenvalues. However, due to the different magnitudes of such variables, the analysis of the eigenvectors alone may incur an error of judgment. The solution for that is the use of the Participation Factors, which is calculated according to:

$$p_{ki} = \frac{\psi_{ik} \cdot \phi_{ki}}{\sum_{m=1}^{n} \psi_{im} \cdot \phi_{mi}} \tag{1.21}$$

where P_{ki} stands for the participation factor of the kth variable at the ith eigenvalue, ψ_{ik} and ϕ_{ki} are the kth element of the left and right eigenvector related to the ith eigenvalue.

The calculation and analysis of the Participation Factors of each eigenvalue of the system may indicate the variables that have more impact over a single eigenvalue. Such information is a valuable measure to determine variables influence over system's stability (Kundur 1994, Abdi 2007, Sauer and Pai 1997).

2.4 Oscillation Modes

As discussed previously, a system can be represented by a set of DAE, which describes the behavior of such system in terms of its dynamic and algebraic variables. Furthermore, for small disturbances, the system can be written by a state-space representation of the linearized equations. Then, the stability can be assessed by analyzing the eigenvectors and eigenvalues associated with the system's matrices.

If an appropriate selection of variables is made, the state-space representation of the Power System can be approximated by an open-loop transfer function. This is made by writing the system by means of its eigenvectors and eigenvalues. In this sense, the eigenvalues can be related to the poles of the transfer function. Then, one can write:

$$\lambda_i = -\xi \cdot \omega_n \pm \omega_d \tag{1.22}$$

$$\lambda_i = -\xi \cdot \omega_n \pm \omega_n \sqrt{1 - \xi} \tag{1.23}$$

where ξ is the damping ratio, ω_n is the undamped natural frequency and ω_d is the damped frequency.

In general, the cause of small-signal stability problem in most power systems is due to insufficient damping of oscillation associated with electromechanical variables or voltage controls. The Small-signal stability can be assessed by analysis of the eigenvector or dynamic modes, most related to such variables, and consequently, the eigenvalue associated.

The instability due to such Oscillation Modes can be classified by (Pal and Chaudhuri 2005):

(a) Local modes: This is associated with the oscillations of a generation unit with respect to rest of the system. It is typically between 1 and 3 Hz.

(b) Interarea modes: This type is observed in large systems, when a group of generators swings against other groups of machines or the whole system. The oscillation range is less than 1 Hz.

(c) Control modes: Associated typically with generators, excitation systems, governors and other controls.

(d) Torsional modes: Associated with the shaft system of the turbine generator. The oscillation range is between 10 and 46 Hz.

2.5 Hopf Bifurcation Point

As stated above, the Small-Signal Stability is the ability of the system to maintain its stability after the occurrence of a small perturbation. On small disturbances over the system's equilibrium points, the linearized analysis of the system's equations can provide valuable information about its stability. Regarding the local stability of the system around a stationary point, the eigenvalues locations indicate the type of singularity of the oscillation modes (Pulgar-Painemal and Sauer 2009, Dobson et al. 1992, Mithulananthan et al. 1999).

The critical oscillation modes seen in power system are usually the ones associated with a complex pair of eigenvalues that cross the imaginary axis from the left half-plane to the right half-plane when the system undergoes a contingency. This chapter follows the basics of Hopf bifurcation as presented in (Seydel 1994). Thus, firstly, a theoretical approach is shown. Consider the set of equations

$$\dot{x}_1 = -x_{2_+} x_1 (\beta - x_1^2 - x_2^2)$$
$$\dot{x}_2 = x_{1_+} x_2 (\beta - x_1^2 - x_2^2)$$

(1.24)

The equilibrium is given by $(x_1, x_2) = (0,0)$. For this equilibrium point the Jacobian matrix is:

$$\begin{bmatrix} \beta & -1 \\ 1 & \beta \end{bmatrix}$$

(1.25)

Whose eigenvalues are $\beta \pm i$. Note that the system is stable if $\beta < 0$ and unstable for $\beta > 0$. For $\beta = 0$ a limit cycle appears. It may be constructed by changing the variables as:

$$x_1 = \rho \cos \theta$$
$$x_2 = \rho \sin \theta$$

(1.26)

After manipulations, one obtains

$$\dot{\rho} = \rho(\beta - \rho^2)$$
$$\dot{\theta} = 1$$

(1.27)

The periodic orbit depends on β. In general, the transversality conditions of Hopf bifurcation may be summarized as:

(1) $f(x_0, \beta_0) = 0$
(2) $f(x_0, \beta_0) = 0$ has a pair of pure imaginary eigenvalues $\delta = \pm _i\gamma$ and no zero-real eigenvalue.

The conditions above, however, are not sufficient to detect a Hopf bifurcation, so a third condition is imposed.

(3) $\dfrac{d(Re\delta(\beta_0))}{d\beta} \neq 0$

When it comes to power systems, one is interested in identifying both Hopf, induced and saddle-node bifurcations. Figure 2 depicts the tracking of the dominant eigenvalues for a nine-bus system. It has been conveniently

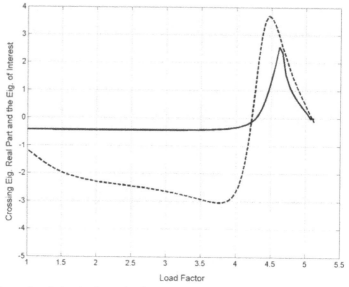

Fig. 2: Eigenvalues behavior for a nine-bus system.

extracted from (Zambroni de Souza et al. 2005). This reference proposes an index to detect both saddle-node and Hopf bifurcations with the help of the tangent vector. Further discussions on bifurcations are found in (Mello et al. 2006), where dynamic and algebraic approaches are employed to understand how bifurcations may take place in power systems.

3. Transient Stability

The linear approach through small-signal stability analysis is not adequate when large disturbances are observed in the system. The system response is evaluated by means of the step-by-step solution of the non linear differential-algebraic equations, which was previously shown by equation (1.1).

The great majority of the nonlinear differential equations do not present analytical solution, that is, they cannot be obtained through elementary algebraic functions. In this sense, it is necessary to seek numerical solutions by using integration methods.

There are several methods available in the literature for solving the set of equations (1.1). These methods are classified according to the following factors (Kundur 1994):

- How the algebraic and differential equations are solved: in a partitioned or simultaneous way;
- The integration method: implicit or explicit;

- The technique used to solve the algebraic equations: Gauss-Seidel, Newton-Raphson, among others.

Two approaches are commonly used in power systems simulation: partitioned solution, based on explicit integration methods, and the simultaneous solution using implicit methods. Both methods are presented in more detail below.

3.1 Integration Methods (Partitioned Explicit/Simultaneous Implicit)

In the partitioned-explicit approach, the algebraic and differential equations are solved separately and alternately. As the solution is explicit, the integral is approximated by values already known in the previous instant. Differential equations are solved using an integration method to find x and the algebraic equations are solved separately for y. If the goal is to determine x, the set y must be known even if it is approximate. Likewise, an estimate of x is required for the calculation of y (Padiyar 2008). This alternating solution can lead to the appearance of interface errors. The most well-known explicit method is Runge-Kutta.

In the simultaneous-implicit method, the differential equations are transformed into algebraic equations by using an implicit method, usually the trapezoidal method. Then, the set of equations, now algebraic, is solved simultaneously with the original algebraic equations composing a unique system of equations. The solution on the set of non-linear algebraic equations is usually obtained by the Newton-Raphson method.

The partitioned-explicit method is more computationally efficient than the simultaneous-implicit method. However, the latter has no interface errors and is numerically more stable.

3.2 Case Study on Angular and Frequency Stability

In order to study the transient stability, the IEEE 9-bus system is used. Synchronous generators are represented by fourth-order models and equipped with IEEE voltage regulators Type DC1. The interconnection between buses is made through six 230 kV transmission lines, (corresponding to the transformers secondary voltage side). Loads are connected to buses 5, 6 and 8. The power flow and dynamic data can be found in (Anderson and Fouad 2003). Appropriate hydro (generator 1) and thermal (generators 2 and 3) speed governors are also considered. The power system employed in the tests is depicted in Fig. 3.

A three-phase fault near Bus 5 during 100 milliseconds is considered such that the transmission line 5–7 is disconnected in order to isolate the fault. The fault starts at t = 2 seconds and the simulation is performed for

40 seconds. Figure 3 shows the rotor angle variation on generators 2 and 3 considering generation 1 as the reference. According to Fig. 4, the system is stable from the angular stability point of view, since it has recovered a state of equilibrium that is feasible after being subjected to a disturbance. In this case, all generating units maintain the synchronism.

In order to evaluate the response of generating units by considering speed governing, the same disturbance is considered. Figures 5 and 6 show the frequency and the rotor angle behavior, respectively, when governors are considered or not. The system is stable only when the governors are present, since the frequency and the rotor angle grow indefinitely without regulators.

Figure 7 shows the rotor angle response after a three-phase fault near Bus 5 during 185 milliseconds. In order to isolate the fault, the transmission line 5–7 is disconnected. The fault starts at t = 2 seconds. Generator 1 is considered again as the reference.

According to Fig. 7, generators angles increase without limits. Thus, the system is unstable when the fault is cleared in 185 milliseconds.

3.3 *Equal Area Criterion*

As described, the transient stability analysis of a power system may require dynamic simulations in order to evaluate the ability of such a system to maintain a normal operation after a contingency takes place. Such simulation involves dynamic modeling and integration methods, which, despite the computational efforts, provide valuable responses.

However, once the focus of the analysis becomes the response of a single machine over the whole system, the transient stability can be

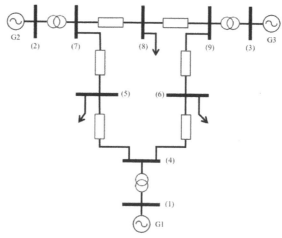

Fig. 3: Power system under analysis.

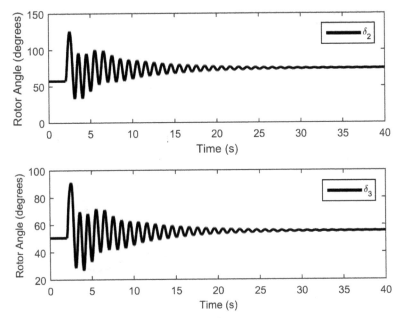

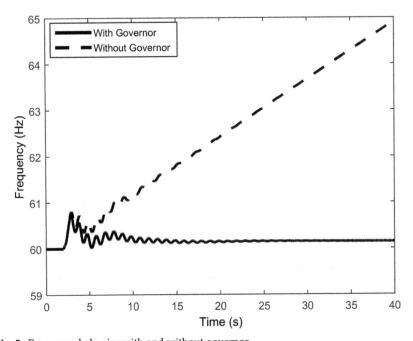

Fig. 4: Rotor angle after a 100 ms three-phase fault.

Fig. 5: Frequency behavior with and without governor.

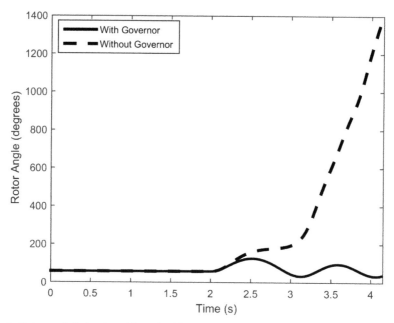

Fig. 6: Rotor angle behavior with and without governor.

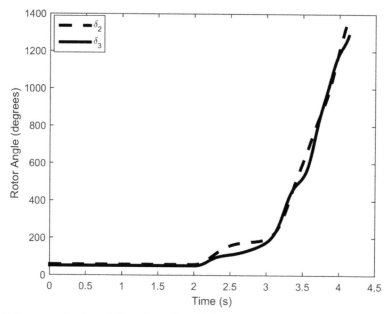

Fig. 7: Power angle after a 185 ms three-phase fault.

assessed through the investigation of the machine's first swing response. Then, considering a contingency, if the angular deviation between the single machine and the system becomes larger than a specific value, the system may face angular instability. Such analysis can be done by using the Equal Area Criterion (Kundur 1992).

The Equal Area Criterion is based on three conditions:

(a) The mechanic power is considered to be a constant.

(b) The machine is represented by a transient reactance and a Voltage Bus.

(c) The damping torque is neglected at the swing equation.

For a given system, the motion equation can be written as:

$$\frac{2H}{\omega_s} \cdot \frac{d^2\delta}{dt^2} = P_m - P_e \tag{1.28}$$

which describes the angular acceleration of the machine.

Multiplying both sides of equation 1.28 by $2d\delta/dt$ and simplifying, one can write:

$$2 \cdot \frac{d\delta}{dt} \cdot \frac{d^2\delta}{dt^2} = \frac{\omega_s}{H} \cdot (P_m - P_e) \cdot \frac{d\delta}{dt} \tag{1.29}$$

$$\frac{d}{dt}\left[\frac{d\delta}{dt}\right]^2 = \frac{\omega_s}{H} \cdot (P_m - P_e) \cdot \frac{d\delta}{dt} \tag{1.30}$$

Now, integrating equation 1.30, one can get:

$$\left[\frac{d\delta}{dt}\right]^2 = \int \frac{\omega_s}{H} \cdot (P_m - P_e)\, d\delta \tag{1.31}$$

For a stable operation, the speed deviation – $d\delta/dt$ – must be zero at some point after the disturbance and the acceleration must be either zero or in opposing direction, i.e. $(Pm - Pe) \leq 0$. At such point, one can assume that the angle δ reaches its maximum value, δ_{max}. Hence:

$$\int_{\delta_0}^{\delta_{max}} (P_m - P_e)\, d\delta = 0 \tag{1.32}$$

Therefore, by equation 1.32, one can imply that the area formed between the mechanic power and the electric power functions from δ_0 to δ_{max} must be zero.

Consider a system comprising a single machine connected by two transmission lines (TL1 and TL2) to an infinite bus, as depicted in Fig. 8.

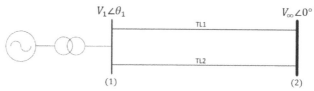

Fig. 8: Single machine infinite bus system.

The Electric Power flow from machine to the infinite bus can be described as:

$$P_{e_{pre}} = \frac{E \cdot V_\infty}{X'_d + X_t + X_{TL}/2} \cdot \sin(\delta) \qquad (1.33)$$

where:

P_e : Electric Power;

E : Internal voltage of the machine;

V_∞ : Voltage at the infinite bus;

X'_d : Transient reactance of the machine;

X_t : Transformer reactance;

X_{TL} : Transmission line reactance;

Considering the loss of one of the transmission lines of the system, the electric power in a post-contingency situation would be:

$$P_{e_{post}} = \frac{E \cdot V_\infty}{X'_d + X_t + X_{LT}} \cdot \sin(\delta) \qquad (1.34)$$

Equations 1.33 and 1.34 provide the electric power function before and after the contingency takes place at the system. Figure 9, depicts both functions with respect to δ, as well as the mechanic power, described by a straight line. The P_{pre} is considering the system at the pre-contingency scenario and, consequently, the P_{post} is the system after the transmission line loss.

At the pre-contingency scenario, the system is in equilibria with δ_0. At the time the contingency takes place, the P_e is reduced and becomes lower than P_m, which accelerates the machine in order to reach the new equilibrium point, δ_1. As the system has no damping, the machine will cross the equilibrium point and, then, P_e becomes higher than P_m, which

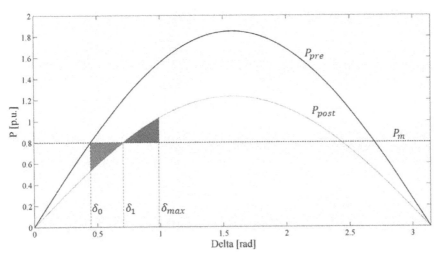

Fig. 9: Equal area criterion analysis.

slows down the machine. The acceleration area is called A_1 while the deceleration area is called A_2. When the acceleration area becomes the same as the deceleration area, i.e., $A_1 = A_2$, then the system may be considered stable.

Substituting equation 1.34 into equation 1.32, one can write:

$$\int_{\delta_0}^{\delta_1} \left(P_m - \frac{E \cdot V_\infty}{X'_d + X_t + X_{LT}} \cdot \sin(\delta) \right) d\delta = A_1 \qquad (1.35)$$

$$\int_{\delta_1}^{\delta_{max}} \left(\frac{E \cdot V_\infty}{X'_d + X_t + X_{LT}} \cdot \sin(\delta) - P_m \right) d\delta = A_2 \qquad (1.36)$$

Consider a system with the parameters given in Table 1.

Table 1: Test system parameters.

$V_1 = 1.05\ pu$	$V_\infty = 1.00\ pu$	$P_m = 0.8\ pu$
$X'_d = 0.2\ pu$	$X_t = 0.1\ pu$	$X_{LT} = 0.6\ pu$

Solving the system for its initial condition:

$$\theta_1 = \text{asin}\left(\frac{P_m \cdot \left(X_t + \frac{X_{LT}}{2}\right)}{V_1 \cdot V_\infty}\right)$$

$$\theta_1 = 17.74°$$

$$\hat{I} = \frac{\left(V_1 \cdot e^{j\theta_1} - V_\infty\right)}{j \cdot \left(X_t + \frac{X_{LT}}{2}\right)}$$

$$\hat{I} = 0.8 \angle 0° \ pu$$

$$\hat{E} = V_1 \cdot e^{j\theta_1} + j(X_d' \cdot \hat{I})$$

$$\hat{E} = 1.109 \angle 25.64° \ pu$$

$$\delta_0 = 25.64°$$

As for the equilibrium point, the machine angle δ_1 can be calculated as:

$$\delta_1 = \text{asin}\left(\frac{P_m \cdot (X_d' + X_t + X_{LT})}{E \cdot V_\infty}\right)$$

$$\delta_1 = 40.48°$$

The acceleration area, A_1, can be calculated by equation 1.31:

$$A_1 = P_m \cdot (\delta_1 - \delta_0) + \left\{\frac{E \cdot V_\infty}{X_d' + X_t + X_{LT}} \cdot [\cos(\delta_1) - \cos(\delta_0)]\right\}$$

$$A_1 = 0.0336$$

As for the δ_{max}, it can be calculated integrating equation 1.32 and solving the following equation, assuming $A_2 = A_1$:

$$P_m \cdot (\delta_{max} - \delta_1) - \left\{\frac{E \cdot V_\infty}{X_d' + X_t + X_{LT}} (\cos(\delta_{max}) - \cos(\delta_1))\right\} = A_1$$

$$\delta_{max} = 56.72°$$

Therefore, the maximum angle deviation that the machine may have is 56.72° in order to maintain its angular stability.

4. Voltage Stability

The voltage stability definition and classification was presented in Section 1. Some of the techniques used for voltage stability evaluation are presented below. The analyzes are carried out from the static and dynamic approaches.

4.1 Static Analysis

Static analyzes are recommended for real-time voltage stability studies, especially when it is necessary to know the system behavior in the face of a great number of disturbances or operating conditions. The PV and QV curves play a key role among the most known static methods since the first one allows determining the load margin and the latter indicates the reactive power margin and the best places for power system reinforcement.

4.1.1 PV Curve

PV curves show the voltage variation at a bus with respect to load variation. The latter also shows the maximum load that a system can supply due to the transmission power limits, which is called the maximum loading point (*MLP*). The additional load that causes voltage collapse is designated as load margin (*LM*). The *LM* can also be defined as the load difference between the base case (P_0) and P_{MLP}. Figure 10 shows a PV curve highlighting the quantities defined above. Note that there is a unique voltage level in MLP, called the critical voltage (V_{cr}) (Marujo et al. 2015).

According to Fig. 10, there are two voltage levels for a given power P_0 less than P_{MLP}. At point V_1 the voltage is high and the current is low, corresponding to a stable operating point. At point V_2, the voltage is low and the current high, which represents an unstable operation point.

If the power system operates with a small load margin, the greater the possibility of instability problems when disturbances occur. In this way, the PV curves are determined for important buses of the system, in order to sustain the voltage levels of the region, and to the critical buses, which are those located in the areas where the instability phenomenon usually starts. From the voltage collapse point of view, the critical bus is that which has the greatest voltage variation when a system parameter changes.

The PV curves can be obtained by performing successive power flow calculations considering the increase in load and/or generation. This process ends when the method diverges. However, this methodology is not very efficient, since some problems of convergence are found near the

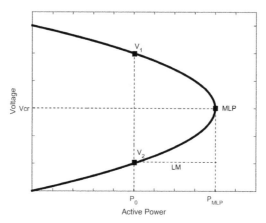

Fig. 10: PV curve.

MLP point. In addition, this process would require manual interventions, which would make the system slower and prone to errors. An alternative approach to overcome such drawbacks is to use the continuation method (Ajjarapu and Christy 1992). The latter is used in power systems to find a continuum of power flow solutions starting at some equilibrium point based on a predictor-corrector scheme including a parameter in the power flow equations, which drives the system from one equilibrium point to another (Marujo et al. 2015). The new set of power flow equations including the system parameter is written as:

$$f(x, \sigma) = 0 \tag{1.37}$$

where x and σ represents the state variables and the system parameter.

The continuation method starts with the predictor step. By using the Tangent Vector (TV), the latter defines the parameter variation, as shown in Equation (1.38) (Zambroni de Souza et al. 1997)

$$TV = \begin{bmatrix} \Delta\theta \\ \Delta V \end{bmatrix} \frac{1}{\Delta\lambda} = J^{-1} \begin{bmatrix} P_0 \\ Q_0 \end{bmatrix} \tag{1.38}$$

where J^{-1} stands for the inverse of system Jacobian matrix. P_0 and Q_0 represent the net power in the nodes. Then, the step size is given as shown in (1.39), where $\|\cdot\|$ indicates the Euclidean norm and k a constant that can accelerate or decelerate the process

$$\Delta\sigma = \frac{k}{\|TV\|} \tag{1.39}$$

Next, the corrector step is executed by using a conventional power flow program.

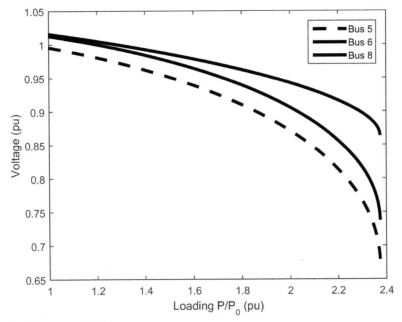

Fig. 11: PV curves of load buses.

By considering the 9 Bus Test System shown in Section 3.2 and using the Continuation method, Fig. 11 shows the PV curves for buses. The load margin is 1.38 p.u. and the critical bus in Bus 5.

4.1.2 QV Curve

The voltage stability depends on how the variation of Q and P affects the voltages levels in load buses. The influence of the reactive power of some equipment is more visible in the QV relationship. QV curves show the sensitivity and variation of bus voltages with respect to reactive power injections or absorptions (Kundur 1994). The latter also depicts the reactive power margin of a bus. This method was initially developed to avoid convergence difficulties in power flow programs (Chowdhury and Taylor 2000). The two main pieces of information provided by the QV curve are highlighted in Fig. 12: Reactive Load Margin (RLM) and the Voltage Stability Limit Point (VSLP).

The reactive load margin is defined as the distance from the minimum point of the QV curve to the voltage axis. A bus has positive RLM when the minimum point of the QV curve is below the voltage axis. If the minimum point is above the voltage axis, the RLM is negative. When RLM is positive, its absolute value indicates the amount of reactive power that this bus can supply to the system. If RLM is negative, this margin shows that the system has a reactive power deficit, requiring an

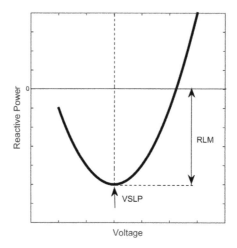

Fig. 12: QV curve.

additional supply of reactive power in order to avoid the voltage collapse (Kundur 1994).

The minimum point of the QV curve, where a derivative dQ/dV is zero, represents the VSLP. If the operating point is on the right side of the QV curve, the system is stable, since an increase the voltage level is accompanied by an increase in reactive power. On the other hand, when the operation point is on the left side, the system is unstable (Kundur 1994).

The specific case in which the system is operated on the unstable side represents an abnormal operating region because of the voltage and reactive power control act in the opposite way. In fact, assuming a load increase pattern in a given system, it can be shown that under certain conditions the system's operating point may be led from the right side on the unstable side.

Figure 13 shows six QV curves for different loading conditions. The dots at the bottom of each curve represent the boundary of the stable and unstable sides. The intersection between each curve and the vertical dashed line represents the operating points. As one can see, at the initial condition point, the system is operated at the right side of the QV curve. However, as the loading is increased, the operation point moves to the unstable side.

Consequently, when the system is on the unstable side of the QV curve, an inadvertently sequence of actions can lead to a Voltage Collapse. A correct adjustment of voltage setpoint at buses, reactive power redispatch or active power redispatch may be performed as corrective actions in such scenarios (Marujo et al. 2015).

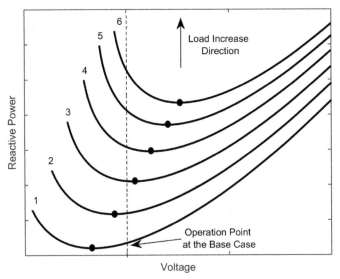

Fig. 13: Effect of operating point overloading increase.

In order to obtain the QV curve the following steps must be considered (Chowdhury and Taylor 2000):

1. Consider the connection of a fictitious synchronous compensator in the bus under analysis, without reactive power limits. This step represents converting the bus into PV. If the bus under analysis is originally a generation bus, it is enough to simply neglect the reactive power limits;

2. Change the terminal voltage level of the bus, preferably in small steps;

3. Run the power flow;

4. Save the terminal voltage (independent variable) and the reactive power supplied by the synchronous compensator (dependent variable) in the bus under analysis;

5. Repeat steps 2 to 4 until enough points have been obtained;

6. Plot the curve from the points obtained in the previous steps.

Another way of obtaining the QV curve with reduced computational effort and with robustness is the QV Continuation Method. The latter is very similar to the Continuation Method used to obtain the PV curve. The idea of this approach is to determine the size of the voltage step in a controlled manner and subsequently to correct the voltage level in order to facilitate the convergence. Further information from this methodology can be found in (Mohn and Zambroni de Souza 2006).

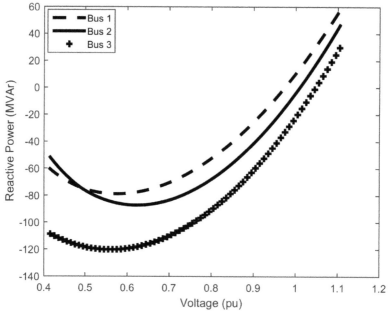

Fig. 14: QV curves of generation buses.

By considering the IEEE 9 Bus shown in Fig. 3, the QV curves of the system generation buses are depicted in Fig. 14. The reactive load margins are 78.9, 87.2 and 120.2 for Bus 1, Bus 2 and Bus 3, respectively. At this operating point, the generators connected to these buses can still provide reactive power to the system without the latter becoming voltage unstable since all the reactive load margins are positive. In (Zambroni et al. 2011), the QV and PV curves are used in the planning scenario, by considering system reinforcements and contingency screening. Furthermore, the long-term voltage security analysis are carried out in (Ferreira et al. 2013), based on information provided by P–V and Q–V curves.

4.2 *Transient and Long Term Voltage Stability Analysis*

The transient and long-term studies provide a more accurate version of the dynamics involved in voltage instability, through a more detailed modeling of generators and loads. This approach is important for studies involving disturbance analysis and coordination of controls and protections.

The transient stability analysis is performed by considering the set of equations (1.1). The devices that have the greatest influence on voltage stability in this period are the generators, induction motors, HVDC and Static Var Compensator. Other elements and processes associated with voltage control are not considered because the simulation time is small when compared to the time of operation of under-load tap changers and over-excitation limiters (OLX), among others. On the other hand, when the study covers the long-term voltage stability, the effect of the discrete changes (z), the evolution of the load over the time w and the operation of devices with slow dynamics should be considered. In this case, the power system model is represented by the equations (1.40)–(1.43) (Cutsem and Vournas 1996, Cutsem and Vournas 2008):

$$\dot{x} = f(x, y, z, w) \tag{1.40}$$

$$0 = g(x, y, z, w) \tag{1.41}$$

$$z(k + 1) = h(x, y, z(k), w) \tag{1.42}$$

$$w = \phi(t) \tag{1.43}$$

Another way to assess the long-term stability is to consider the Quasi-Steady State approach (QSS) (Zambroni de Souza and Lopes 2009). The latter is computationally faster than the transient simulation. In comparison with the set (1.40)–(1.41), the QSS analysis consider $\dot{x} = 0$, in equation (1.40), since the devices with fast dynamics have already reached the steady state. Then, the set of equations to be solved becomes purely algebraic, such that the solution is affected by the actuation of the elements associated with equations (1.42) and (1.43). Reference (Cutsem and Vournas 2008) presents the original approach to QSS simulation.

In the long-term approach, the equations of discrete nature (1.42) capture the transition imposed by controls, protection and limiters according to well-known time scales.

In order to demonstrate the transient and long-term simulation of the voltage stability phenomena, the IEEE 9 Bus system is used again. The transformers between lines 1–4 and 2–7 are now equipped with Load Tap Changing, with the objective of maintaining the voltage at Bus 4 and Bus 7 in 1.025 pu. The following sequence of events is considered:

o The load at Bus 5 is increased in 45% at t = 10 seconds;

o A 5MVAr capacitor is switched on at Bus 5 in t = 70 seconds;

o Line 7–8 trips out at t = 100 seconds.

o The load at Bus 5 is shed in 30% at t = 160 seconds.

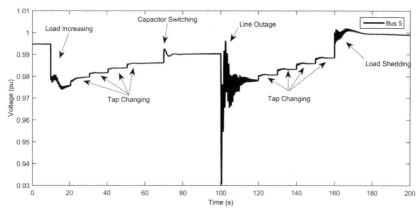

Fig. 15: Effect of successive voltage control actions.

The simulation is performed for 200 seconds. Figure 15 shows the voltage at Bus 5, such that the major events that occur over time are highlighted. The voltage is held constant up to 10 seconds when the load at Bus 5 increases by 45%. Consequently, the voltage levels of the load buses decrease. In this way, the voltage in the buses controlled by the LTCs leaves the reference value. The LTC acts in order to recover the voltage level of the controlled buses, such that the effect of this control action reflects in the other load buses of the system. The capacitor switching occurs in 70 seconds, further raising the voltage level. At t = 100 seconds, the line 7–8 is tripped. Again, the voltage is reduced and successive LTC actions are required to restore the voltage level. Finally, load shedding is performed at Bus 5 in t = 160 seconds, such that the voltage level is increased.

References

Adbi, H. 2007. The eigen-decomposition: Eigenvalues and eigenvectors. pp. 304–308. *In*: Neil Salkind (ed.). Encyclopedia of Measurement and Statistics. Thousand Oaks, CA: Sage.

Ajjarapu, V. and C. Christy. 1992. The continuation power flow: A tool for steady state voltage stability analysis. IEEE Transactions on Power Systems 7: 416–423.

Anderson, P.M. and A.A. Fouad. 2003. Power System Control and Stability. John Wiley & Sons, USA.

Chowdhury, B.H. and C.W. Taylor. 2000. Voltage stability analysis: Power flow simulation versus dynamic simulation. IEEE Transactions on Power Systems 15: 1354–1359.

Cutsem, T.V. and C.D. Vournas. 1996. Voltage stability analysis in transient and mid-term time scales. IEEE Transactions on Power Systems 11: 146–154.

Cutsem, T.V. and C. Vournas. 2008. Voltage Stability of Electric Power Systems. Springer.

Dobson, I., F. Alvarado and C.L. Demarco. 1992. Sensitivity of Hopf bifurcations to power system parameters. Proceedings of the 31st Conference on Decision and Control, Tucson, Arizona.

Ferreira, L.C.A., J.A. Passos Filho, A.C. Zambroni de Souza and J.C. Stacchini de Souza. 2013. Enhanced long-term voltage security assessment through a methodology that combines

P–V and Q–V curves analyses. Journal of Control, Automation and Electrical Systems 24: 702–713.

Kundur, P. 1994. Power System Stability and Control. McGraw-Hill, New York-USA.

Kundur, P., J. Paserba, V. Ajjarapu, G. Andersson, A. Bose, C. Canizares, N. Hatziargyriou, D. Hill, A. Stankovic, C. Taylor, T. Van Cutsem and V. Vittal. 2004. Definition and classification of power system stability: IEEE/CIGRE joint task force on stability terms and definitions. IEEE Transactions on Power Systems 19: 1387–1401.

Marujo, D., A.C. Zambroni De Souza, B.I.L. Lopes, M.V. Santos and K.L. Lo. 2015. On control actions effects by using QV curves. IEEE Transactions on Power Systems 30: 1298–1305.

Mello, L.F., A.C. Zambroni de Souza, G.H. Yoshinari, Jr. and C.V. Schneider. 2006. Voltage collapse in power systems: Dynamical studies from a static formulation. Mathematical Problems in Engineering, Article ID 91367, http://dx.doi.org/10.1155/MPE/2006/91367.

Mithulananthan, N., C.A. Cañizares and J. Reeve. 1999. Hopf bifurcation control in power systems using power system stabilizers and static var compensators. North American Power Symposium (NAPS), San Luis Obispo, California.

Mohn, F.W. and A.C. Zambroni De Souza. 2006. Tracing PV and QV curves with the help of a CRIC continuation method. IEEE Transaction on Power Systems 21: 1115–1122.

Padiyar, P.K. 2008. Power System Dynamics: Stability and Control. BS Publications, Hyderabad-India.

Pal, B. and B. Chaudhuri. 2005. Robust Control in Power Systems. New York, NY: Springer.

Pulgar-Painemal, H.A. and P.W. Sauer. 2009. Bifurcations and loadability in power system. IEEE Bucharest Power Tech Conference, Bucharest, Romania.

Sauer, P.W. and M.A. Pai. 1997. Power System Dynamics and Stability. Campaign, IL:Stipes.

Seydel, R. 1994. Practical Bifurcation and Stability Analysis. From Bifurcation to Chaos. Second Edition, Springer-Verlag.

Zambroni de Souza, A.C., C.A. Cañizares and V.H. Quintana. 1997. New techniques to speed up voltage collapse computations using tangent vectors. IEEE Transaction on Power Systems 12: 1380–1387.

Zambroni de Souza, A.C., B.I.L. Lopes, R.B.L. Guedes, N.G. Brettas, A.C.P. Martins and L.F. Mello. 2005. Saddle-node index as bounding value for Hopf bifurcations detection. IEE Proceedings - Generation, Transmission and Distribution 9: 737–742.

Zambroni de Souza, A.C. and B.I.L. Lopes. 2009. Unified computational tool for transient and long-term stability studies. IET Electric Power Applications 3: 173–181.

Zambroni de Souza, A.C., F.W. Mohn, I.F. Borges and T.R. Ocariz. 2011. Using PV and QV curves with the meaning of static contingency screening and planning. Electric Power Systems Research 81: 1491–1498.

CHAPTER 8

Electricity Markets Operation with Renewable Energy Sources

George Cristian Lazaroiu[1,2,*] and *Virgil Dumbrava*[1]

1. Introduction

In order to comply with the international and national procedures (European Union directives, the electrical energy laws, network codes), the system operators must ensure the access of all users, including producers holding renewable energy sources, to the public electrical network (Romanian Electricity Market Operator 2018). The important benefits, which these electrical energy generators that use renewable sources offer regarding environment protection, diversification of energy sources, technology development and innovation, reduced cost of primary energy (especially for wind and solar), as well as the inexhaustible character, makes them very attractive for a society's sustainable development. However, the constraints related to safe operation of the electrical power system must be carefully considered (Romanian Transmission System Operator 2018).

The constraints on renewable energy based plants depend on a series of factors like location, voltage level of the public network at point of common coupling, the characteristics of the network where the plant is interconnected, as well as the power system's characteristics. The constraints on local injected powers are determined by the transfer

[1] University POLITEHNICA of Bucharest—Department of Power Systems, Splaiul Independentei 313 – 060042 Bucharest – Romania.
[2] University Maritima of Constanta, Mircea cel Bătrîn 104 – 900663 Constanta – Romania.
* Corresponding author: cristian.lazaroiu@upb.ro

capacity of the area network lines, and the capacity to evacuate/use the generated power. In addition, constraints on the admitted power quality levels at the point of common coupling can exist in practice.

At the power system level, the constraints can be determined by the generation adequacy, by the insertion in the production profile, as well as the existence within the network area of sufficient energy reserve. In addition, the typology and power of fossil fuelled power plants, their flexibility (the nuclear power plants have low flexibility), the number of interconnection lines, and the voltage and frequency control systems are all considered when evaluating the maximum power of renewable sources to be connected to the network.

Another constraint imposed by the renewable sources production, considering their participation to the generation schedule profile and for the safe operation of power system, is that the classical power plants must provide the frequency regulation and are to operate at partial capacity or even to be turned off. This leads to increased operation costs and lifetime reduction of classical power plants' equipment. By increasing the installed power in renewable energy sources, a supplementary power in classical power plants and cross-border interconnections must be guaranteed.

The procedures used for eliminating the congestions may impact the share of produced energy by renewable energy sources. Thus, in some power systems, the renewable energy sources can be turned off during outages or congestions that can jeopardize the safe operation of the power system.

The constraints presented above highlight the necessity that, for the interconnection of renewable energy sources to the power system, an in-depth analysis regarding the safe operation of the network must be carried on. The capability of the power system to balance the demand, at any moment, considering the outages of system components (planned or not-planned with their associated probability) and the variability of the renewable energy sources, has to be investigated for different areas, as well as on the entire power system.

The stochastic optimization applied to energy system was discussed in detail in Powell and Meisel (2015, 2011), from the point of view of characteristics and implementation. The security constrained unit commitment modelled as a stochastic optimization problem was conducted in (Papavasiliou and Neill 2011, Li and Shahidehpour 2005, Khazali and Kalanta 2015). The analysis of power systems in presence of intermittent renewable sources and the necessary reserve for safe operation were discussed in (Pappala et al. 2009, Ortega-Vazquez and Kirschen 2010, 2009). The spinning reserve requirements, considering errors in the wind power generation and load forecasts, were determined through an optimization model minimizing the operating and the socio-economic costs associated with load shedding. The elaboration of the stochastic optimization model for market clearing, under demand forecast uncertainties, was conducted in Bouffard et al. 2005a. The proposed model was solved on an IEEE RTS test network in Bouffard et al. 2005b, the results revealing that the unintended load shedding value has high impact on reserve requirements. The reserve requirements in power systems with high wind power integration, computed based on a probabilistic method, such that to ensure a specific level of reliability was carried on in Meibom et al. 2011. A stochastic optimization problem is formulated for spot market clearing and reserve requirement determination in power systems with high penetration of wind generation (Morales et al. 2009, Jafari et al. 2014, Ahmadi-Khatir et al. 2013). The proposed model optimizes the day-ahead and intra-day market operation, and determines the reserve requirements for safe operation of the power system. The optimization model was built considering in the power system thermal power plants and wind generators.

With respect to the previous researches, the elaborated stochastic optimization model accounts for the spot market clearing and real-time balancing in a real case power network. The analysed power system contains thermal generators and hydro power plants with limited flexibility, a nuclear power plant is practically inflexible, 7 wind generators each of them having different production scenarios, and real practice inflexible demand. The analysis is carried on over a 24-h horizon, the analysis being conducted on each base interval with duration of 1 hour. The obtained results reveal that a 37% higher cost for power system operation, compared to the case when the wind generation is perfectly matching the production value notified to the transmission system operator, can occur.

2. Mathematical Model

The mathematical model is elaborated considering the operation of the electricity market facing various regimes of operation, with high variability, of renewable energy sources and the availability of fossil fuelled power plants and hydro power plants for ensuring safe power system functioning. In agreement with the Romanian electricity market operation, the participants (producers and consumers) to the day-ahead market submit bids of pairs energy-price for buying or selling energy in this market and for reserve capacity (Romanian Electricity Market Operator 2018). The electricity market operator clears the market and establishes the day-ahead schedule of dispatched generators and required reserve, considering the notified productions received from the renewable energy sources. The procedure is carried on for each one hour base interval of the 24 hours of the operation day. The central system operator can revise the decisions function of power system safe operation and reliability requirements. Even if the renewable energy producers receive production forecasts from several different sources with some accuracy, during real-time operation, the production of renewable energy sources can largely deviate from the notified values (Dumbrava et al. 2011). In this case, the system operator has to take real-time balancing decisions by operating additional classical power plants or hydro power plants, or adopting other countermeasures like wind spillage and load shedding. This action can incur important supplementary costs for the system operator.

The stochastic mathematical model is formulated for minimizing the objective function *FOB*, i.e., the power system total costs, of operating the day-ahead market (dispatching power and reserve) and the real-time balancing market, without renewable energy sources and under different scenarios (probabilities π_s) of production from renewable energy sources (King and Wallace 2014, Shapiro et al. 2014). Within the model, thermal power plants and hydro power plants are considered to provide power and reserve services, i.e., upward and downward reserve. As in Romania, the installed power in wind power plants is relatively high (about 24% of the entire generation portfolio), wind is considered within this chapter as the only renewable energy source and the only source of uncertainty modelled through production scenarios probabilities (Morales et al. 2010, Baringo and Conejo 2013, Oskouei and Yazdankhah 2010). The owners of wind power plants sell all their notified wind power production. The load is considered to be not-flexible, for safe operation of power system load shedding is implemented. The network power flow is elaborated using a DC power flow approximation, being limited by the maximum transmission lines capacity and neglecting power losses.

The mathematical model is formulated as follows (all variable definitions are given in the Nomenclature at the end of this chapter):

$$[\text{MIN}]\ FOB = \sum_{t=1}^{24} \left\{ \begin{array}{l} \left[\begin{array}{l} \sum\limits_{g \in G} \left(C_{g,t} \cdot P_{g,t} + C_{g,t}^{up} \cdot R_{g,t}^{up} + C_{g,t}^{dwn} \cdot R_{g,t}^{dwn} \right) + \\[2mm] \sum\limits_{h \in H} \left(C_{h,t} \cdot P_{h,t} + C_{h,t}^{up} \cdot R_{h,t}^{up} + C_{h,t}^{dwn} \cdot R_{h,t}^{dwn} \right) \end{array} \right] + \\[8mm] \sum\limits_{s \in S} \pi_s \cdot \left[\begin{array}{l} \sum\limits_{g \in G} \left(Cr_{g,t}^{up} \cdot r_{g,t,s}^{up} - Cr_{g,t}^{dwn} \cdot r_{g,t,s}^{dwn} \right) + \\[2mm] \sum\limits_{h \in H} \left(Cr_{h,t}^{up} \cdot r_{h,t,s}^{up} - Cr_{h,t}^{dwn} \cdot r_{h,t,s}^{dwn} \right) + \\[2mm] + \sum\limits_{l \in L} \lambda_{l,t}^{shed} \cdot L_{l,t,s}^{shed} + \sum\limits_{y \in Y} \left(\lambda_t^{DAM} + \lambda_t^{GCM} \right) \cdot P_{y,t,s}^{spill} \end{array} \right] \end{array} \right\}$$

(1)

subject to the following constraints:
- power system balance constraint on the day-ahead market:

$$\sum_{g \in G} P_{g,t} + \sum_{h \in H} P_{h,t} + \sum_{y \in Y} P_{y,t}^d = \sum_{l \in L} L_{l,t} + \sum_{i,j \in N} \frac{\left(\delta_{i,t} - \delta_{j,t}\right)}{X_{i,j}} \qquad (2)$$

- constraints related to day-ahead market operation:
  - dispatched power plus scheduled reserve to be between the power plant capacity limits:

$$P_{g,t} - R_{g,t}^{dwn} \geq P_g^{min}, \qquad P_{g,t} + R_{g,t}^{up} \leq P_g^{max}, \qquad g \in G, t = 1,..,24$$
$$P_{h,t} - R_{h,t}^{dwn} \geq P_h^{min}, \qquad P_{h,t} + R_{h,t}^{up} \leq P_h^{max}, \qquad h \in G, t = 1,..,24 \qquad (3)$$

  - upward/downward scheduled reserve to be smaller than the reserve capacity offers:

$$R_{g,t}^{up} \leq R_g^{up,\,max}, \qquad R_{g,t}^{dwn} \leq R_g^{dwn,\,max}, \qquad g \in G, t = 1,..,24$$
$$R_{h,t}^{up} \leq R_h^{up,\,max}, \qquad R_{h,t}^{dwn} \leq R_h^{dwn,\,max}, \qquad h \in H, t = 1,..,24 \qquad (4)$$

  - wind power plant schedule to be below the notified power:

$$P_{y,t}^d \leq P_{y,t}^{nfd} \qquad y \in Y, \ t = 1,..,24 \qquad (5)$$

  - production of hydro power plant h on 24 hours to be below maximum available energy:

$$\sum_{t=1}^{24} P_{h,t} \leq E_h^{max} \qquad h \in H \qquad (6)$$

  - DC power flow F between buses i and j smaller than the maximum line flow capacity:

$$\frac{\left(\delta_{i,t} - \delta_{j,t}\right)}{X_{i,j}} \leq F_{i,j}^{max} \qquad \forall i, j \in N, i \neq j \qquad (7)$$

  - real-time market balance constraint for a given realized scenario s:

$$\sum_{g \in G} \left(r_{g,t,s}^{up} - r_{g,t,s}^{dwn}\right) + \sum_{h \in H} \left(r_{h,t,s}^{up} - r_{h,t,s}^{dwn}\right) + \sum_{l \in L} L_{l,t,s}^{shed} +$$
$$+ \sum_{y \in Y} \left(P_{y,s,t} - P_{y,t}^d - P_{y,t,s}^{spill}\right) = \sum_{i,j \in N} \frac{\left(\delta_{i,t} - \delta_{i,t,s} - \delta_{j,t} - \delta_{j,t,s}\right)}{X_{i,j}} \qquad (8)$$

- constraints related to real-time balancing market operation:
- upward/downward deployed reserve to be smaller than the scheduled reserve:

$$0 \le r_{g,t,s}^{up} \le R_{g,t}^{up}, \quad 0 \le r_{g,t,s}^{dwn} \le R_{g,t}^{dwn}, \quad g \in G, t = 1,..,24, s \in S$$
$$0 \le r_{h,t,s}^{up} \le R_{h,t}^{up}, \quad 0 \le r_{h,t,s}^{dwn} \le R_{h,t}^{dwn}, \quad h \in H, t = 1,..,24, s \in S \tag{9}$$

- wind power plant spillage limit:

$$0 \le P_{y,t,s}^{spill} \le P_{y,t,s} \quad y \in Y, \ t = 1,..,24, s \in S \tag{10}$$

- load shedding limit:

$$0 \le L_{l,t,s}^{shed} \le L_{l,t} \quad l \in L, \ t = 1,..,24, s \in S \tag{11}$$

- hydro power plant h availability on 24 hours:

$$\sum_{t=1}^{24} \left(P_{h,t} + r_{h,t,s}^{up} - r_{h,t,s}^{dwn} \right) \le E_h^{max} \quad h \in H, s \in S \tag{12}$$

- power flow on the line $i - j$ during real time operation of the power system:

$$\frac{\left(\delta_{i,t,s} - \delta_{j,t,s} \right)}{X_{i,j}} \le F_{i,j}^{max} \quad \forall i, j \in N, i \ne j, s \in S \tag{13}$$

The mathematical model is composed of the objective function (1), the day-ahead market operation constraints (2)–(7), and real time balancing market constraints (7)–(13). The first two terms in (1) represents the cost of the day-ahead market decisions for energy generation by thermal power plants and hydro power plants, respectively. The third and fourth terms in (1) are the costs associated with real-time balancing market operation, corresponding to the effective deployment of procured reserves from thermal power plants and hydro power plants, for each wind power production scenario. The fifth and sixth terms in (1) correspond to the costs associated with load shedding, and wind power spillage.

The results of the stochastic optimization model (1)–(13) are the day-ahead dispatched powers and reserve capacities (the variables that do not depend on scenario s realization), and the real-time balancing market deployed reserve capacities and curtailment actions (load shedding and wind spillage) characteristic to each scenario realization.

3. Case Studies

The mathematical model (1)–(13) is initially applied on a small scale illustrative network, and afterwards on a larger real-case network.

3.1 Test Network

A 6 bus test system with 5 thermal power plants, 1 hydro power plant, 2 wind turbines and 2 aggregated loads is illustrated in Fig. 1. The characteristic data of the thermal and hydro power plants are reported in Table 1, the hydro power plant being located at bus 4. The reactance of the transmission lines is 0.10 p.u. on a 100 MVA base, with a maximum power flow capacity of 100 MW on each line. The load demand at each bus is reported in Table 2. The wind power production scenarios, for the wind turbines placed at buses 1 and 6, are reported in Table 3 using real

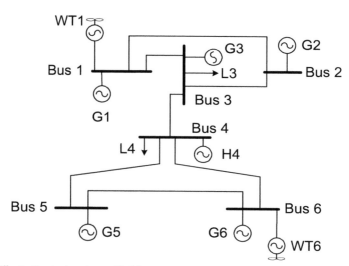

Fig. 1: Illustrative test system with 6 buses.

Table 1: Characteristic data of thermal and hydro power plants.

Generator	G1	G2	G3	H4	G5	G6
P^{max} (MW)	50	50	50	55	62	62
P^{min} (MW)	0	0	0	0	0	0
C (Euro/MWh)	40.88	42	55.55	32	78.51	81
C^{up} (Euro/MWh)	14	14.2	15.55	7	17.8	18
C^{dwn} (Euro/MWh)	14	14.2	15.55	7	17.8	18
Cr^{up} (Euro/MWh)	40.88	42	55.55	32	78.51	81
Cr^{dwn} (Euro/MWh)	40.88	42	55.55	32	78.51	81

Table 2: Load data.

Time period	1	2	3	4	5	6
L_3 (MW)	137.6	133.3	131.1	130.1	131	133.8
L_4 (MW)	102.5	97.39	93.54	90.18	88.35	87.43

Table 3: Wind power production scenarios.

Time period (hours)	WPP bus 1 (MW)				WPP bus 6 (MW)			
	s_1	s_2	s_3	s_4	s_1	s_2	s_3	s_4
t1 (1 hr)	7.09	9.45	25.98	4.72	27.00	25.44	20.00	30.00
t2 (1 hr)	6.29	8.39	23.07	4.19	9.27	30.00	21.43	10.72
t3 (1 hr)	5.85	7.80	21.45	3.90	9.83	25.14	30.00	10.72
t4 (1 hr)	5.39	7.19	19.78	3.60	17.92	23.90	30.00	11.95
t5 (1 hr)	4.94	6.59	18.11	3.29	30.00	30.00	24.00	20.62
t6 (1 hr)	4.77	6.36	17.49	3.18	20.41	27.21	30.00	13.60

production data from a 26 MW 30 MW wind turbine and respectively. For the four wind power production scenarios, the scenario production probabilities are 0.15, 0.2, 0.55, and 0.1 respectively. All the terms used in Tables 1–3 are defined in the Nomenclature. For the case when the wind power production is exactly equal with the notified production, no reserve is scheduled on the day-ahead market. As there is no wind power production deviation, no reserve is deployed on the real time balancing market in either scenario. The total cost associated with this operation is 40438.80 Euros.

The system operation under the 4 wind power production scenarios is illustrated in Fig. 2. The hydro power plant, being the cheapest classical energy source within the network, is operated at its full capacity of 55 MW and it is not contributing to up/down reserve. Some thermal power plants are operated below their maximum capacity, contributing to reserve requirements.

In the real-time balancing market, under the 4 production scenarios of the 2 wind power plants reported in Table 3, the deployed reserve and operation results are shown in Fig. 3. As illustrated, in conditions of low (scenario 1) or high (scenario 3) wind power productions, upward reserve r^{up}_{s1} =83.79 MW and, downward reserve r^{dwn}_{s3} = 50.74 MW, are deployed respectively. The total cost associated with this case of operation is 48744.28 Euros, which is 20.54% higher with respect to the case when the wind power production is considered as perfectly accurate.

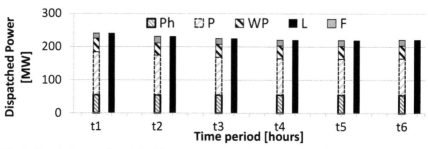

Fig. 2: Simulation results of the 6 bus system operation on the day-ahead market, under 4 different wind power production scenarios.

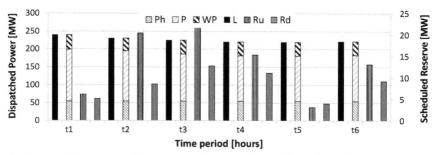

Fig. 3: Simulation results of the 6 bus system operation on the real-time balancing market, under 4 different wind power production scenarios.

3.2 Large Case Network

A real case large system with 25 buses, 10 classical thermal power plants, 1 hydro power plant, 1 nuclear power plant, 7 wind power plants, and 16 important loads is illustrated in Fig. 4. In this case study, one nuclear power plant is added to the system and maintained in operation during the whole analysed period and constantly supplying 1400 MW. Since the nuclear power plant usually operates at constant power, it is not considered in the objective function (hence it is not included in the scheduling calculations nor can provide the power reserve for the power system).

However, the nuclear power plant is not neglected in the constraints of the mathematical model for not disturbing the influence of wind power production variability on the thermal and hydro power plants operation.

The operation is conducted in 24 hours, the analysis horizon being composed of 24 periods of 1 hour each. The reactance of transmission lines is illustrated in Fig. 4.

The characteristic data of the thermal and hydro power plants are reported in Table 4. Load curtailment is penalized with a factor of 10 times the average price of day-ahead market price, and a penalty factor of

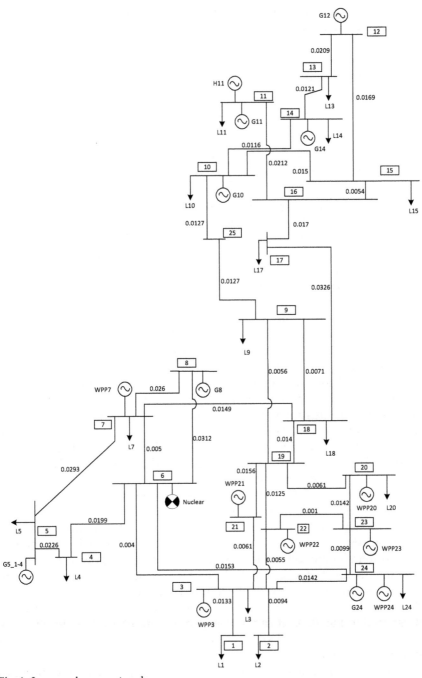

Fig. 4: Large scale case network.

Table 4: Characteristic data of thermal and hydro power plants.

Generator bus	G12	G14	G11	G10	G8	G24	G5_1	G5_2	G5_3	G5_4	H11
P^{max} (MW)	50	100	200	420	160	50	155	155	155	155	210
P^{min} (MW)	0	0	0	0	0	0	0	0	0	0	0
C (Euro/MWh)	40.88	40.88	42	42	40.88	55.55	40.88	55.55	78.51	81	32
C^{up} (Euro/MWh)	12	14	14	14	14.2	15.55	14	15.55	17.8	18	7
C^{dwn} (Euro/MWh)	12	14	14	14	14.2	15.55	14	15.55	17.8	18	7
Cr^{up} (Euro/MWh)	40.88	40.88	42	42	40.88	55.55	40.88	55.55	78.51	81	32
Cr^{dwn} (Euro/MWh)	40.88	40.88	42	42	40.88	55.55	40.88	55.55	78.51	81	32

1000 Euro/MWh, respectively. For modelling the possible output production of the 7 wind power plants during the analysis horizon, 92 scenarios with equal probability were generated based on real-outputs of wind power plants located in south-eastern region of Romania. In order to release the computation burden associated with this large number of scenarios for 7 wind power plants, these scenarios were reduced to the most 5 representative scenarios (Dupacova et al. 2003). A number of 5 scenarios for each of the 7 wind power plants, located in this real-case network was considered sufficient to test the operation of power system. The wind curtailment is penalized with a factor of 1650 Euro/MWh.

The variation of wind power production, during the scheduling horizon for all the wind power plants, generated through scenarios and the expected value of the aggregate power for the entire wind power generation are illustrated in Fig. 5. In Fig. 6, the resulting 5 scenarios for each wind power plant, and the expected value variation during the 24 hours analysis horizon are shown. The obtained probabilities of the reduced 5 scenarios are 0.1, 0.11, 0.34, 0.29, and 0.16 respectively.

The nuclear power plant production is illustrated in Fig. 7 at the base of the generation profile. In this case, no reserve is scheduled on the day-ahead market. As there is no wind power production deviation, no reserve is deployed on the real time balancing market in either scenario. The total cost associated with this operation is 507868.089 Euros. The real case system operation under the reduced 5 wind power production scenarios is illustrated in Fig. 8. The total thermal power and the hydro power dispatched on the day-ahead market, as well as the total upward/downward scheduled reserve from thermal and hydro units, are illustrated in Fig. 8. As it can be seen, an important active power reserve is scheduled for upward or downward regulation, depending on wind power production.

In the real-time balancing market, under the 5 production scenarios of each of the 7 wind power plants illustrated in Fig. 6, the deployed reserve and operation results for one significant scenario is shown in Fig. 9, for

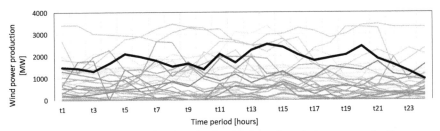

Fig. 5: Generated scenarios and expected value for the entire wind power production.

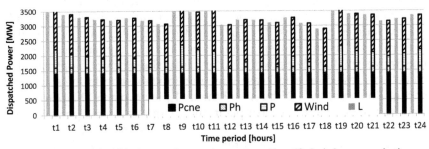

Fig. 6: Reduced scenarios and expected value wind power production for some wind power plant.

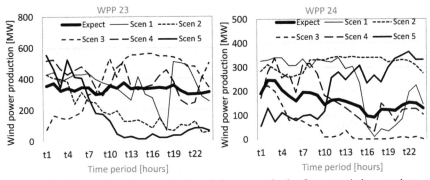

Fig. 7: Simulation results of the large-scale network operation with notified wind power production.

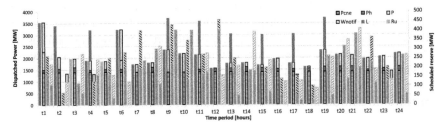

Fig. 8: Simulation results of the large scale network operation on the day-ahead market, under 5 different wind power production scenarios.

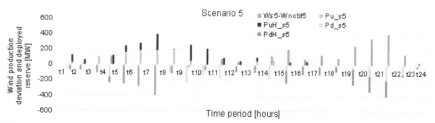

Fig. 9: Simulation results of the large scale network operation on the real-time balancing market, under the most relevant wind power production scenarios.

scenario 5. We illustrate here the case when the first half-day deviation between the wind production and notified wind production is negative, while in the second half-day this deviation is positive. The obtained results reveal the deployed upward and downward reserve, respectively, from all thermal power plants and all hydro power plants.

5. Results Interpretation

This chapter deals with the power system's operation with renewable energy sources. The stochastic optimization model proposed minimizes the costs associated with the day-ahead dispatch and scheduled reserve and the real-time balancing actions, in case of different wind production scenarios deviating from the notified value submitted by the renewable energy sources operators. The elaborated mathematical model including thermal and hydro power plants was initially implemented on an illustrative small scale network. Afterwards, the stochastic optimization model was tested on a real-case large network, existing in the south-east part of Romania, with 25 busses, 9 thermal power plants, 1 hydro plant, 1 nuclear power plant, 1 nuclear plant, and 7 large scale wind power plants. Based on real output wind power productions, 92 scenarios were generated for each wind power plant, and through scenario reduction the five most relevant scenarios for each wind production were used in the simulations carried on in GAMS. For the large scale real network, the obtained cost for the case without reserve deployment (perfectly accurate wind production) is 507868 Euros. For the case with 5 wind production scenarios, the total cost is 698933 Euros, which is 37.62% higher with respect to the case when the wind power production is considered as perfectly accurate. Further analysis will be carried on considering the use of intra-day market for balancing the large mismatches raised by the intermittent wind production.

6. Nomenclature

T	index of time periods running from 1 to 24, each equal with 1 hour;
G	index of classical thermal power plants, $g = 1,.., G$
H	index of hydro power plants, $h = 1,.., H$
L	index of loads, $l = 1,.., L$
Y	index of wind power plants, $y = 1,.., Y$
S	index of wind power production scenarios, $s = 1,.., S$
i, j	index of buses within the set N of area buses, $i, j = 1,.., N$
$C_{g,t}$	offer cost of thermal power plant g in period t (Euro/MWh)
$C_{h,t}$	offer cost of hydro power plant h in period t (Euro/MWh)
$C_{g,t}^{up}, C_{g,t}^{dwn}$	offer cost for upward/downward reserve, deployed by unit g in period t (Euro/MWh)
$R_{g,t}^{up}, R_{g,t}^{dwn}$	scheduled upward/downward reserve capacity, for thermal unit g in period t (MW)
$R_{g,t}^{up,max}, R_{g,t}^{dwn,max}$	maximum scheduled upward reserve capacity, respectively downward reserve capacity, for thermal unit g in period t (MW/h)
$C_{h,t}^{up}, C_{h,t}^{dwn}$	offer cost for upward reserve, respectively downward reserve, deployed by unit h in period t (Euro/MWh)
$R_{h,t}^{up}, R_{h,t}^{dwn}$	scheduled upward reserve capacity, respectively downward reserve capacity, for hydro unit h in period t (MW)
$R_{h,t}^{up,max}, R_{h,t}^{dwn,max}$	maximum scheduled upward reserve capacity, respectively downward reserve capacity, for hydro unit h in period t (MW/h)
$Cr_{g,t}^{up}, Cr_{g,t}^{dwn}$	offer cost for upward reserve, respectively downward reserve, deployed by unit g in period t (Euro/MWh)
$r_{g,t,s}^{up}, r_{g,t,s}^{dwn}$	upward reserve, respectively downward reserve, deployed by thermal unit g, in period t, and scenario s (MW)
$Cr_{h,t}^{up}, Cr_{h,t}^{dwn}$	offer cost for upward reserve, respectively downward reserve, deployed by unit h in period t (Euro/MWh)
$r_{h,t,s}^{up}, r_{h,t,s}^{dwn}$	upward reserve, respectively downward reserve, deployed by hydro unit h, in period t, and scenario s (MW)
$\lambda_{l,t}^{shed}$	penalty cost of load shedding in period t (Euro/MWh)

$L_{l,t,s}^{shed}$	load shedding imposed to load l, in period t, and scenario s (MW)
λ_t^{DAM}	price of electrical energy on day-ahead market (DAM) (Euro/MWh)
λ_t^{GCM}	price of 1 green certificates (for each 1 MWh produced from renewable sources) on the green certificate market (GCM) (Euro/MWh)
$P_{y,t,s}^{spill}$	wind power spillage to plant y, in period t, and scenario s (MW)
$P_{g,t}$	scheduled output power of thermal unit g in period t (MW)
$P_{h,t}$	scheduled output power of hydro unit h in period t (MW)
$P_{y,t}^d$	scheduled power of wind power plant y in period t (MW)
$\delta_{i,t}$	scheduled phase angle of bus i in period t (rad)
$\delta_{i,t,s}$	phase angle of bus i in period t and scenario s (rad)
$P_{y,t}^d$	scheduled wind power of plant y in period t (MW)
E_h^{max}	maximum hydro energy available on 24 hours (MWh)
$X_{i,t}$	reactance of the transmission line between buses i and j (p.u.)

Acknowledgement

This work was supported in part by a grant of the Romanian Ministry of Research and Innovation, CCCDI – UEFISCDI, project number PN-III-P1-1.2-PCCDI-2017-0404/31PCCD/2018, within PNCDI III.

References

Ahmadi-Khatir, A., A.J. Conejo and R. Cherkaoui. 2013. Multi-area energy and reserve dispatch under wind uncertainty and equipment failures. IEEE Trans. Power Syst. 28: 4373–4383.

Baringo, L. and A.J. Conejo. 2013. Correlated wind-power production and electric load scenarios for investment decisions. Appl. Energy 101: 475–482.

Bouffard, F., F. Galiana and A. Conejo. 2005a. Market-clearing with stochastic security—Part I: Formulation. IEEE Trans. Power Syst. 20: 1818–1826.

Bouffard, F., F. Galiana and A. Conejo. 2005b. Market-clearing with stochastic security—Part II: Case studies. IEEE Trans. Power Syst. 20: 1818–1826.

Dumbrava, V., P. Ulmeanu, M. Scutariu and G.C. Lazaroiu. 2011. Analysis of reliability aspects of wind power generation in Romania using Markov models. Proc. IET RPG.

Dupacova, J., N. Growe-Kuska and W. Römisch. 2003. Scenario reduction in stochastic programming: An approach using probability metrics. Math Program. 95: 493–511.

Jafari, A.M., H. Zareipour, A. Schellenberg and N. Amjady. 2014. The value of intra-day markets in power systems with high wind power penetration. IEEE Trans. Power Syst. 29: 1121–1132.

Khazali, A. and M. Kalantar. 2015. Spinning reserve quantification by a stochastic–probabilistic scheme for smart power systems with high wind penetration. Energy Convers Manage 96: 242–257.

King, A.J. and S. Wallace. 2012. Modeling With Stochastic Programming. Springer New York, NY.

Li, Z. and M. Shahidehpour. 2005. Security-constrained unit commitment for simultaneous clearing of energy and ancillary services markets. IEEE Trans. Power Syst. 20: 1079–1088.

Meibom, P., R. Barth, B. Hasche, H. Brand, C. Weber and M. O'Malley. 2011. Stochastic optimization model to study the operational impacts of high wind penetrations in Ireland. IEEE Trans. Power Syst. 26: 1367–1379.

Morales, J.M., A.J. Conejo and J. Pérez-Ruiz. 2009. Economic valuation of reserves in power systems with high penetration of wind power. IEEE Trans. Power Syst. 24: 900–910.

Morales, J.M., A.J. Conejo and R. Minguez. 2010. A methodology to generate statistically dependent wind speed scenarios. Appl. Energy 87: 843–855.

Ortega-Vazquez, M. and D. Kirschen. 2009. Estimating the spinning reserve requirements in systems with significant wind power generation penetration. IEEE Trans. Power Syst. 24: 2009.

Ortega-Vazquez, M. and D. Kirschen. 2010. Assessing the impact of wind power generation on operating costs. IEEE Trans. Smart Grid 1: 295–301.

Oskouei, M.Z. and A.S. Yazdankhah. 2010. Scenario-based stochastic optimal operation of wind, photovoltaic, pump-storage hybrid system in frequency-based pricing. Energy Convers Manage 105: 1105–1114.

Papavasiliou, A., S.S. Oren and R.P.O. Neill. 2011. Reserve requirements for wind power integration: A scenario-based stochastic programming framework. IEEE Trans. Power Syst. 26: 2197–2206.

Pappala, V.S., I. Erlich, K. Rohrig and J. Dobschinski. 2009. A stochastic model for the optimal operation of a wind-thermal power system. IEEE Trans. Power Syst. 24: 940–950.

Powell, W.B. and S. Meisel. 2011. Tutorial on stochastic optimization in energy—part II: an energy storage illustration. IEEE Trans. Power Syst. 31: 1468–1475.

Powell, W.B. and S. Meisel. 2015. Tutorial on stochastic optimization in energy—part I: modeling and policies. IEEE Trans. Power Syst. 31: 1459–1467.

Romanian electricity market operator. 2018. [Online]. Available: http://www.opcom.ro

Romanian transmission system operator. 2018. [Online]. Available: http://www.transelectrica.ro/en.

Shapiro, A., D. Dentcheva and A. Ruszczyski. 2014. Lectures on Stochastic Programming: Modeling and Theory, 2nd ed. SIAM-Society for Industrial and Applied Mathematics, Philadelphia.

CHAPTER 9

Integrating Community Resilience in Power System Planning

Lamine Mili,[1,*] *Konstantinos Triantis*[2] and *Alex Greer*[3]

1. Introduction

The Executive Office of the President Report 2013 (EOP 2013) revealed that electric power outages caused by severe weather conditions over the period 2003–2012 resulted in an average financial loss of between $18 billion and $33 billion adjusted for inflation. Uninterrupted power supply has, in fact, become a necessity, particularly during extreme events when it impacts a community's functioning and ability to provide emergency services. These services rely on the continuing operation of the communications, transportation, and water infrastructures, which, in turn, depend on the availability of electric power supply during and after a disaster. To overcome problems due to long lasting power outages (Nigg 1995, Webb et al. 2002, Webb 2000), this chapter investigates the deployment of micro-grids (MGs)—a relatively new technology that can enhance the utilization of scarce resources, save lives and property, and aid in a quick recovery during the aftermath of a disaster making communities more resilient (Abbey et al. 2014, Yang et al. 2009, Hirose et al. 2013).

[1] Bradley Department of Electrical and Computer Engineering, Virginia Tech, Falls Church, VA 22043.

[2] Grado Department of Industrial Systems Engineering, Virginia Tech, Falls Church, VA 22043. Emails: triantis@vt.edu; triantis@vt.edu

[2] Department of Political Science, Oklahoma State University, Stillwater, OK 74078. Email: alex.greer@okstate.edu

* Corresponding author: lmili@vt.edu

Now, what exactly are MGs and how could they improve community resilience? A MG consists of a collection of interconnected small-scale electric generation units, storage devices, and power electronic interfaces that deliver electric energy and heat to local or remote customers. It may operate in a stand-alone or grid-connected mode. It may be owned by households, the so-called prosumers (those who produce and consume electric energy), or by businesses, municipalities, emergency services, or utilities. Because MGs are situated within the communities they serve, they can provide, when endowed with the appropriate technologies, continuing services during and after a disaster. For example, the MG in Fukushima, Japan, continued to supply electricity to critical facilities during and after the tsunami disaster, because of its reliance on gas engines, fuel cells, and photovoltaic panels (Strickland 2011). The MG located in Verizon's Garden City, NY (Abbey et al. 2014) is another successful example. These are in stark contrast to conventional, remotely located, large power plants that deliver electric power through transmission and distribution systems that are prone to outages or destruction (Abbey et al. 2014). Thus, the focus of our study is to understand how MGs can be used to address infrastructure and community resilience.

2. Defining Robustness, Reliability and Resilience for Power Systems

2.1 The Five Operating States of an Electric Power System

In the wake of the 1965 Northeast blackout in the US (UFOS 1965), the electric power system community has embarked on a journey of fruitful research and developments aimed at modernizing the monitoring, protection, and control of electric power generation, transmission, and distribution systems. This endeavor has resulted in many methodological and technological advancements, which include the following: (i) the provision of control centers with mainframes and, later on, with distributed computers; (ii) the initiation of a theoretical framework along with an ensemble of computer-aided functions to partially automate the operation of the transmission system; and (iii) the development of fast algorithms based on sparse matrix techniques for modeling the steady-state and the dynamic operating conditions of the system. The theoretical framework was developed by Dy Liacco (1978) and then Fink and Carlsen (1978) with the definitions of five operating states of a power system which, as depicted in Fig. 1, consist of the normal, alert, emergency, in extremis, and restorative states. These definitions are motivated by the characteristics of the dynamics of a power system, which are governed by a set of nonlinear differential equations and two nonlinear algebraic equations. The latter

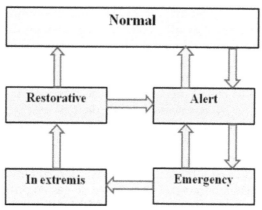

Fig. 1: The five states of an electric power system and their transitions as defined by Fink and Carlsen (1978).

consist of two ensembles of equality and inequality constraints. It is worth recalling these definitions:

Normal state: It is a state characterized by the satisfaction of all the equality and inequality constraints and by a sufficient level of stability margins in transmission and generation so that the system can withstand a single contingency, be it a loss of a transmission line, a transformer, or a generator. In this case the system state is deemed to be secure; consequently, no action is taken.

Alert state: It is a state typified by the satisfaction of all the equality and inequality constraints and by an insufficient level of stability margins, which is an indication that the system is dangerously vulnerable to failures. This means that in the event of a contingency, at least one inequality constraint will be violated such as, for example, that due to the overload of a transmission line or a transformer. To bring the system to a normal state, *preventive actions* have to be taken, typically by increasing the structural redundancy in the transmission system.

Emergency state: It is a state where all the equality constraints are satisfied and at least one inequality constraint is violated, indicating that the system is experiencing overloads. Obviously, the system calls for the immediate implementation of *corrective actions* to remove the overloads, prevent the damage of equipment, and mitigate the risk of cascading failures that may result in a blackout. These actions consist of load shedding, transmission line tripping, transformer outages, or generating unit disconnections.

In extremis state: It is characterized by the violation of both equality and inequality constraints that stem from the chain of actions taken at a previous emergency state, while the transmission network remains interconnected. At this stage, *heroic actions* are implemented to either reconnect the disconnected load and generation if this is at all possible or to perform additional outages to protect the overloaded equipment, which may result in the breakup of the network.

Restorative state: It is a state where the equality and inequality constraints are violated, while the system is breaking up into pieces, resulting in the formation of islands that may be energized or not. Here, restorative actions need to be implemented to bring the system to a normal or alert state.

In view of these definitions, an interesting question arises: when a power system is in a normal state, can the monitoring of its state be relaxed? The answer is negative because of the ever-changing conditions of its loads and structure. Due to these changes, which are stochastic in nature, its operating point may be driven dangerously close to the stability boundaries of the basin of attraction of its current stable equilibrium point. In other words, the safety margins of a power system may quickly erode with time as the internal and external conditions evolve. Consequently, a continuous online assessment of the stability margins of the system has to be executed to check whether it is still in a normal state.

2.2 *Defining Robustness and Resilience for Power Systems*

In contrast to the definitions of robustness and resilience given in ecology (Holling 1973, 1996), or in biology (Kitano 2004), or in complex systems (Carlson and Doyle 2002, Jen 2005), which are inclusive to each other in that robustness includes resilience or vice versa, we argue that for designed systems such as infrastructures, the definitions of these two concepts should be distinct from each other so that they may become useful tools during the design process. Furthermore, they must integrate the two principal characteristics of a system, namely, its function and its structure. For example, the function of an electric power system is to deliver electric energy at contractual voltage magnitudes and frequency within specified ranges. Thus,

Definition 1: The robustness of a system to a *given* class of perturbations is defined as the ability of this system to maintain its function when it is subject to a set of perturbations of this class, which may induce changes in its structure.

Definition 2: The resilience of a system to a class of *unexpected* extreme perturbations is defined as the ability of this system to (i) gracefully degrade its function by altering its structure in an agile way when it is subject to a set of perturbations of this class and (ii) quickly recover it once the perturbations have ceased.

Regarding electric power systems, robustness applies to the normal and alert states and their transitions whereas resilience applies to the emergency and in extremis states and their transitions. As for the restorative state and its transitions to the other states, they call for the definitions of more advanced properties such as *self-reconfiguration* and *self-sustainability*. These properties are outside the realm of resilience or robustness since the system needs to self-restore its integrity.

The definitions given earlier require further elaboration. First, it is important to notice that robustness and resilience are not general properties of a system but are relative to specific classes of perturbations. As pinpointed by Carlson and Doyle (2002), the more an engineering system is designed to be robust to a specific class of failures, the more it will become fragile and vulnerable to another class of failures. Hence, robustness comes intertwined with fragility. The example that they give to illustrate this duality is enlightening. They stress that the control systems of a modern airplane are made of an intricate hierarchical network of thousands of computerized feedback loops that provide high robustness to a variety of environmental perturbations but by the same token, expose the aircraft to dangerous vulnerabilities to major power failures. This duality, robustness/fragility, also applies to critical infrastructures, including electric power systems. For instance, a bug in the software program carrying out a key monitoring function or a vital control scheme can play havoc with the normal operation of these infrastructures. See for instance the 2003 US-Canada Northeast blackout or the more recent deadly accident of the Metrorail system of Washington DC. In the former event, the origin of the blackout can be traced back to a topology error in the database of the state estimator (USCPSOTF 2004), while in the latter, the accident was due to errors in the software programs that automatically drive the Metrorail trains and control their emergency stop mechanism. Similarly, a system that is resilient to a certain type of failure may be fragile to another one. This point will be further discussed later on.

An important additional elaboration is that robustness is a predefined characteristic of a system embedded in the original system blueprint envisioned by the designer or the planner whereas resilience is an *emergent* property of distributed agents that control a system of weakly coupled modules. Let us elaborate on this point. In resisting perturbations, robustness requires strong coupling between the various components of the system. It is usually achieved either via an enhancement of the redundancy in the system by providing multiple paths to the input/output relationships or via conventional feedforward or feedback loops that have well-defined responses to perturbations. In contrast, resilience requires flexibility, agility, and adaptability to a changing environment that has not been entirely envisioned during the design process. In other words, the control actions are to be specific to each situation that is new and unique. This is possible only if the controllers are agents endowed with a certain degree of self-learning, self-innovating, and self-improvising abilities to generate the novel actions required in response to unexpected events whose occurrences come as a surprise.

2.3 *Robustness, Stability, Reliability, and Homeostasis*

Let us now discuss how robustness differs from the concepts of stability, reliability, and homeostasis.

a) Defining the Concept of Stability

In a broad sense, stability is synonymous with insensitivity to small perturbations. These perturbations can be defined in many different ways depending on the problem at hand, including approximations in the assumptions, uncertainties of the model being used, or uncertainties in the measurements, to cite a few. Specifically, we can define the stability of an equilibrium point in linear or nonlinear system theory (Nayfeh and Balachandran 1995), the stability of the bias and variance of an estimator in robust estimation theory (Huber and Ronchetti 2009, Hampel et al. 2005), and the stability of the probability of the first and second kind in hypothesis testing (Huber and Ronchetti 2009). A general definition of this concept is as follows:

Definition 3: A system is said to be stable if small perturbations induce small responses in its outputs.

In view of this definition, we may say that the robustness of a system is broader than the stability of an equilibrium point since it may involve changes in the structure of the system (Kitano 2004). For instance in power systems, robustness refers to the ability of a system to transition smoothly from a pre-fault to a post-fault stable equilibrium point that may be different from each other if permanent line outages have occurred to clear the faults.

b) Defining the Concept of Reliability

Reliability is a concept that together with a collection of theories introduced in statistics and in industrial and systems engineering is used to analyze and characterize the performance of engineering systems when they are subject to random failures (Yang 2007). Its general definition is as follows:

Definition 4: The reliability of a system is defined as the probability that this system is able to retain, over a given time period, its intended function under given conditions when it is subject to internal or external failures.

What is apparent from this definition is that unlike robustness, reliability is a probabilistic concept. Furthermore, reliability requires a certain degree of robustness to occur. Indeed, a non-robust system is not reliable since it cannot maintain its function when subject to failures. Reliability also deals with average failures, not extreme events. This is true in power systems where the usual reliability indices are the loss of load probability, which is a probability index when it is expressed in days per day, the loss of load expectation, and the loss of energy expectation. In a non-composite reliability analysis, these indices account for neither the transmission line constraints nor the transmission line outages, which are modeled in the second step of a two-step procedure via the so-called N-1 contingency analysis. The latter assesses the ability of the power system to withstand the loss of a single piece of equipment, including a generating unit, a transmission line, or a transformer. However, extreme events other than natural hazards are typically local failures that cascade through the system and possibly to other interconnected systems. Their modeling is essential in the evaluation of the risk of blackouts and the derivation of reliability indices for extreme events (Mili and Dooley 2010).

c) Defining the Concept of Homeostasis

Homeostasis is a concept introduced by Cannon in 1932 to describe the ability of the physiological processes of an organism to maintain in a coordinated manner its steady state conditions (Kitano 2007). In a more general setting, we may give the following definition:

Definition 5: Homeostasis is defined as the ability of a system to retain its stability by regulating its internal processes when it is subject to internal and external perturbations.

Since this definition involves the stability rather than the function of a system, we may say that homeostasis is a special case of robustness.

Homeostasis was introduced into power systems by Fred Schweppe in a visionary paper published in 1980 (Schweppe et al. 1980). He envisioned the forthcoming revolution in power systems that will be made possible by the development of computers and communication technologies. This revolution represents a paradigm shift where both the load and the generation of a power system will participate in a coordinated manner to maintain its stability via appropriate market mechanisms. Currently, demand response has this potential if made to respond to the frequency changes of a power system. Recall that demand response is the ability of end-users' appliances to switch on and off in response to electric energy price signals transmitted via the Internet.

2.4 *Sustainability*

Recently, the concept of sustainability (McDonough and Braungart 2002) has received growing attention by researchers from many disciplines, including ecology, the environmental sciences, biology, economics, the political and social sciences, architecture, physics, mathematics, computer science, and engineering. This comes as no surprise since this concept is at the intersection of all scientific and engineering fields. Indeed, it concerns all human activities and their impact on the environment and the earth in general. We will give a definition that specifically applies to engineering systems:

Definition 6: Sustainability is defined as the ability of an engineering system to restore its intended function following a major breakdown, while subjecting the earth's ecosystem to a minimum of damage in terms of resource usage and waste disposal during the system's entire life cycle.

What transpires from this definition is that a sustainable engineering system should (i) maintain its functionality during its expected normal operating life span and (ii) induce a minimum perturbation to the environment for its construction, operation, and disposal. Let us further discuss these two points. The first point concerns the ability of the system to recover after a breakdown, which is an evident property needed for sustainability. This recovery can be done thanks to human intervention, the current state of affair for critical infrastructures, or performed automatically via an intrinsic ability of the system to self-reconfigure. Current computer-based technologies have the potential to achieve the latter. Note that according to the definitions given by Bossel (1998), self-sustaining systems are less complex than self-organizing systems. Indeed, the latter have the ability to adapt to unexpected events in a novel way and to self-organize into new forms able to survive and recover from harsh conditions. As of today, robots do not have these capabilities; only biological systems exhibit these characteristics. Regarding the second point, which concerns the assessment of the impact on the environment of a designed system, the analysis should account for the energy required for the extraction, transportation, and transformation of the raw materials involved in its construction. It should also account for the energy involved in its operation, maintenance, and cleaning of the produced wastes and pollutants. Finally, after its retirement, it should account for the resources needed for its disposal and the recycling of the materials of which it is made. These are the objectives of a life cycle analysis described as "from cradle to grave" when all materials used are disposed of or "from cradle to cradle" when they are recycled into another system.

3. Resistance/Robustness and Resilience of the Community

It is important to also consider the resistance and resilience of the community serviced by the MG. While there are a number of ways to conceptualize the relationship between resistance and resilience, we adopt the following definitions:

Definition 7: Resistance (i.e., robustness) is defined as the degree to which a community can use their existing resources, or resources they acquire rapidly, to avoid the impacts of a potential disaster (EOP 2013). Resistance is effective when resources are robust enough or are rapidly provided to prevent a disaster from interrupting community functioning.

An example of this would be having adequate mitigation measures in place that prevent excessive rainfall from becoming a damaging flood, or a MG that prevents electricity interruption. With that in mind, our definition of resilience of the community is in line with definition two provided above.

Definition 8: Resilience of a community is defined as the ability to recover from disaster impacts (Nigg 1995, Webb et al. 2002). Recovery, then, is achieved when the community reaches a new normal, and measures of resilience often consider the latent characteristics and resources a community has pre-event that enable or hinder their recovery post-event and the adaptive processes communities undertake during recovery that increases their resistance to future events.

In this same way, resilience is effective when resources are robust enough or rapidly provided in adequate measure that the community can return to a new normal level of functioning post-impact. When considered this way, resources, provided for either resistance or resilience, must counteract the vulnerability in the system and the impacts of the disaster.

As noted by Cutter (2015) and Aldrich (Aldrich and Meyer 2015), social capital theory provides meaningful insights as to what makes social units resilient. Social capital is defined as the social norms, trust, and networks that exist between and modify human interactions (Nakagawa and Shaw 2004). In contrast to other types of capital, like economic capital that exists at an individual level, social capital exists in relationships between people and across bonds that exist within communities (Aldrich and Meyer 2015). This social capital allows for collective emergent action and, in times of need, this capital is actualized to acquire resources for individuals (Coleman 1988). In relation to the resiliency contributed to a community via MGs, social capital offers us both a proxy for the measurement of social resilience and insights on what might influence sharing behaviors and access to the electricity post-event. When considered in relation to MG, social capital offers explanations as to how communities could reach a consensus to adopt microgrids and how post-event power sharing may unfold.

4. Metrics of Robustness, Resilience, and Sustainability

4.1 Metrics of Stability and Robustness

All the metrics of stability of a dynamic system can be considered as metrics of robustness as well. The usual metric of stability is based on the Lyapunov theory. When the Lyapunov function reduces to the system's energy function, as is the case in power systems, the system stability margin is defined as the difference between two potential energies, namely, the minimum potential energy among those of all the unstable equilibrium points (u.e.p.'s) located on the basin of attraction boundary of the post-fault stable equilibrium point (s.e.p.) and the potential energy of that s.e.p. Because this metric can be too conservative for multi-machine systems, the u.e.p. with the smallest potential energy is replaced with the closest u.e.p. along the fault-on trajectory of the system, termed the controlling u.e.p. (Chang et al. 1988).

Due to its strong nonlinearities, a power system exhibits a rich dynamical behavior that ranges from stable and unstable limit cycles to period doubling and chaos. Consequently, other metrics of stability have been defined. The most popular ones are the distance to a saddle-node bifurcation point, termed point of voltage collapse, and the distance to a Hopf bifurcation point.

4.2 Metrics of Resilience of the Power System

Metrics of resilience can be defined as *adaptive flexibility* and *adaptive capacity* (Hollnagel et al. 2008). They are measures of the ability of a system to smoothly transit from one stable equilibrium point to another one, to reshape and stretch the basins of attraction of these s.e.p.'s, and to increase its stability margins so that its function is maintained during and after major perturbations. In power systems, this system agility can be achieved via transmission line switching, the fast and adaptive control actions of High Voltage Direct Current (HVDC) converters and MGs, and demand response to frequency changes. However, quantifying the adaptive flexibility and capacity of a system is a challenge.

The most common resiliency metrics proposed in the literature are: (1) the system/customer average interruption duration/frequency indices assessed only for extreme events (Lo Petre et al., 2010, Cano-Andrade et al. 2015) and (2) metrics based on the time of partial or total recovery to the normal state from the impact of the event (Norris et al. 2008, Adger 2000). We propose these metrics in our characterization but argue that they do not measure some important resiliency characteristics provided in our definition stated in Section 3. To bridge this gap, we define some new resiliency metrics of an agile power system with MGs:

a) Customers-Based Resilience Metrics

(1) Time of recovery of the critical load on outage (partial recovery);
(2) Time of recovery of all the load on outage (total recovery);
(3) Amount of critical electric energy on outage;
(4) Total amount of electric energy on outage;
(5) Number of critical customers on outage;
(6) Total number of customers on outage;

b) Infrastructure-Based Resilience Metrics

(7) Number of clustered stable equilibrium points (s.e.p.) that exist while the operating voltage, current and frequency constraints are satisfied (i.e., that are associated with a normal sate);
(8) Degree of flexibility of the potential energy function associated with a normal sate; we propose to calculate the derivatives of the potential energy function with respect to some relevant state/control variables, e.g., state variables of FACTS devices that have the ability to change the shape of the potential well of a s.e.p.;
(9) Amount of potential energy that can be transferred to a u.e.p. along the trajectory of the operating point to absorb the injected kinetic energy during a short-circuit;
(10) Amount of power that can be transferred via the tie-lines of one area that is experiencing severe instabilities to damp its oscillations and help its operating point to reach an acceptable stable equilibrium point.

4.3 Metrics of Resilience of the Community

Given what we know about community resilience, we suggest that the "measurement core of resilience indicators" proposed by (Cutter 2015) serve as the state of science and a starting point to explore both inherent (pre-existing, baseline indicators) and adaptive (learning after an event) resilience indicators at a community level of communities. Specifically, Cutter recommends researchers measure: (1) economic assets (e.g., per capita income), (2) social capital (e.g., number of civic organizations), (3) institutional attributes (e.g., percentage of population covered by mitigation plans), (4) information and communication (e.g., experience with hazards in the past 10 years), (5) infrastructural assets (e.g., spending on mitigation over the last ten years), (6) emergency management assets (e.g., sheltering capacity and the mapping of evacuation routes), and (7) additional critical social assets (e.g., percentage of the population with less than a high school diploma).

4.4 *Metrics of Sustainability*

Metrics of sustainability of an electric power system include various indices that assess the exergy efficiency of the power plants (Lo Petre et al. 2010); the amount of waste, pollutants, and effluents released into the environment; the impact of power plants and power networks on the fauna and flora of various ecosystems such as lakes, rivers, forests, and seas; the amount of energy needed for the extraction, processing and transportation of the fuel and the raw materials involved in the construction, operation and disposal of the power equipment.

5. Theoretical Service Efficiency and Effectiveness Measurement Framework

Not only do we need to design for robust and/or resilient infrastructure systems, but we are also concerned with the ability of their underlying systems to provide the appropriate services ("effectiveness") with the least amount of resources ("efficiency"). These services are provided both during normal and extreme events. Given our concern with the efficiency and effectiveness with which the power and emergency service infrastructures deliver services, we provide a framework for decision makers that can describe their processes, their associated services, and their interdependencies. Our approach starts with the assumption that service "performance" is a socio–technical phenomenon requiring us to define the basic processes and phenomena associated with electric power generation, transmission, and distribution as well as with the provision of emergency services. We consider five stakeholder groups (Fig. 2) and study them during each of the four time periods around a disaster. We assume that the five stakeholder groups are: utilities, prosumers, and businesses working with emergency management agencies (including police and fire departments) and the media to communicate electricity sharing plans and provide essential emergency management services for all time periods of interest so as to satisfy community needs. We assume that all of these stakeholders monitor what transpires during these four time periods and make adjustments according to what they perceive to be "good outcomes" (effectiveness). Nevertheless, they have limited resources available for the provision of services and are mindful of how well these are used (efficiency).

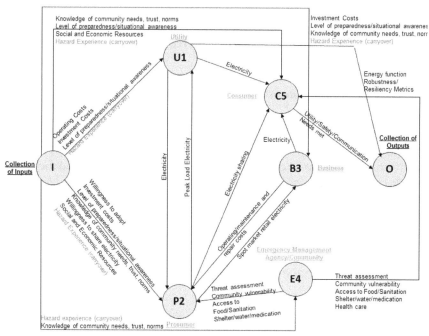

Fig. 2: Network of the five stakeholder groups and their interrelations (pre-disaster period).

Our conceptualization of how the five stakeholder groups define what constitutes efficient and effective services is unique in many respects. Firstly, determining particular stakeholder behaviors or activities allows us to assess high or low performance via different measures aimed at each and all of the groups collectively. Secondly, since the services are collectively produced through the decisions and actions of all stakeholder groups, considering their interdependencies is critical. Thirdly, to analyze these interdependencies, we link to mathematical models that will help each stakeholder group measure and analyze the lessons learned from previous disasters and make necessary adjustments to regulations and policies. This is achieved by extending the dynamic network Data Envelopment Analysis (DEA) (Charnes et al. 1978) approach proposed in (Herrera-Restrepo et al. 2015). Here, the stakeholder groups are regarded as a "system" that collectively provides services.

In summary, our approach offers an integrated mechanism to: (1) evaluate MG design configurations including types of generating units, storage and heating devices, and control systems that when varied could affect service performance; (2) identify key determinants of service performance for all five points of view and explore the concepts of "efficiency" related to resource use and "effectiveness" related to outcome achievement, especially in the context of community needs; (3) recognize the physical and social resources needed to achieve goals for all points of view; (4) identify the interdependencies among these points of view; and (5) provide dynamic performance measures for each point of view and for the "system" as a whole since the timing of electricity sharing decisions by prosumers significantly impacts community resilience.

As noted earlier, our initial normative abstraction is represented as a network (Fig. 2) where we represent the five stakeholder groups as nodes for a given time period. The assumption is that underlying each node are specific processes. The physical infrastructure is distributed among the utility, the prosumer, and business stakeholder groups, whereas the emergency services infrastructure is associated with all five stakeholder groups. The consideration of infrastructure systems as processes is one of the key points of our approach (Herrera-Restrepo et al. 2015).

Our approach allows for the computation of efficiency-effectiveness targets for each of the nodes and for the network as a whole. Specifically, it allows exploration of whether the provision of services improves if people choose to use MGs. It also addresses this premise by consideration of the knowledge and commitment to share before, during, and after an extreme event that will determine the benefits of a MG for community resilience. This second premise can then be explored via an understanding of the role of social capital that is manifested in social networks, norms, and the trust, which plays out in the community.

6. Least Cost Power System Planning Subject to Reliability and Resilience Constraints

6.1 *Planning for Enhanced Community and Infrastructure Resilience*

One question that arises is how to define in a unified manner the resilience for social communities and infrastructures and how to differentiate it from the well-established concept of robustness in systems theory. While acknowledging the Department of Homeland Security (DHS) (DHS 2000) definition of resilience as the ability to adapt to changing conditions and to withstand and rapidly recover from disruption due to emergencies, we focus on an in-depth representation of infrastructure and community resilience and how it is aided by the penetration of MGs. We borrow from (Bruneau et al. 2003, Wilbanks 2008, Peacock et al. 2008) to define community resilience as its ability to survive the damage inflicted by extreme events by maintaining subsistence ensured by critical infrastructure services and by preserving the bio-physical systems upon which it depends, while being engaged in relationships that are fair, legitimate, and adaptive. We contrast community resilience with infrastructure resilience. We define the resilience of coupled physical critical infrastructures to a class of *unexpected* extreme disturbances as the ability of this system to degrade gracefully and to recover its function once the disturbances cease (Mili 2011). Underpinning these concepts are multi-time-scale considerations from decisions that are realized almost instantaneously to ones that span hours or days. One important time scale is the window of opportunity that opens up just before the disaster where critical actions are taken based on community and infrastructure situational awareness and vulnerabilities. In this research, we emphasize four critical time periods: normal, the window of opportunity, the disaster, and the recovery time periods. Our assumption is that behaviors such as the development of social capital bonds, etc., during the normal time period will affect community resilience.

6.2 Robustness/Resilience Tradeoff

Robustness and resilience are obviously complementary but not antinomic properties. This can be illustrated via a metaphor conveyed by a fable of Jean de La Fontaine (1621–1695), which is entitled "The Oak and the Reed". The moral of the fable is that to survive strong storms, it is better to be flexible but weak like a reed, which bends under any breeze, than to be robust but fragile like an oak, which can be uprooted if the wind gust is stronger than its capacity to resist. Of course, de la Fontaine did not have properties of infrastructures in mind when he was writing this fable. Rather, he was giving advice to the subjects of his majesty, the king of France, on how to survive stringent but harsh conditions. From a material science view point, the trunk of an oak consists of layers of hard cellulose made of tightly coupled molecules, hence its strength and its brittleness, while the stem of a reed is made of elastically coupled molecules, hence its flexibility and its weakness.

From a philosophical viewpoint, this fable contrasts robustness with resilience and shows a preference for the latter when facing adversity. Let us further explore this claim. Consider a table made of rubber instead of wood. Obviously, unlike the latter, it is highly resilient to shocks. However, it cannot be used as a desk since it will undergo too large a dip under the weight of office supplies and books. For this application, a robust table is preferred. In fact, the desk can be made of composite materials that exhibit a good level of robustness and some degree of resilience to shocks. Mammal bones provide a good example of such a tradeoff with the addition of a self-healing ability (Taylor 2010).

Let us now discuss how tradeoffs between robustness and resilience can be achieved in an electric power system. Resilience requires structural modularity and hierarchies together with adaptive and intelligent control. As depicted in Fig. 3, this can be achieved via two interesting

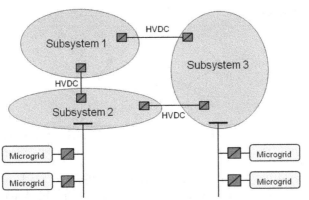

Fig. 3: Segmentation of a power system into subsystems weakly interconnected via power electronic interfaces represented as red boxes.

proposals made by the Electric Power Research Institute (EPRI), which are (i) a segmentation of the transmission network into subnetworks connected through HVDC links to confine a cascading failure to the originating subsystem (Clark et al. 2008) and (ii) the partial restructuring of the distribution systems into MGs (Lasseter et al. 2002). A MG is a small-scale power system that can either be operated as a standalone system or interconnected to a distribution feeder or a substation. Typically, it consists of generating units; loads; storage devices; circuit breakers with transmitters and receivers, power electronic interfaces such as converters, inverters and boosters; and communications networks.

Evidently, a power distribution system with interconnected MGs and distributed generation (DG) needs to be provided with a large number of sensors and actuators that exchange data with distribution control centers via dedicated communications channels or the Internet. When supervised by a distributed and coordinated control scheme, for example, based on multi-agent technologies, MGs and loads will be able to significantly contribute to peak shaving, ancillary services, and the dampening of the transients induced by faults on the power system (Hatziargyriou et al. 2007, Djiapic et al. 2007, Strbac et al. 2005, Schuler 2007, Goulding et al. 1999, Clark and Bradshaw 2004, Reites et al. 2002). However, that control scheme has to act rapidly and robustly under stringent conditions that are characterized by, for example, incomplete and partially erroneous data sets when part of the communications infrastructure fails. Another potential benefit of MGs is their ability to supply, in the aftermath of natural disasters such as hurricanes, the needed electric energy to sustain customers and emergency services during long-lasting power outages since some segments of the utility distribution networks typically experience physical damage.

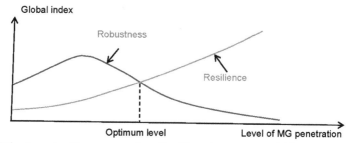

Fig. 4: Robustness/resilience curves for a fixed investment cost vs. MG penetration.

An interesting question arises here: is there a trade-off between robustness/reliability and resilience? As observed in Fig. 4, we conjecture that for a given investment cost and for an increasing level of penetration of the distributed resources, including MGs, the robustness/reliability of the system increases, reaches a maximum and then decreases while the resilience of the system steadily increases. Therefore, there is a point where the robustness and the resilience curves intersect; this point achieves an optimal trade-off between system robustness and resilience. In this characterization, we propose that this be tested using a co-simulation approach. It is worth noting that the potential negative impact of a large penetration of distributed resources on power system reliability has recently been documented by NERC in the US (NERC 2011) and by SINTEF in Norway (Tande 2006).

7. Conclusions

The mapping between the concepts of power system resilience and robustness with community resilience and robustness provides a unique opportunity to consider important behavioral and community considerations in future power system infrastructure design. There are multiple challenges that this research faces. Originating from the power system side, can one find equivalent conceptual and operational representations of the energy function notion to describe and measure community resilience? On the other hand, originating from the community side, in what way do conceptual and operational representations of social capital speak to the agility of the community's resilience and in a direct or indirect way affect the power system resilience?

The expansion of the framework of Section 3 to include the underlying mapping between the resilience of the power system to the community resilience will inform the further evolution of the theoretical considerations associated with the power system and disaster management critical infrastructures. The last but not the least issue, pertains to the appropriate consideration of dynamics. While in the power system literature, the consideration of dynamics is ingrained in the power system design and planning, in the disaster management literature, dynamic representations are not the norm when one considers the well accepted conceptual/ theoretical representations as well as the methods that are available to capture the dynamics. This last point offers a research challenge and opportunity when considering the integration of the power system and resilience concepts as we attempt to achieve in Section 3, Fig. 2, of this chapter.

References

Abbey, C., D. Cornforth, H. Hatziargyriou, K. Hirose, K. Kwasinski, E. Kyriakides, G. Platt, L. Reyes and S. Suryanarayanan. 2014. Powering through the storm: Microgrids operation for more efficient disaster recovery. IEEE Power and Energy Magazine 12: 67–76.

Adger, W.N. 2000. Social and ecological resilience: are they related? Prog. Hum. Geogr. 24: 347–364.

Aldrich, D.P. and M.A. Meyer. 2015. Social capital and community resilience. Am. Behav. Sci. 59: 254–269.

Bossel, H. 1998. Earth at a Crossroad: Paths to a Sustainable Future. Cambridge University Press, Cambridge,UK.

Bruneau, M., S.E. Chang, R.T. Eguchi, G.C. Lee, T.D. O'Rourke, A.M. Reinhorn, M. Shinozuka, K. Tierney, W.A. Wallace and D. Von Winterfeldt. 2003. A framework to quantitatively assess and enhance the seismic resilience of communities. Earthquake Spectra 19: 733–752.

Cano-Andrade, S., M.R. von Spakovsky, A. Fuentes, C. Lo Prete and L. Mili. 2015. Upper-level of a sustainability assessment framework for power system planning. Journal of Energy Resources Technology 137: 041601–11.

Carlson, J.M. and J. Doyle. 2002. Complexity and robustness. Proceedings of the National Academic of Science (PNAS) 99: 2538–2545.

Chang, H.-D., F.F. Wu and P.P. Varaiya. 1988. Foundation of potential energy boundary surface boundary method for power systems transient stability analysis. IEEE Trans. on Circuits and Systems 35: 712–728.

Charnes, A., W.W. Cooper and E. Rhodes. 1978. Measuring the efficiency of decision making units. Eur. J. Oper. Res. 2: 429–444.

Clark, II, W.W. and T.K. Bradshaw. 2004. Agile Energy Systems: Global Lessons from the California Energy Crisis. Elsevier, Amsterdam, NL.

Clark, H., A.-A. Edris, M. El-Ghasseir, K. Epp, A. Isaacs and D. Woodford. 2008. Softening the blow of disturbances: segmentation with grid-shock absorbers for reliability of large transmission interconnections. IEEE Power and Energy Magazine 6: 30–41.

Coleman, J.S. 1988. Social capital in the creation of human capital. Am. J. Sociol. 94: 95–120.

Cutter, S.L. 2015. The landscape of disaster resilience indicators in the USA. Natural Hazards 80(2): 741–758.

DHS. 2000. DHS definition of resilience. DHS Report. https://www.dhs.gov/topic/resilience.

Djapic, P., C. Ramsay, D. Pudjianto, G. Strbac, J. Mutale, N. Jenkins and R. Allan. 2007. Taking and active approach. IEEE Power & Energy Magazine 5: 68–77.

Dy Liacco, T.E. 1978. Systems security: The computer's role. IEEE Spectrum 15: 43–50.

EOP. 2013. Executive Office of the President. Economic Benefits of Increasing Electric Grid Resilience to Weather Outages. Report, The White House, Washington DC, August 2013. https://energy.gov/sites/prod/files/2013/08/f2/Grid%20Resiliency%20Report_FINAL.pdf.

Fink, L.H. and K. Carlsen. 1978. Operating under stress and strain. IEEE Spectrum 15: 48–53.

Goulding, A.J., C. Rufin and G. Swinand. 1999. The role of vibrant retail electricity markets in assuring that wholesale power markets operate effectively. The Electricity Journal 12: 61–73.

Hampel, F.R., E.M. Ronchetti, P.J. Rousseeuw and W.A. Stahel. 2005. Robust Statistics: The Approach Based on Influence Functions. John Wiley, New Jersey, USA.

Hatziargyriou, N., H. Asano, R. Iravani and C. Marnay. 2007. Microgrids. IEEE Power & Energy Magazine 5: 78–94.

Herrera-Restrepo, O., K. Triantis, J. Trainor, P. Murray-Tuite and P. Edara. 2015. A multiperspective dynamic network performance efficiency measurement of an evacuation: A dynamic network-dea approach. Omega 60: 45–59.

Hirose, K., T. Shimakage, J.T. Reilly and H. Irie. 2013. The sendai microgrid operational experience in the aftermath of the tohoku earthquake: A case study. NEDO Microgrid Case Study, New Energy and Industrial Technology Development Org.

Holling, C.S. 1973. Resilience and stability of ecological systems. Annual Review of Ecological Systems 4: 1–23.

Holling, C.S. 1996. Engineering resilience versus ecological resilience. pp. 31–43. In: Schulze, P.C. (ed.). Engineering with Ecological Constraints. National Academy of Engineering.

Hollnagel, E., D.D. Woods and N. Leveson. 2008. Resilience Engineering: Concepts and Precepts. Ashgate, Aldershot, UK.

Huber, P. J. and E.M. Ronchetti. 2009. Robust Statistics. John Wiley, 2nd Edition., New Jersey, USA.

Jen, E. 2005. Stable or robust? What is the difference? pp. 7–20. In: Jen, E. (ed.). Robust Design: A Repertoire of Biological, Ecological, and Engineering Case Studies, Oxford University Press.

Kitano, H. 2004. Biological robustness. Nature Reviews 5: 826–837.

Kitano, H. 2007. Towards a Theory of Biological Robustness. Molecular System Biology 3: 1–7.

Lasseter, R., A. Akhil, C. Marnay, J. Stephens, J. Dagle, R. Guttromson, A.S. Meliopoulos, R. Yinger and J. Eto. 2002. Integration of Distributed Energy Resources: The CERTS MicroGrid Concept. White Paper, CERTS.

Lo Prete, C., B. Hobbs, C.S. Norman, S. Can-Andrade, A. Fuentes, M.R. von Spakovsky and L. Mili. 2010. Sustainability Assessment of Microgrids in the Northwestern European Electricity Market. Proceedings of the 23rd International Conference on Efficiency, Cost, Optimization, Simulation and Environmental Impact of Energy Systems (ECOS 2010), Lausanne, Switzerland.

McDonough, W. and M. Braungart. 2002. Cradle to Cradle: Remaking the Way we Make Things. North Point Press.

Mili, L. and K. Dooley. 2010. Risk-based power system planning integrating social and economic direct and indirect costs. In: Momoh, J. and L. Mili (eds.). Economic Market Design and Planning for Electric Power Systems. John Wiley, New Jersey, USA.

Mili, L. 2011. Taxonomy of the characteristics of power system operating states. Proceedings of the 2nd NSF-RESIN Workshop, Tucson, AZ.

Nakagawa, Y. and R. Shaw. 2004. Social capital: a missing link to disaster recovery. Int. J. Mass Emerg. Disasters 22: 5–34.

Nayfeh, A.H. and B. Balachandran. 1995. Applied Nonlinear Dynamics: Analytical, Computational, and Experimental Methods. John Wiley, New Jersey, USA.

NERC. 2011. North American Electric Reliability Co. Potential Bulk System Reliability Impacts of Distributed Resources. NERC Special Report.

Nigg, J.M. 1995. Disaster Recovery as a Social Process. Disaster Research Center, Newark, DE.

Norris, F.H., S.P. Stevens, B. Pfefferbaum, K.F. Wyche and R.L. Pfefferbaum. 2008. Community resilience as a metaphor, theory, set of capacities, and strategy for disaster readiness. Am. J. Community Psychol. 41: 127–150.

Peacock, W.G., H. Kunreuther, W. Hooke, S.L. Cutter, S.E. Chang and P.R. Berke. 2008. Toward a Resiliency and Vulnerability Observatory Network : RAVON. HRRC Reports: 08-02R. archone.tamu.edu/hrrc/publications/researchreports/RAVON.

Reitzes, J.D., L.V. Wood, J. Arnold Quinn and K.L. Sheran. 2002. Designing Standard-Offer Service to facilitate Electric Retail Restructuring. The Electricity Journal 15: 34–51.

Schuler, R.E. 2007. Two-sided electricity markets: Self-healing systems. *In*: Richardson, H., P. Gordon and J. Moore II (eds.). Economic Costs and Consequences of Terrorist Attack. E. Elgar, Cheltenham, UK.

Schweppe, F.C., R.D. Tabors, J.L. Kirtley, H.R. Outhred, F.H. Pickel and A.J. Cox. 1980. Homeostatic utility control. IEEE Trans. on Power Apparatus and Systems 99: 1151–1163.

Strbac, G., J. Mutale and D. Pudjianto. 2005. Pricing of Distribution networks with distributed generation. Proceedings of the IEEE 2005 International Conference on Future Power Systems.

Strickland, E. 2011. A Microgrid that Wouldn't Quit. IEEE Spectrum Newsletter, 26th October 2011 http://spectrum.ieee.org/energy/the-smarter-grid/a-microgrid-that-wouldnt-quit/0.

Tande, J.O. 2006. Impact of integrating wind power in the norwegian power system. SINTEF Energy Research Technical Report.

Taylor, D. 2010. Why are your Bones not Made of Steel? Materialstoday.com 13: 6–7.

UFOS. 1965. NICAP's report of UFO Sightings during the November 9, 1965 Blackout. http://www.nicap.org/nyne.htm.

USCPSOTF. 2004. U.S.-Canada Power System Outage Task Force. Final Report on the August 14, 2003 Blackout in the United States and Canada: Causes and Recommendation. April. https://energy.gov/sites/prod/files/oeprod/DocumentsandMedia/BlackoutFinal-Web.pdf.

Webb, G.R., K.J. Tierney and J.M. Dahlhamer. 2000. Businesses and disasters: Empirical patterns and unanswered questions. Natural Hazards Review 1: 83–90.

Webb, G.R., K.J. Tierney and J.M. Dahlhamer. 2002. Predicting long-term business recovery from disaster: A comparison of the Loma prieta earthquake and hurricane andrew. Global Environmental Change Part B: Environmental Hazards 4: 45–58.

Wilbanks, T. 2008. Enhancing the resilience of communities to natural and other hazards: What we know and what we can do. Nat. hazards Obs. 41: 283–295.

Yang, G. 2007. Life Cycle Reliability Engineering. John Wiley, New Jersey, USA.

Yang, Y., L. Wu, W. Song and Z. Jiang. 2009. Collaborative control of microgrid for emergency response and disaster relief. Proceedings of the International Conference on Sustainable Power Generation and Supply 1–5.

CHAPTER 10

Developments in Power System Measurement and Instrumentation

Carlos Augusto Duque,[1,*] *Leandro Rodrigues Manso Silva,*[1] *Danton Diego Ferreira*[2] and *Paulo Fernando Ribeiro*[3]

1. Introduction

Measurement and instrumentation in power systems are often concerned with the analysis of voltage and current waveforms. Harmonic analysis and phasor estimation play an important role in several power system areas, from protection to power quality (PQ). PQ analyzers estimate the harmonic content using the Discrete Fourier Transform (DFT) and the Phasor Measurement Units (PMU) computes the time-synchronized phasor using modified DFT algorithm, or other signal processing tools. Independent of the power system application, a measurement system is composed of four main parts as illustrated in Fig. 1. The first block is the instrument transformers, which are used for "stepping down" the voltage or current of the system to measurable values. Typical values are 110V for voltage and 5A for current. The second block is the signal conditioner,

---
[1] Federal University of Juiz de Fora, Brazil.
  Emails: leandro.manso@ufjf.edu.br
[2] Federal University of Lavras, Brazil.
  Email: danton@deg.ufla.br
[3] Federal University of Itajuba, Brazil.
  Email: pfribeiro@ieee.org
* Corresponding author: carlos.duque@ufjf.edu.br

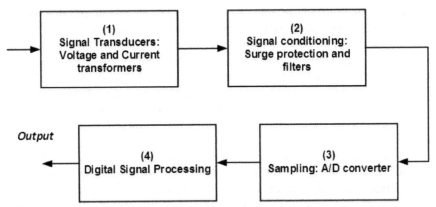

Fig. 1: The block diagram of the measurement and instrumentation system for power system.

composed of surge suppress and analog filters. The third block is the sampling block, composed of a sample and hold (S/H) circuit and an Analog-to-Digital converter (ADC). The last block is the digital signal processing where signal processing-based algorithms are performed to produce the desired output that can be information to be recorded and analyzed off-line or to be used in real time, such as a trip for opening a recloser in protection application.

This chapter discusses some current topics regarding power system measurement and the instrumentation system.

2. Instrumentation Transformer

The main function of the instrument transformers is "stepping down" the voltage or current of the system to measurable values. Furthermore, the instrument transformers are used for decoupling the primary circuit from the secondary, so that there is no connection between the primary and secondary circuits, providing electrical isolation for the electronic circuits and human being. The physical principle used to transfer information between the primary and the secondary circuits is, nowadays, mainly based on the electromagnetic theory, although the use of transducers based on optical principles has gained space in measurement systems, but the high cost of these transducers is a limiting factor for their proliferation. As a consequence, the vast majority of electronic measuring and instrumentation systems are connected to the traditional electromagnetic voltage and current transformers, which are designed for a high accuracy at rated frequency and their frequency response for other frequencies are often not known. This fact directly affects the accuracy of the harmonic content measurements in power systems, and, consequently, the results presented by the power quality monitors become unreliable.

Stiegler et al. (2016) presented an extensive study on the accuracy of the voltage transformers used for the measurement of harmonics. Figure 2, extracted from their work, presents the voltage harmonics in a 330 kV substation measured using three different voltage transformers (VTs), connected at the same point and at the same time. As it can be seen, differences up to 200% can be observed, which puts the data in doubt for further processing.

While the accuracy of the measurement equipment is addressed in the IEC 61000-4-30 standard (IEC 61000-4-30, 2008), nothing is said regarding the instrument transformer accuracy. In the IEC 61869-3 standard, the accuracy of instrument transformers is defined only for the nominal frequency (IEC 61869-3, 2011). The IEC is currently working on the future version of the standard that should include the frequency response of the transformers. However, this topic is not easy to address as can be seen from the few technical reports found in literature regarding frequency response of transformers (Seljeseth et al. 1998, Zhao et al. 2014, Stiegler et al. 2016). The authors have identified the main source of errors as the resonance frequency within the VTs that are dependent on the manufactures design. In their work, they have presented interesting comparative results that led to the following important conclusions:

- Inductive VTs for extra high voltage (EHV) are able to reach enough accuracy up to the 15th harmonic using a combination of invasive and noninvasive calibration methods;

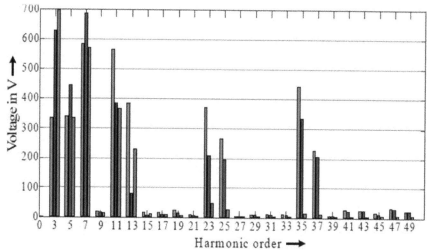

Fig. 2: Harmonic voltage of the same busbar, using three different voltage transformers (Extracted from Stiegler et al. 2016).

- Capacitive VTs are usually not suitable for harmonic measurements;
- If accurate measurements are required in the full frequency range up to 2.5 kHz, other technologies like RC dividers and optical transformers have to be used.

2. The Sampling Rate and Resolution

Block 3 of Fig. 1 deals with two important problems in power system measurement and instrumentation: the sampling rate and resolution. The crucial question is: what are the sampling rate and the ADC resolution to be used? Obviously, the answering of this question is application dependent. For protection application, typical sampling rates are low: 8, 16, 32 and 64 samples/cycle, reaching 128 samples/cycle in the latest relay that includes fault recorder functionality. For PQ monitoring it is possible to finding three main frequency bands and the corresponding sampling rate:

(i) Classic low frequency band that comprises harmonics up to 50th order can be accessed using sampling rates of 128, 256 and 512 samples per cycle;

(ii) The 2 kHz–150 kHz band is becoming important due to the growing of power-electronic device and power line communication (PLC).

(iii) High frequency impulse recorder uses sampling rates from 1 MHz up to 4 MHz.

The resolution of the ADC should be chosen according to the accepted error for the conversion that is inversely proportional to the number of bits of the ADC. Commonly, 10 or 12 bits for voltage signals, 16 bits for current signals for protection application, and 16 bits for both voltage and current in PQ are used.

Table 1 summarizes the application, the corresponding sampling rates and the A/D converters resolution.

Table 1: Sampling rate and A/D resolution.

Application	Sampling rate	A/D resolution
Protection/Fault Recorder	8, 16, 32 and 64 samples/cycle	Voltage: 12 bits Current: 16 bits
PQ: Classic frequency band (up 50th harmonic)	128, 256 and 512 samples/ cycle	16 bits
PQ: 2 k–150 kHz	1 MHz	16 bits
PQ: High frequency impulse recorder	1 or 4 MHz	8, 10, 12 bits

2.1 Subsampling Application

The Nyquist theorem states that a signal with maximum frequency F_M (Hz) can be represented by its samples since the sampling rate (F_s) obeys the Nyquist criterion, $Fs > 2F_M$. In order to guarantee the Nyquist criterion, an analog low-pass filter must be used in the signal conditioning section (Block 2 of Fig. 1). This filter is commonly known as anti-aliasing or guard filter. Figure 3 shows the basic specification for the guard filter. In this figure, f_p and f_s are the passing and stopping frequency, A is the minimum attenuation in the stopband.

The Nyquist criterion can be extended to bandpass signal. If the *bandwidth* of the signal is supposed to be B_W, then the theorem can be rewritten as:

"if a signal is a bandpass signal, with bandwidth B_W, then the signal can be represented by their samples since $Fs > 2B_W$".

Figure 4a illustrates a bandpass signal, with B_W = 50 kHz, centered at 975 kHz. Using the extended definition of the Nyquist criterion, this signal can be sampled using a sampling rate higher than 100 kHz. Figure 4b shows the spectrum of the sampling signal for F_s = 100 kHz. Note the aliasing does not mix the spectrum, so it is possible to work with the baseband signal (highlighted band) and after that, the signal can be modulated to the original frequency.

The subsampling idea can be used for spectrum analysis of high frequency harmonics, such as the *supraharmonics* as presented in Meyer et al. (2014). Using this methodology, it is possible to preserve the main

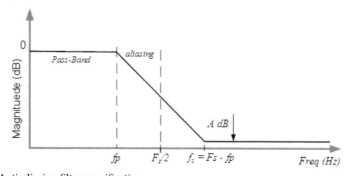

Fig. 3: Anti-aliasing filter specification.

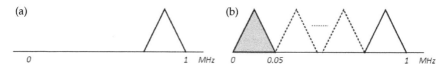

Fig. 4: Extended Nyquist criterion: (a) Band-limited signal; (b) Subsampling signal.

architecture of the present day PQ monitors. To do this, changes have to be made in the conditioning block (block 2—Fig. 1) and in the algorithms used to estimate the spectrum, as described below.

Figure 5 illustrates the basic modification to be made at conditioning block.

From this figure an analog filter bank, with each filter centered at specific spectrum, must be implemented, jointly with an analog multiplexer used to choose the desired band where harmonics or the spectrum will be estimated. The number of filters in the bank corresponds to the subsampling factor M. In this figure, the bottom filter selects the *supraharmonic* spectrum and $M = 4$. The ADC converter is not able to work at the high frequency required by this band, but it works with a sampling rate of just a quarter. In this case, the subsampling will be produced at the ADC output. The bandpass filter prevents destructive aliasing and the corresponding spectrum is like the one shown in Fig. 4b. The *supraharmonics* frequencies are shifted to the lowest band. The Fast Fourier Transform (FFT) applied to the subsampling signal will estimate the correct magnitude, however the frequencies presented by the FFT need to be corrected.

A simple example of a subsampling FFT is presented here. Consider a signal described by,

$$x(t) = \sum_{h=1}^{50} A_h \cos(h\omega_0 t + \theta_h) + A_{sh_1} \cos(sh_1 \omega_0 t + \theta_{sh_1}) + A_{sh_2} \cos(sh_2 \omega_0 t + \theta_{sh_2}) \ (1)$$

This signal is composed by conventional harmonic and two *supraharmonics* of orders $sh_1 = 965$ and $sh_2 = 1000$ respectively. The

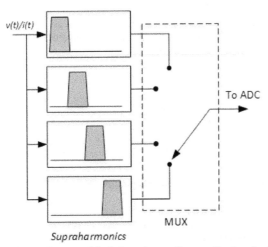

Suprahamonics

Fig. 5: Modification at conditioning block for subsampling application for $M = 4$.

amplitudes are 1 and 0.8 respectively and the phases are zero for both. The signal is first filtered by a bandpass filter whose bandpass frequencies are 960 × 60 Hz and 1024 × 60 Hz. The subsampling factor used is $M = 16$. The filter is assumed ideal in this example, in order to avoid amplitude or phase corrections. The sampling frequency is $F_s = 128 \times 60$ Hz.

Figure 6a presents the FFT of the subsampled signal. The x-axis is labeled with the apparent frequency f_A normalized from 0 to 1. Figure 6b presents the subsampled spectrum with the x-axis in real frequency f_R in Hz.

As observed in Fig. 6, the subsampling can invert the position of the *Supraharmonics*, however after applying the correction to obtain the real frequency the inversion is corrected. The algorithm shown in Fig. 7 must be used to obtain the real frequency, for this case. In Fig. 7, M is the subsampling factor and F_s is the sampling frequency:

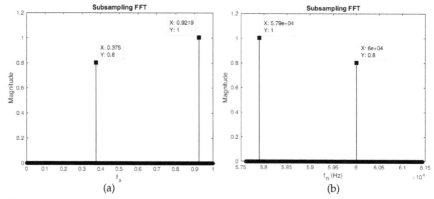

Fig. 6: Supraharmonics estimation using subsampling FFT ($M = 16$): (a) normalized Frequency; (b) real frequency.

$$f_A \text{ is initialized with normalized frequency } [0,1]$$

$$f_A = (f_A + 2 \times fix(M/2))/M$$

$$if \ f_A > 1$$
$$f_A = 2 - f_A$$
$$end$$

$$f_R = f_A \times M \times F_s / 2$$

Fig. 7: Algorithm used for correcting the frequency axis of the subsampling FFT for frequencies yield in the last band.

The above algorithm corrects the apparent frequency if the real frequency is in the last band defined by the filter banks. There is an apparent frequency correction algorithm for each band where the actual frequency is found.

3. Noise Reduction

Noise reduction is an important issue in signal processing. There are several situations in power system measurement and instrumentation that are crucial to reduce the amount of noise in the sampled signal. There are two ways to deal with noisy signal: (a) The first one is commonly known as denoising; and (b) the second one is to develop algorithms that are robust to noise. The former option consists in filtering the noise from the signal, before applying the desired algorithms to extract the information. In the last option the algorithms are able to extract the desired information even in presence of a certain amount of noise.

This section will firstly discuss how to denoise a signal using the Wavelet transform, its advantages and limitation. Then a noise robust algorithm will be presented to compute the derivative of a noisy signal using the Savisky Golay filter.

3.1 Noise in Power Signal

In the context of power system measurement and instrumentation, noise, or interference, can be defined as undesirable electrical signals that distort the desired signal from those features and information that need to be extracted. There are several sources of noise; the main ones are caused by: (a) large motors being switched on; (b) power electronic devices, such as converters and drives; (c) lightning strikes; (d) large nonlinear load, such as arc furnace; (e) quantization noise caused in analogue to digital converter, etc. An example of noisy signal is presented in Fig. 8. This signal represents a typical partial discharge (PD) in a Generator Stator (Cunha et al. 2015). The Signal-to-Noise Ratio (SNR) of this signal is 20dB. The goal is to identify the decaying pattern inside this noisy signal.

Mathematical model for noise is traditionally represented as WGN (White Gaussian Noise) with zero mean and standard deviation σ. This fact is supported by the Central Limit Theorem that tells, in simple words, that the summation of several noise sources with arbitrary probability density function lead to a signal with normal distribution. Thus a discrete time power signal can be expressed as

$$x[n] = s[n] + w[n] \qquad (2)$$

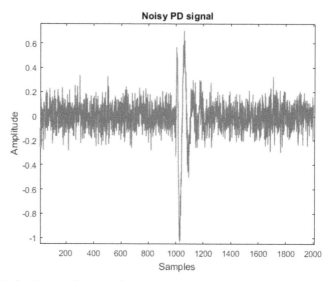

Fig. 8: Typical noisy signal: partial discharge in a generator stator.

where $x[n]$ is the observed noisy signal, $w[n]$ is the noise and $s[n]$ is the desired signal. The aim of denoising algorithms is to eliminate or reduce the noise from the observed signal.

There are in literature several methods to reduce noise present in the measured signal; however, the Wavelet Transform (WT) has been widely used as a denoising method. In Liao et al. (2011), the authors used WT for noise suppression in voltage signal for better detecting and localizing the occurrence of the power disturbances. In Cunha et al. (2015), WT based denoising is applied for identifying the pattern of partial discharge and consequently evaluates the equipment condition.

3.2 Wavelet Denoise

The block diagram shown in Fig. 9 presents the steps for noise reduction through wavelet transform. In block 1 the samples of the raw signal are acquired to be processed. In block 2 a discrete Wavelet Transform (DWT) is applied to generate the details and scales components. Blocks 3 and 4 are the core of the denoising process and will be detailed next. Block 5 is the inverse DWT that reconstructs the signal with reduced noise.

There are several parameters to be chosen in order to apply WT for denoising, among them: the number of levels in DWT, the mother wavelet used, the limiarization function and the threshold value. Some of these points are addressed in this section.

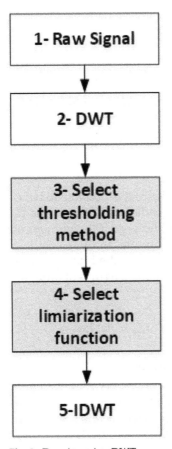

Fig. 9: Denoise using DWT.

3.2.1 The Maximum Number of Levels in DWT

The choice of the number of levels of DWT decomposition and consequently in the inverse discrete Wavelet Transform (IDWT) reconstruction is an important parameter seldom addressed in literature in formal way. Based on their experience, it is left choose. In fact, the number of levels in decomposition is application dependent, so no general theory is available and users must use some empirical process to define this number. However, the maximum number of decomposition levels (J_{max}) can be defined from the number of samples inside the window to be processed (N),

$$J_{max} = fix(\log_2 N) \tag{3}$$

where function $fix(x)$ returns the integer part of x. This rule is very simple to understand. Consider a three level DWT as shown in Fig. 10. If the

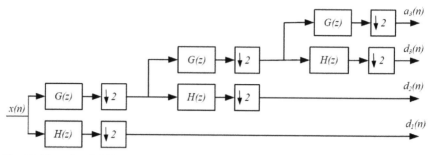

Fig. 10: Three levels DWT.

length of $x[n]$ is 16, using (3) it is possible to have up to 4 levels in the tree, so the approximation coefficient $a_3[n]$ and the detail coefficient $d_3[3]$ will have 2 samples each. Otherwise if the number of samples in $x[n]$ is 8, J_{max} = 3, and the number of samples in those coefficients will be just one. This is a general rule that supposes the filter length equal to one, which is not feasible. If the wavelet filters' length is of order M, then a better expression is given by,

$$J_{max} = fix\left(\log_2\left(\frac{N}{M-1}\right)\right) \qquad (4)$$

3.2.2 The Denoising Problem

After the decomposition process, noise spreads in all detail and approximation coefficients. The next and important step is to choose a threshold value in order to reduce the noise. Figure 11 presents a sinusoid signal corrupted by WGN and its DWT, using a tree of 5 levels and Daubechies 5 (dB5) as mother wavelet. At the top plot is the original signal. The second plot is the approximation a5 (note that this signal is almost free of noise). The other plots show the details coefficient where the noise is the main component.

The previous signal is very easy to denoise, because all relevant information is in the approximation coefficient, so all the detail coefficients can be discarded and the approximation coefficients can be used as a denoised signal. However, if part of the information is spread along the details coefficients it is important to define a threshold limit and a threshold function to separate signal from noise. Figure 12 shows a DWT of a signal given by the following equation,

$$x[n] = \cos\left(\omega_0 n\right) + 0.03\cos\left(5\omega_0 n\right) + 0.04\cos\left(7\omega_0 n\right) + 0.02\cos\left(764\pi n\right) + w[n] \,(5)$$

where $\omega_0 = 120\pi$ and $w[n]$ is a WGN with zero mean and standard deviation such as the SNR in (5) is *20dB*.

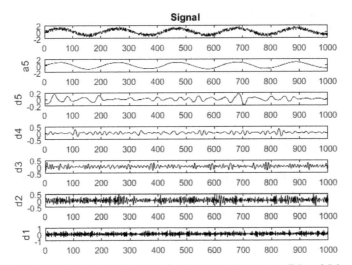

Fig. 11: Noisy sinusoid signal and its wavelet decomposition using db5 and 5 levels. The details coefficients carry only noise.

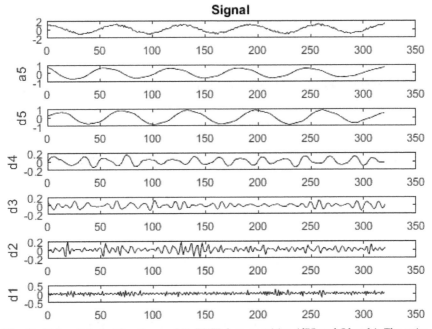

Fig. 12: Noisy distorted function and its DWT decomposition (*d*B5 and 5 levels). The noise and relevant information is mixed in the details coefficients.

From Fig. 12 it is possible to note that noise and relevant information are mixed in the details coefficients, so it is not possible simply to discard these coefficients without losing relevant information.

3.2.3 Linearization Function

The basic idea of the wavelet denoise process is that the energy of a signal will often be concentrated in a few coefficients in the wavelet domain while the noise energy spreads to all coefficients. In this way, a *thresholding* function tends to maintain the highest energy coefficients that represent the signal of interest, while noise is mitigated.

There are in literature basically two thresholding function: *hard* and *soft thresholding* (Ribeiro et al. 2014, Burrus et al. 1998). *Hard thresholding* is given by the following equation

$$\hat{X} = T_{hard}(Y,\tau) = \begin{cases} Y, & |Y| \geq \tau \\ 0, & |Y| < \tau \end{cases} \tag{6}$$

where τ is the *thresholding value*.

Soft thresholding is given by,

$$\hat{X} = T_{soft}(Y,\tau) = \begin{cases} \mathrm{sgn}(Y)(|Y|-t), & |Y| \geq \tau \\ 0, & |Y| < \tau \end{cases} \tag{7}$$

where $sgn(x)$ is the signal function, such as

$$\mathrm{sgn}(x) = \begin{cases} 1, & x > 0 \\ -1, & x < 0 \end{cases} \tag{8}$$

Figure 13 illustrates the *hard* and *soft thresholding* function for $\tau = 0.25$.

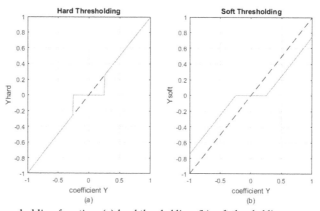

Fig. 13: The resholding function: (a) *hard thresholding*; (b) *soft thresholding*.

Hard thresholding better preserves the amplitude characteristics of coefficients than *soft threshold* because it does not reduce the coefficients above threshold values. *Hard thresholding* is indicated for application where discontinuity needs to be detected, such as the beginning and ending of sag disturbance, impulsive signals, etc. (Nasri and Nezamabadi-Pour 2009). Denoise applications generally uses *soft thresholding* (Donoho 1995).

3.3.4 Threshold Value

The frequent question about threshold is how to choose its value. Donoho and Johnstone (1994), suggest some answers for this question. The most single choice is the universal threshold, to be used in all coefficients, given by following equation:

$$\tau = \sqrt{2 \log N} \qquad (9)$$

where N is the number of samples in the original signal.

The second one suggested by Donoho is defining one threshold for each scale,

$$\tau_j = \sqrt{2 \log N_j} \qquad (10)$$

where N_j is the number of samples at scale j and τ_j is the threshold for the j scale.

Rescaling method can be incorporated in (9) and (10). According with Zhang et al. (2007), (9) and (10) can be rewritten as,

$$\tau = \frac{MAD|d_1|}{0.6745} \sqrt{2 \log N} \qquad (11)$$

And

$$\tau_j = \frac{MAD|d_j|}{0.6745} \sqrt{2 \log N_j} \qquad (12)$$

where MAD is the median absolute deviation of the detail coefficient.

The *threshold SURE* (Stein's Unbiased Risk Estimate) is used to define a threshold for *soft threshold* function. The threshold selection rule is based on Stein's Unbiased Estimate of Risk (quadratic loss function) (Stein 1981). One gets an estimate of the risk for a particular threshold value (τ). Minimizing the risks in (τ) gives a selection of the threshold value.

Figure 14 shows a denoised signal using SURE criterion and four level DWT when applied to the signal given by (5). The mother wavelet used was Daubechies of order four ("*db4*"). As can be seen, despite the noise reduction, part of relevant information is lost from the denoising operation. In fact, wavelet based denoise is not appropriate for denoising signal composed by components that spread their information along all scales and all time, such as harmonic distorted signals. On the other hand, denoise is very helpful to deal with noisy signal with localized information, such as partial discharge (PD) signal or signal containing sag disturbance, transient disturbance, etc. Figure 15 shows denoising being applied to a PD signal presented in Fig. 7. The mother wavelet used was the Symlet wavelet, specifically the "*sym6*" and a 6 level DWT was used

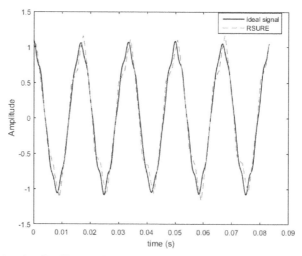

Fig. 14: Denoise signal in (5) using SURE soft threshold, "db4" and 4 levels of DWT. Despite noise reduction part of information from the original signal is lost.

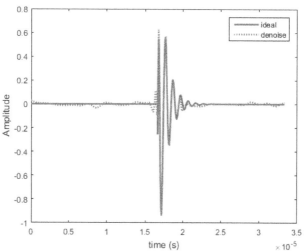

Fig. 15: Denoise of PD signal using Penalize Low hard threshold, Symlet mother wavelet "sym6" and 6 levels DWT.

for signal decomposition. The thresholding method used was *"Penalize low"* and *hard threshold*. Note that the denoised signal carries the main information regarding decaying component, so an identification algorithm may estimate this component very close to the real one.

4. The Savitzky-Golay Filters

In the last section the Savitzky-Golay filter was defined as a useful tool to reduce noise from power system signal. Once noise is reduced further algorithms may be used to extract desired information from the signal. However, if the final algorithm is made robust to noise, the preprocessing step can be eliminated. Designing an algorithm robust to noise, ever possible, is the best solution, but also is a challenge. This section will focus on an algorithm used to compute the derivative of a signal. Derivative is a very common application in power system, and very troubling one as well because it is very sensitive to noise. The Savitzky-Golay filter will be presented next, as a good solution to compute derivative of a noisy signal.

Savitzky and Golay presented a smoothing technique based in the concept of least squares (LS) interpolation in 1964. The initial intention of the authors was to present a data analysis tool in the field of analytical chemistry destined to the implementation in digital computers. Although the work of Savitzky and Golay (SG) was considered a "watershed" in the field of computational analytical chemistry and was elected by the editors of Analytical Chemistry, as the fifth most important article ever published in the journal (Riordon et al. 2000), its application in other areas of digital signal processing, according to Schafer (2011), is still surprisingly incipient. Practically nothing has yet been produced in the areas of energy quality, protection and instrumentation related with Electric Power Systems, other than the article proposed by Schettino (Schettino et al. 2016). The SG method can be used as a smoothing filter or as a derivative approach for high noisy signal. The formalization of the SG filter is presented below.

Let y a vector containing $2N + 1$ samples of a noisy signal, and let $f(x)$ a polynomial of degree m ($m < 2N + 1$) with $m + 1$ coefficients c_i ($i = 0,1,...,m$). The goal is to find c_i in order to minimize the error equation,

$$\varepsilon\,(c_0, c_1, ..., c_m) = \sum_{j=-N}^{N} \left(f(j) - y_j \right)^2, \tag{13}$$

where the polynomial $f(x)$ is,

$$f(x) = c_0 + c_1 x + c_2 x^2 + ... + c_m x^m \tag{14}$$

The LS solution of (13) is obtained by equalizing to zero the partial derivatives of the cost function in relation to each of the coefficients of the polynomial,

$$\frac{\partial \varepsilon}{\partial c_i} = \frac{\partial}{\partial c_i}\left[\sum_{j=-N}^{N}\left(f(j)-y_j\right)^2\right]=0\,,\quad i=0,1,...,m. \tag{15}$$

Substituting (14) in (15) and after some algebraic manipulation,

$$\frac{\partial \varepsilon}{\partial c_i} = \frac{\partial}{\partial c_i}\left[\sum_{j=-N}^{N}\left(f(j)-y_j\right)^2\right]$$

$$=2\sum_{j=-N}^{N}\left[\left(\sum_{k=0}^{m}c_k j^k - y_j\right)j^i\right]=0\,,\quad i=0,1,...,m. \tag{16}$$

Interchanging the summation order, yields

$$\sum_{k=0}^{m}\left[\left(\sum_{j=-N}^{N}j^{i+k}\right).c_k\right]=\sum_{j=-N}^{N}\left(j^i.y_j\right),\quad i=0,1,...,m. \tag{17}$$

Equation (17) is known as normal equation of the LS problem (Kay 2010). In order to obtain the coefficients, the matrix A is first built,

$$\mathbf{A}=\begin{bmatrix} (-N)^0 & \cdots & (-1)^0 & 1 & 1^0 & \cdots & N^0 \\ (-N)^1 & \cdots & (-1)^1 & 0 & 1^1 & \cdots & N^1 \\ (-N)^2 & \cdots & (-1)^2 & 0 & 1^2 & \cdots & N^2 \\ \vdots & \vdots & \vdots & \vdots & \vdots & \vdots & \vdots \\ (-N)^m & \cdots & (-1)^m & 0 & 1^m & \cdots & N^m \end{bmatrix}^{\mathrm{T}} \tag{18}$$

Using (18) it is possible to rewritten the right side of (17) as

$$\sum_{j=-N}^{N}\left(j^i.y_j\right)=\left\{\mathbf{A}^{\mathrm{T}}\right\}_{\text{Linha } i+1}.\mathbf{y}\quad i=0,1,...,m. \tag{19}$$

where,

$$\mathbf{y}=\begin{bmatrix} y_{-N} & y_{-N+1} & \cdots & y_0 & y_1 \cdots y_N \end{bmatrix}^T \tag{20}$$

On the hand, it is possible to observe that,

$$\mathbf{A}^{\mathrm{T}}.\mathbf{A}=\begin{bmatrix} \sum_{j=-N}^{N}j^{0+0} & \sum_{j=-N}^{N}j^{0+1} & \cdots & \sum_{j=-N}^{N}j^{0+m} \\ \sum_{j=-N}^{N}j^{1+0} & \sum_{j=-N}^{N}j^{1+1} & \cdots & \sum_{j=-N}^{N}j^{1+m} \\ \vdots & \vdots & \ddots & \vdots \\ \sum_{j=-N}^{N}j^{m+0} & \sum_{j=-N}^{N}j^{m+1} & \cdots & \sum_{j=-N}^{N}j^{m+m} \end{bmatrix}, \tag{21}$$

Therefore, the left side of (17) is giving by

$$\sum_{k=0}^{m}\left(\sum_{j=-N}^{N}j^{i+k}c_k\right)=\left\{\mathbf{A}^{\mathrm{T}}.\mathbf{A}\right\}_{\mathrm{Linha}\,i+1}.\mathbf{c}\qquad i=0,1,...,m. \tag{22}$$

Using (19) and (22) the normal equation can be rewritten in matrix form,

$$\left(\mathbf{A}^{\mathrm{T}}\cdot\mathbf{A}\right)\cdot\mathbf{c}=\mathbf{A}^{\mathrm{T}}\cdot\mathbf{y}. \tag{23}$$

Finally, the LS solution is given by,

$$\mathbf{c}=\left(\mathbf{A}^{\mathrm{T}}\cdot\mathbf{A}\right)^{-1}\cdot\mathbf{A}^{\mathrm{T}}\cdot\mathbf{y}=\mathbf{A}^{+}\cdot\mathbf{y} \tag{24}$$

From (24), each coefficient is the linear combination of the data vector $\mathbf{y}$ and the respective row of the matrix $\mathbf{A}^+$. For example,

$$c_0=\mathbf{A}^{+}(1,:)\mathbf{y} \tag{25a}$$

$$c_i=\mathbf{A}^{+}(i,:)\mathbf{y} \tag{25b}$$

where, $\mathbf{A}^+(i,:)$ is a row vector that corresponds to i^{th} row of the matrix $\mathbf{A}^+$.

The first line of $\mathbf{A}^+$ corresponds the coefficients of the smoothing SG filter,

$$H_{SG}(z)=\sum_{i=1}^{2N+1}a_{1,i}z^{(-N-1+i)} \tag{26}$$

where $a_{1,i}$ corresponds to the i-th column of the first line.
The difference equation represented by (26) is given by,

$$c_0[n]=\sum_{i=1}^{2N+1}a_{1,i}y(n-i-1-N) \tag{27}$$

$H_{SG}(z)$ is a noncausal FIR filter and to be implemented in real time, a N delay buffer must be included. Figure 16 illustrates SG filter as a smoothing filter. An observation vector $\mathbf{y}$ given by the following samples:

$$\mathbf{y}=\begin{bmatrix}1 & 2 & 1.5 & \underset{\uparrow}{1} & 0.5 & 1 & 2\end{bmatrix}^{T},$$

where the arrow under number 1 indicates the index $n = 0$ of the sequence. The central sample, $y[0]$, will be modified by the smoothing filter of (27). In this example the polynomial order is $m = 3$ and the number of samples in the buffer is 7, i.e., $N = 3$. The filter coefficients of (26) is given by,

$$\mathbf{a}_1=\frac{1}{21}\cdot\begin{bmatrix}-2 & 3 & 6 & 7 & 6 & 3 & -2\end{bmatrix} \tag{28}$$

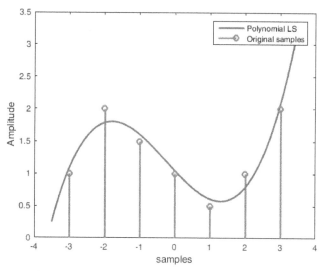

Fig. 16: The smoothing SG filter of order three. The central sample is smoothed.

Figure 16 shows that the smoothed samples at $n = 0$ is the one obtained by the polynomial (14). In this case, x represents the sample index n. For smoothing the next sample, i.e., $n = 1$, the buffer must be updated and (27) performed again. Note that the coefficients of the filter remain constant, while the buffer of the filter is updated sample by sample.

A first order SG differentiator can be obtained using the following equation

$$c_1[n] = \sum_{i=1}^{2N+1} a_{2,i} y(n-i-1-N) \tag{29}$$

The advantage of SG differentiator is the ability to reduce noise effect in derivative calculation, once it is well known; the noise is amplified by using simple numerical derivative approaches, such as Euler Backward approximation.

The reduction of the noise effect can be understood by the magnitude response of the derivative filter. Figure 17 shows the magnitude response for ideal differentiator and four SG differentiator filters, of orders $N = 2, 3, 4$ and 5. At low frequency, all SG differentiator filters follow the frequency response of the ideal differentiator. However, while the ideal differentiator amplifies noise, the SG differentiators attenuate the high frequencies and consequently the noise. The differentiators that have similar behavior in frequency domain, are named narrow band differentiators, and work very well if the input signal to be differentiated has low frequency content compared with the sampling rate.

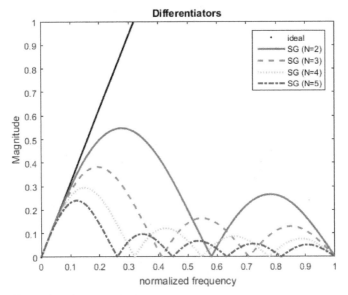

Fig. 17: Differentiators frequency response.

The above characteristics of differentiator will be useful in several power system applications, mainly in detection application. For example, in Schettino et al. (2016), the second order SG differentiator filter is used to detect the Current Transformer (CT) Saturation. When the CT saturates, the secondary current (i_s) can be segmented into two types of regions: (i) the unsaturated regions, where the secondary current is a scaled version of the primary current (i_p), and (ii) the saturated regions, where there is not linear relationship between i_p and i_s. Figure 18(a) shows typical signals of primary current (dashed line) and secondary current (solid line) of a saturated CT. Then the goal is to detect the abrupt changes in the secondary current as can be seen by the green points that delimitates the saturated and unsaturated regions. The abrupt changes can be detected by using narrow band differentiator filters, such as SG differentiator filters.

The second-order SG differentiator filter using a third-order polynomial and a small seven-samples data window is proposed in Schettino et al. (2016), for detecting CT saturation when applied to the sampled signal. The filter coefficients can be obtained by taking the third column of (25b), yielding in the causal second-order derivative SG filter:

$$del2sg[n] = \{5i_s[n] - 3i_s[n-2] - 4i_s[n-3] - 3i_s[n-4] + 5i_s[n-6]\}/42, \quad (30)$$

where n is the sample index.

The secondary distorted current signal generated by the RTDS shown in Figure 18(a), henceforth called the clean signal, was embedded in

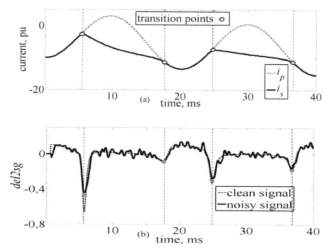

Fig. 18: Detection of transition points: (a) Typical signals of primary current and secondary distorted current of a saturated CT, (b) Second-order derivative estimation using SG filter.

white Gaussian noise (WGN) with a signal-to-noise-rate (SNR) of 40 dB. Next, the noisy signal was pre-filtered with an anti-aliasing second-order Butterworth low-pass filter having a cutoff frequency of 300 Hz (for 50 Hz system frequency), which is commonly found in protective relays. Finally, the pre-filtered noisy signal was differentiated by the SG filter presented in (30). The result for the clean and noisy signal is shown in Figure 18(b). As can be seen with the aid of the dash-dotted vertical lines in Fig. 18(a) and (b), the transition points (green points) match the del2sg negative peaks (red points). Therefore, the choice of a suitable threshold allows precise location of the transition points and achieves CT saturation detection.

5. Conclusions

This chapter discusses current issues regarding power system measurement and instrumentation that range from instrumentation transformers (IT) to signal processing. In order to obtain an accurate measure and instrumentation system it is crucial to have accurate IT. This chapter has shown that IT is an important topic of investigation since the current magnetic transformers do not meet the repeatability and accuracy criteria when it comes to measuring harmonics and other high frequency components. The second focus of this chapter was on sampling and resolution with emphasis on the sub-sampling technique that can be used for sampling supra-harmonics. The chapter showed how it is possible to take advantage of sub-sampling to estimate high order harmonics without the need to increase the sampling rate that is commonly used in present day power quality monitors. Finally, two signal processing application were presented, for power system measurement and instrumentation related with signal embedded in white Gaussian noise. The first focused on reducing the noise of the original signal as a preprocessing tool, using wavelet transform for denoising. The second application showed very useful filters named Savitzky-Golay filter that can be used for denoising or derivative application, such as saturation on current transformer detection. The entire chapter sought to address some important topics that deserve attention from engineers and researchers working in the field of power system measurement and instrumentation.

References

Burrus, C.S., R.A. Gopinath and H. Guo. 1998. Introduction to Wavelets and Wavelets Transforms, A primer; Prentice Hall, New Jersey.

Cunha, C.F., A.T. Carvalho, M.R. Petraglia and A.C. Lima. 2015. A new wavelet selection method for partial discharge denoising. Electric Power Systems Research 125: 184–195.

Donoho, D.L. and I.M. Johnstone. 1994. Threshold selection for wavelet shrinkage of noisy data. Proceedings of 16th Annual International Conference of the IEEE Engineering in Medicine and Biology Society, Baltimore, MD 1: A24–A25.

Donoho, L. 1995. De-noising by soft-thresholding. IEEE Trans. on Information Theory 41(3): 613–627.

IEC 61000-4-30:2008, Electromagnetic compatibility (EMC)—Part 4-30: Testing and measurement techniques – power quality measurement methods, 2008.

Kay, S.M. 2010. Fundamentals of Statistical Signal Processing, Volume 1: Estimation Theory. Prentice-Hall. ISSN 1050–2769.

Liao, C.C., H.T. Yang and H.H. Chang. 2011. Denoising techniques with a spatial noise-suppression method for wavelet-based power quality monitoring. IEEE Transactions on Instrumentation and Measurement 60(6): 1986–1996.

Meyer, J., M. Bollen, H. Amaris, A.M. Blanco, A.G. de Castro, J. Desmet, M. Klatt, L. Kocewiak, S. Sarah Rönnberg and K. Yang. 2014. Future work on harmonics - some expert opinions Part II - supraharmonics, standards and measurements. 16th International Conference on Harmonics and Quality of Power (ICHQP), Bucharest, pp. 909–913. doi: 10.1109/ICHQP.2014.6842871.

Nasri, M. and H. Nezamabadi-Pour. 2009. Image denoising in the wavelet domain using a new adaptive thresholding function. Neurocomputing 72(46): 1012–1025.

Ribeiro, P.F., C.A. Duque, P.M. Silveira and A.S. Cerquira. 2014. Power Systems Signal Processing for Smart Grids, Wiley.

Riordon, J., E. Zubritsky and A. Newman. 2000. Top 10 Articles. Analytical Chemistry 72(9): 324 A–329 A.

Savitzky, A. and M.J.E. Golay. 1964. Smoothing and differentiation of data by simplified least squares procedures. Anal. Chem. J. 36: 1627–1639.

Schafer, R.W. 2011. What is a savitzky-golay filter? [lecture notes]. IEEE Signal Processing Magazine 28(4): 111–117.

Schettino, B.A., C.A. Duque and P.M. Silveira. 2016. Current-transformer saturation detection using savitzky-golay filter. *In*: IEEE Transactions on Power Delivery 31(3): 1400–1401.

Seljeseth, H., E.A. Saethre, T. Ohnstad and I. Lien. 1998. Voltage transformer frequency response. Measuring harmonics in Norwegian 300 kV and 132 kV power systems. *In*: Proc. 8th International Conference on Harmonics and Quality of Power 2: 820–824.

Stein, C.M. 1981. Estimation of the mean of a multivariate normal distribution. Ann. Statist. 9(6): 1135–1151.

Stiegler, R., J. Meyer, J. Kilter and S. Konzelmann. 2016. Assessment of voltage instrument transformers accuracy for harmonic measurements in transmission systems. 17th International Conference on Harmonics and Quality of Power (ICHQP), Belo Horizonte, 2016, pp. 152–157. doi: 10.1109/ICHQP.2016.7783472.

Zhang, H., T. Blackburn, B. Phung and D. Sen. 2007. A novel wavelet transform technique for on-line partial discharge measurements: Part 1. WT de-noising algorithm. IEEE Trans. Dielectr. Electr. Insul. 14(1): 3–14.

Zhao, S., H. Li, P. Crossley and F. Ghassemi. 2014. Test and analysis of harmonic responses of high voltage instrument voltage transformers. 12th IET International Conference on Developments in Power System Protection (DPSP), Copenhagen.

CHAPTER 11

High Capacity Consumers Impact on Power Systems

Harish Sharma

1. Introduction

High capacity consumers are key customers for electric service providers, accounting for dependable base loads and revenue streams. Examples include steel mills employing electric arc furnaces (EAFs), metals processing plants using induction furnaces, facilities employing large motors with fluctuating loads, facilities employing large HP rated Variable Frequency Drives (VFD), and more recently, data centers with large numbers of computers. It is in the best interest of service providers to ensure that the quality of the supply voltage to any customer is not affected by the loads in these high capacity facilities. However, many of these facilities have a potential of being a source of power quality (PQ) issues as they can be sources of significant harmonic current injections and/or excessive fluctuations in supply voltage. The adverse impacts on the system can be quite pronounced if the supply system is weak given the large size of the facility. An example would be a large industrial plant located at the end of a weak distribution feeder (Dugan et al. 2012). Therefore, system planners should also include contingency scenarios (e.g., line outages) in the system impact studies when evaluating addition of high capacity consumers with a potential of introducing or aggravating PQ issues.

In this chapter, some example facilities are used for defining the PQ issues that the high capacity consumers can introduce. The key PQ issues

Southern Company Services, 600 North 18th Street, Birmingham, AL 35203, USA.
Email: hsharma@southernco.com

are explained and the methodologies that may be used to avoid and mitigate the same have been described.

2. Electric Arc Furnaces

One example of a high capacity consumer that can cause significant PQ issues is a steel mill employing an EAF. The concerns are mainly related to operation of open arc type furnaces in which arcing between electrodes and charge happens in open air. It may be noted that these can be served by either an AC or DC supply. To minimize system impacts, such facilities are usually connected to the transmission system to provide a strong source capable of delivering thousands of short circuit MVA. These furnaces are normally sized between 100 and 150 MW and the typical setup for an AC EAF is illustrated in Fig. 1. A main transformer is used to step down the voltage at the point of common coupling (PCC) to a medium voltage. A furnace transformer is then used to step the service voltage down to supply the electrodes of the furnace. The furnace transformer typically has several taps to vary the output voltage during different stages of a melt cycle. The furnace itself consists of a refractory lined vessel that is water cooled. The charge in the form of metal scrap is added to the vessel through the slag door. There are three electrodes (one for each phase) through a retractable roof and these can be moved up or down. The arcing happens between the electrodes and the scrap and the resultant heat melts the scrap. At the end of the cycle, a molten steel can be tapped into a ladle which is then taken to a ladle furnace.

The nature of EAFs produces a widely fluctuating and unbalanced load due to the random nature of arcing and possibility of electrodes breaking during a melt cycle. The power factor is usually poor at values as low as 0.7. These factors result in fluctuations in system voltage which happen to be most severe during the initial portion of the melting cycle known as boredown. During this period, the electrodes are lowered into

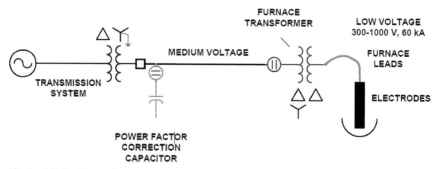

Fig. 1: AC electric arc furnace overview.

the scrap, an arc is struck and the electrodes are bored into the scrap. The arcing during the period is very erratic. After some time, the electrodes are raised, lengthening the arc and the power is increased. During this period the arcing continues to be very fierce. A melt cycle may include multiple charges with each instance followed by boring and melting periods. Eventually, a flat bath stage is reached when the molten pool has formed and the arc has stabilized. The greatest power system impact is related to each initial boredown period as the arcing is highly random and fierce.

2.1 Flicker Impact

Sample EAF current measurements during the initial boredown period are illustrated in Fig. 2, clearly showing the highly unbalanced and erratic nature of this type of load. The resultant fluctuations in the system voltage are apparent in the voltage measurements at the furnace bus as illustrated in Fig. 3. These fluctuations get reflected to the PCC and become a source of flicker to other customers served by the same system in the vicinity.

An example trend of flicker measurements at the PCC of an EAF facility is shown in Fig. 4. It shows the trend of P_{st} that is one of the indices used to characterize flicker and can be obtained by connecting a flickermeter that is compliant with IEC 61000-4-15. P_{st} is basically a measure of short-term

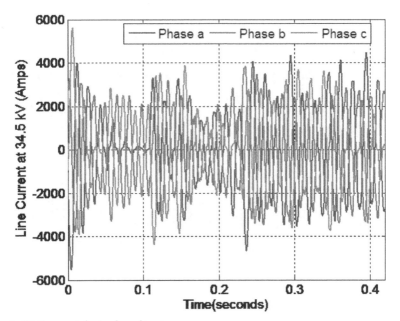

Fig. 2: EAF current during boredown.

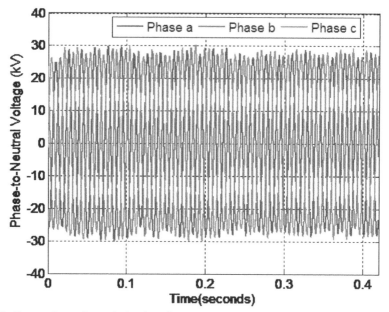

Fig. 3: Furnace bus voltage during boredown.

Pst Trend at PCC

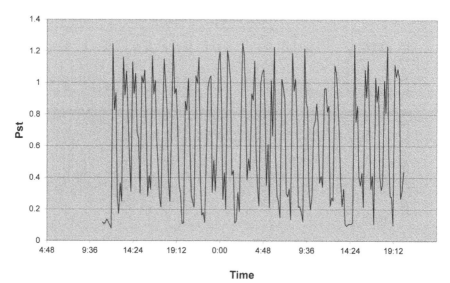

Fig. 4: Example flicker introduced by EAF facility.

perception of flicker that is obtained in ten minute intervals. A Pst value of 1.0 in the supply voltage at lighting service voltages signifies that 50% of the population served by that supply can be expected to perceive the resultant fluctuations in the light output of the lamps.

The voltage fluctuations responsible for Pst in the transmission system are typically somewhat attenuated as they propagate towards the distribution system. This attenuation phenomenon can be predicted by computing flicker transfer coefficients using various methods summarized in IEEE 1453 2015 (IEEE Standards Association 2015). Practically speaking, customer complaints have not been reported to be received by service providers until the Pst values in the HV system start exceeding 1.3 as reported in (CIGRE 2011, WG C4.108 2011). To avoid flicker complaints, industry standards have recommended planning levels for flicker (See Table 1 and Table 2). It may be noted that planning levels are assessed at the PCC and should consider the aggregate impact of all the fluctuating loads at that PCC. This means that emission limits of individual loads should be set so that the combined effects do not exceed the planning levels. It is a prudent practice for service providers to identify and then specify the appropriate emission limits in the contract with the customer. IEEE Std 1453-2015 (IEEE Standards Association 2015) provides detailed procedures for determining appropriate emission limits for customers that produce voltage fluctuations.

For a new EAF application, the need for a mitigation solution can be identified early in the process by estimating its future flicker contribution using easy hand calculations that can be found in IEEE Std 1453-2015 (IEEE Standards Association 2015) and that are reproduced here. The 95th percentile value of flicker index P_{st} can be computed using equation (1)

$$P_{st95\%} = K_{st} \frac{scc_f}{scc_n} \qquad (1)$$

Table 1: Recommended planning levels (IEEE Standards Association 2015).

	Planning levels	
	MV	HV-EHV
Pst	0.9	0.8
Plt	0.7	0.6

Table 2: System voltage levels (IEEE Standards Association 2015).

System	System voltage (UN)
LV	UN ≤ 1 kV
MV	1 kV < UN ≤ 35 kV
HV	35 kV < UN ≤ 230 kV
EHV	UN > 230 kV

Where,

Scc_f = Short circuit capacity of the furnace at the PCC (electrodes shorted)

Scc_n = Short circuit capacity of the system at the PCC (short circuit at the PCC)

K_{st} is a measure of the arc furnace flicker causing characteristics independent of the effect of the short circuit strength. For 120 V systems, a value ranging between 58 and 70 for K_{st} is recommended for estimating purposes. The relationship between 120 V and 230 V systems with regards to K_{st} values is shown in (2). This can be interpreted as the 230 V systems being more susceptible to flicker than 120 V systems.

$$K_{st(120\ V)} = 0.8 K_{st(230\ V)} \qquad (2)$$

It may be possible to obtain estimated K_{st} values from the arc furnace supplier. DC arc furnaces operate similar to AC furnaces except that the DC arc is more stable and usually causes less voltage fluctuations for the same size furnace. It is typically assumed that DC arc furnaces will have about 50–75% of the flicker levels associated with a similar size AC furnace.

The submerged arc furnaces (SAF) usually draw an arc on or in a significant amount of slag. The resistance of this slag prevents the true short circuits seen during an EAF operation. Consequently, the K_{st} value of a submerged arc furnace is much smaller than a similarly rated electric arc furnace. Typical values for SAF are in the 5–30 range.

Electric ladle furnaces are used for secondary refining of the metals exiting the main furnace. They operate in a much more stable mode than the electrical arc furnaces, since in their operation the arc is established on top of a molten metal which is usually covered by a thin slag layer. Typical K_{st} ranges for such furnaces are in the 15 to 30 range.

If higher accuracy is desired, detailed time domain modeling analysis may be performed. The modeling of EAF operation due to its random nature is quite challenging, but has been implemented and validated in (Horton et al. 2009).

2.2 Harmonics Impacts

Three-phase AC arc furnaces have an electrode for each phase which produce a waveform that is almost a square wave voltage across the arc. Such a voltage waveform can be decomposed into the 1/N harmonic sources. The third harmonic content is the largest component and can easily be in 25–35% range. Therefore, a 3-phase electric arc furnace is nearly always fed by a delta/delta or ungrounded-wye/delta transformer to block the flow of zero-sequence currents. However, this blocking of

Table 3: Harmonic Spectrum – arc furnace initial melt stage.

Harmonic	Magnitude (% of Fundamental)	
	IEEE Std 519-1992	Measurements for actual EAF
2	7.7	8.9
3	5.8	5.7
4	2.5	3.0
5	4.2	3.7
7	3.1	2.1

triplen harmonics occurs only during reasonably balanced operating conditions. The typical harmonic spectrum of an arc furnace in the initial melt stage, which is quite unbalanced, is shown in Table 3. It is evident that the spectrums are quite rich in both odd and even low order harmonics and thus need to be mitigated so that the current harmonic injections into the supply system at the PCC are within acceptable limits. In addition, EAFs are known to be a rich source of interharmonics.

2.3 Inter-harmonics Impacts

Currently there are no IEEE limits on interharmonic currents but caution is necessary to avoid potentially catastrophic torsional issues with any nearby conventional generation plants. The interharmonic currents injected into the system by EAFs can travel towards the source and flow into the stator of nearby generators. These interharmonic components can excite mechanical resonances of turbines that can be damaging. Studies required to evaluate this phenomenon are very complicated and the high-level description is provided here:

Step 1—Modal analysis of the turbine generator is performed to determine the mechanical resonant modes of the turbine. Eigen analysis may be performed to determine the modal frequencies of the system and the corresponding mode shapes. Example modal frequencies and normalized mode shapes for a unit are provided in Table 4 and Fig. 5.

Step 2—Current gain calculations need to be performed to determine how much of the interharmonic current injected into the system at the PCC will flow in the stator of the generator. A full detailed model of the system needs to be created in a time domain simulation platform in order to compute these current gains in the interharmonic frequency range of interest (1–119 Hz for 60 Hz and 1–99 Hz for 50 Hz systems). Interharmonic currents that are being injected into the facility need to be measured or estimated and the computed current gain values used to estimate the interharmonic

Table 4: Example modal frequencies.

Mode	Frequency (Hz)
1	12.3
2	21.3
3	25.7
4	39.7

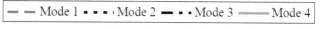

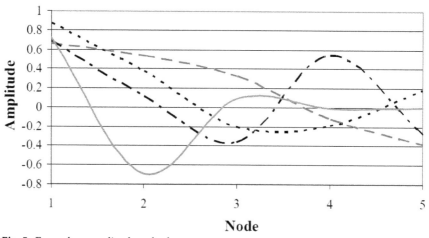

Fig. 5: Example normalized mode shapes.

current flow in the stator for the modal frequencies calculated in the previous step.

Step 3—The detailed time-domain model may be used to compute changes in the torque between shaft sections assuming the worst case of simultaneous excitation of all the modal frequencies. It is not expected to be an issue if the change in torque is below the screening criteria of 0.5 pu (EPRI 2006).

To avoid these torsional issues, EAFs should be located 2–3 buses away from the nearest generation but still ensuring adequate short circuit strength at the facility.

2.4 Mitigation Solutions

One mitigation solution for both harmonics and flicker issues is a Static Var Compensator (SVC) connected in parallel with the EAF. The SVC provides a fixed reactive power compensation component using a combination of tuned filter banks that serve the dual purpose of absorbing harmonic injections from the furnace. Reactive power is compensated but still fluctuating, so a thyristor controlled reactor (TCR) branch is employed. It is basically a reactor branch (sized on the order of 1.5–2.0 times the furnace rating) in series with back-to-back thyristors that provides the variable inductive reactive power component. By controlling the firing angle of these thyristors every half cycle, the desired Q can be drawn from the system to compensate for the variable reactive power due to the arc furnace operation. The SVC has a dynamic response time of 8–10 ms which is fast enough to practically reduce the flicker at the PCC by a factor of up to two. A formula that can be used to size the SVC, (S_{SVC}) based on the desired flicker mitigation is shown in (3). Note that a flicker reduction factor in excess of 2.2 is difficult to achieve in practice.

$$S_{SVC} = \frac{(R-1)}{0.75} S_f \qquad (3)$$

Where,

R = Desired flicker reduction factor

Sf = Size of the furnace (MVA)

Because EAFs produces a wide spectrum of harmonics and inter-harmonics, and typically use multiple capacitor banks, many possible resonant frequencies exist and the application of conventional notch filters becomes difficult. For this reason, a 2nd harmonic filter branch is typically designed as a C-filter (Fig. 6). It is essentially a high-pass filter with the addition of a capacitor, Ca, that is intended to prevent fundamental frequency current from flowing in the damping resistor, R. This filter provides very effective filtering over a wide bandwidth (Dugan et al. 2012).

In cases where much higher flicker mitigation is needed that can't be achieved by a SVC, a device called a STATCOM is an option. Like the SVC, it also uses harmonic filters as source of fixed reactive power. However, it employs IGBT based voltage source converters that can supply both inductive and capacitive reactive power. Higher switching frequencies (~ 1 kHz) means the dynamic response to reactive power variations is much quicker than for an SVC. The STATCOM responds in the order of a couple of ms and the flicker mitigation is improved significantly with reported flicker improvement factors as high as six (ABB 2012). It may be noted that improved flicker mitigation performance of STATCOM comes at a price, as the cost is roughly twice that of an SVC.

With devices such as SVCs and STATCOMs, one needs to be mindful of another torsional interaction called Device Dependent Subsynchronous Oscillations (DDSO) that occur between the controls of these devices and turbine-generators. The phenomenon is due to control interactions, and

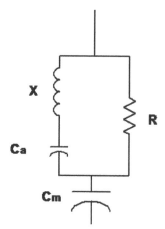

Fig. 6: C filter configuration.

is not a resonance phenomenon. Such interaction results in a negative damping torque phenomenon which can degrade or totally eliminate the inherent mechanical damping of a mechanical torsional mode of oscillation. If this is found to be a problem, the solution is in re-tuning of the control system.

Usage of a series reactor with taps is also an effective way to mitigate the flicker to some extent as it serves to stabilize the arc during the melting phase and also contributes to a reduction in the short circuit capacity of the furnace. Such a reactor is inserted between the main transformer and the furnace transformer. However, it requires making frequent changes in the furnace transformer taps in order to maintain the furnace productivity (Montanari et al. 1994).

Some facilities may employ induction furnaces for heating. In addition to harmonics and line-voltage notching, such furnaces specifically employing current-fed power supply designs have the potential to introduce additional interharmonics over a wide range of frequencies (EPRI 1999). If the melt shop is integrated with a rolling mill, adjustable speed motor drives are frequently employed which can lead to additional harmonic and flicker concerns.

3. Facilities with Large Motors

There are several facilities that employ large sized motors for their operations. Normal motor operation is generally not a PQ concern but the starting of large induction motors can be of concern. During starting, induction motors draw a large amount of current for several seconds at a very poor displacement factor—usually in the range of 15 to 30%. IEC

Standard 61000-3-7 classifies the voltage changes due to motor starting as rapid voltage changes (See Fig. 7). If such changes happen on a frequent basis, customer complaints related to lamp flicker can happen. To avoid complaints, the indicative planning levels in Table 5 may be used to determine emission limits for such changes for individual customers based on the number of changes.

In case the levels above are found to be exceeded, the mitigation solutions could be adopted that include the use of reduced voltage starters, soft-starters and SVCs.

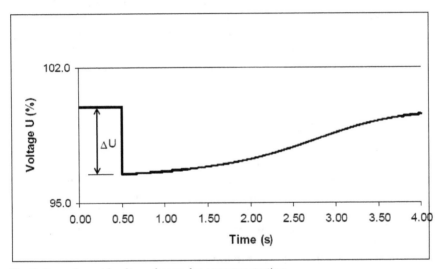

Fig. 7: Example rapid voltage change due to motor starting.

Table 5: Indicative planning levels for rapid voltage changes (International Electrotechnical Commission 2008).

Number of changes N	ΔU/Un (%)	
	MV	HV-EHV
N ≤ 4 per day	5–6	3–5
N ≤ 2 per hour	4	3
2 < N ≤ 10 per hour	3	2.5

4. Facilities with Large Drives

Adjustable speed drives (ASDs) find numerous applications in commercial and industrial systems (e.g., process control in oil and paper industry). These drives can be a significant source of harmonics at their characteristic frequencies. The harmonic distortion in these currents can vary over a wide range but can be classified based on the inverter type (Dugan et al. 2012). If distortion is found to be excessive, input chokes are found to be very effective in reducing harmonic distortion. The inductance in choke will make input current less distorted by slowing down the rate of charging of the dc cap. Input chokes provide additional benefit of damping the transients that may show up at drive input due to switching of capacitor banks upstream and prevent unexpected drive trips. Typical size of input chokes is between three and five percent of the impedance on the drive base.

DC drives can be a significant percentage of plant loading in many industrial facilities. These drives are still the most common type of motor speed control for applications requiring very fine control over wide speed ranges with high torques. Most dc drives use the 6-pulse rectifier. The two largest harmonic currents for the 6-pulse drive are at the 5th and the 7th but they are also the most troublesome in terms of system response.

Another effective method to control harmonics from the drives employing power converters is to use higher pulse number configurations (e.g., 12 pulse or 18 pulse). A 12-pulse configuration can be achieved by supplying one drive through a delta-wye transformer and another drive through a delta-delta transformer. This design causes the overall facility to approach 12-pulse operation with cancellation of the 5th and 7th harmonic components if there is nearly equal loading for the two connections. For a q-pulse converter, IEEE 519-2014 (IEEE Standards Association 2014) limits for characteristic harmonics are increased by a factor equal to the square root of q divided by 6 provided that the amplitudes of non-characteristic harmonics are less than 25% of the limits.

5. Facilities with Large Transformers

Energizing a transformer can cause it to draw significant initial inrush currents that are unbalanced in nature and which decay over time to much smaller steady state magnetizing currents. The main factors affecting the inrush current magnitude and duration include transformer design, initial conditions and network factors. Initial conditions affecting the magnitude of inrush current are residual flux and the point-on-wave at energization. The residual flux is the flux that remains trapped in the core due to a previous de-energization of the transformer. Energization at a voltage zero crossing results in the most severe inrush current for a transformer as it induces a flux-linkage of theoretically up to 2.0 pu (with 1.0 pu DC offset); the residual flux adds on top of that giving a maximum possible flux-linkage of almost 3.0 pu.

These inrush currents that can take several seconds to decay must be supplied by the system. This inrush current flow through the network impedance results in system voltage drops. It may be noted that these voltage fluctuations fall in the category of rapid voltage changes (RVCs) if rms reduction is within 10%. In energizing large facility transformers from weak grids, there is increased likelihood of resultant system voltage drops exceeding the allowable limits in Table 5. In some cases, transformer energization may even result in more than 10% rms reduction and thus fall under the voltage sag category. Voltage sag is rms reduction (to between 0.1 and 0.9 pu) in the AC voltage, at the power frequency, for durations from a half cycle to one minute (IEEE Standards Association 2009, 1159-2009). The preferred terminology when describing voltage sag is remaining voltage. For example, 80% sag refers to a disturbance that resulted in a voltage of 0.8 pu. An example voltage sag in a utility system due to the energization of a large facility transformer is shown in Fig. 8. For nearby facilities with sensitive loads, voltage drops may be compared against the SEMI F47 voltage sag ride through curve shown in Fig. 9.

The solutions that can be implemented to address the above issues include the following:

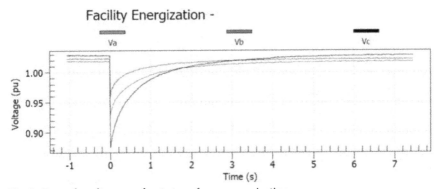

Fig. 8: Example voltage sag due to transformer energization.

5.1 Pre-insertion Switching

This solution involves use of a transformer switching device that is equipped with a pre-insertion resistor. It is a very effective solution to mitigate the transient magnetizing currents and resultant system voltage drops and the scheme itself is illustrated in Fig. 10.

Upon receiving a close command by the energizing breaker, the scheme works by closing the pre-insertion contact that will insert a resistor in series with the transformer to be energized. The voltage drop across the resistor produced by the inrush current will decrease the voltage on the transformer windings, which in turn decreases the magnetic over flux in the core. Consequently, the magnitude of the transient magnetizing currents will be reduced as well. Subsequently, the main contact is assumed

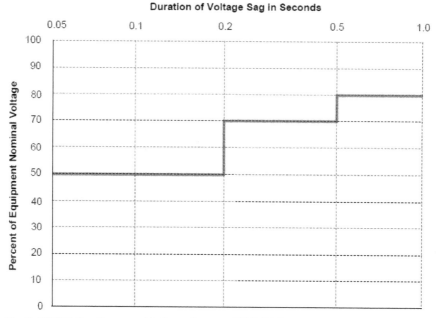

Fig. 9: SEMI F47 voltage sag ride through curve (EPRI 2004).

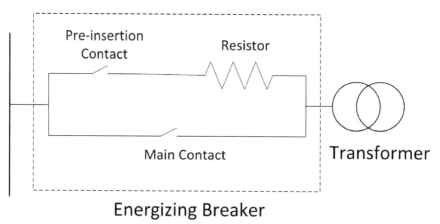

Fig. 10: Pre-insertion closing scheme illustration.

to close about 0.7 cycles or 12 ms later to bypass the pre-insertion resistor. It may be noted that a comprehensive EMT study is recommended for each location to arrive at an optimal value for the pre-insertion resistor and the insertion time.

5.2 Point-on-wave (POW) Control Switching

This solution involves using segregated pole breakers with special purpose point-on-wave control. The principle is to close the first phase at its voltage peak and delaying the closing of the other two phases by several half-cycles. Using such strategy, up to a 50% reduction in the inrush current has been achieved (CIGRE 2014, WG C4.307). Even better performance can be achieved by determining optimal closing times by incorporating residual flux information but it is more complicated to implement in practice.

5.3 Energizing Facility Transformers in Stages

Energizing the main transformer along with all the facility transformers will yield the worst case. Therefore, energizing the main transformer first followed by various facility transformers in stages may also be considered for mitigating these system impacts.

6. Summary

High capacity consumers have the potential to cause several PQ related issues to the service provider and its other customers in the vicinity. This chapter has identified such issues for various types of high capacity consumers and appropriate mitigation options have been presented. It is a good practice for service providers to be proactive in performing necessary studies to anticipate any potential issues and take corrective actions before a large customer is interconnected to the system.

References

ABB. 2012. SVC Light for Powerful Flicker Reduction from EAF Operation, https://library.e.abb.com/public/499fa9721d1e41d0482579f20066622c/1JNS012168%20LR.pdf.

CIGRE. 2011. CIGRE WG C4.108 2011. Review of Flicker Objectives for LV, MV and HV Systems.

CIGRE. 2014. WG C4.307. Transformer Energization in Power Systems: A study guide, published in February, 2014.

Dugan, R.C., M.F. McGranaghan, S. Santoso and H.W. Beaty. 2012. Electrical Power Systems Quality – Third Edition, McGraw-Hill New York.

EPRI. 1999. Power quality for induction melting in metals production. Techcommentary. International Electrotechnical Commission. 2008. IEC 61000-3-7. Electromagnetic Compatibility (EMC) – Part 3-7: Limits – Assessment of Emission Limits for the Connection of Fluctuating Installations to MV, HV and EHV Power Systems.

EPRI. 2004. Impact of SEMI F47 on Utilities and Their Customers. EPRI, Palo Alto, CA: 2004. 1002284.

EPRI. 2006. Torsional Interaction Between Electrical Network Phenomena and Turbine-Generator Shafts: Plant Vulnerability. EPRI, Palo Alto, CA: 2006, 1013460.

Horton, R., T. Haskew and R. Burch. 2009. A time domain AC electric arc furnace model for flicker planning studies. IEEE Transactions on Power Delivery 24(2): 1450–1457.

IEEE Standards Association. 2009. IEEE Std 1159-2009. IEEE Recommended Practice for Monitoring Electric Power Quality.

IEEE Standards Association. 2014. IEEE Std 519-2014. IEEE Recommended Practice and requirements for harmonic Control in Electric Power Systems.

IEEE Standards Association. 2015. IEEE 1453-2015. IEEE Recommended Practice for the Analysis of Fluctuating Installations on Power Systems.

International Electrotechnical Commission. 2010. IEC Standard 61000-4-15. Flickermeter - Functional and Design Specifications.

Montanari, G.C., M. Loggini, A. Cavallini, L. Pitti and D. Zaninelli. 1994. Arc-furnace model for the study of flicker compensation in electrical networks. IEEE Transactions on Power Delivery 9(4).

CHAPTER **12**

Grid-Edge Voltage Control in
Utility Distribution Circuits

Harsha V. Padullaparti,[1]
Pisitpol Chirapongsananurak[2] *and Surya Santoso*[1,*]

1. Introduction

One of the key responsibilities of a distribution utility is voltage regulation.
Voltage regulation refers to maintaining acceptable voltages at all points
along the distribution feeder including secondary service circuit under
all loading conditions. Most of the utilities in the U.S. follow ANSI C84.1
voltage standards (NEMA 2016) for maintaining acceptable levels of service
voltages. The service voltage is the point where the electrical systems of
the supplier and the user are interconnected which is normally at the
meter. According to ANSI C84.1, the service voltages should be within
±5% of their nominal voltage level which translates to 0.95 to 1.05 per unit
(pu) voltage. When a service voltage goes below 0.95 pu, it is considered
as an undervoltage violation and if the service voltage exceeds 1.05 pu,
then it is an overvoltage violation. For voltage regulation, traditionally,
utilities employ the load tap changer (LTC), line voltage regulators, and

---

[1] Electrical and Computer Engineering, The University of Texas at Austin, 2501 Speedway,
Austin, TX, USA-78712.
  Emails: harshap@utexas.edu
[2] Department of Electrical Engineering, Chulalongkorn University, 254 Phayathai Road,
Pathumwan, Bangkok 10330, Thailand.
  Email: pisitpol.c@chula.ac.th
* Corresponding author: ssantoso@mail.utexas.edu

fixed or switched capacitors. These electromechanical control devices are installed on the primary feeder and are controlled either in centralized or decentralized fashion to perform the voltage regulation.

There are few key issues associated with the use of traditional voltage regulation equipment in the present distribution systems with growing penetration of distributed energy resources (DER). One issue is the inherent mechanical nature of the primary voltage regulation devices. These devices can perform a limited number of tap operations per day, significantly limiting their voltage regulation capability especially in the presence of DER and fast-charging electric vehicle loads. Furthermore, with the discrete tap positions in case of voltage regulators and on/off status in case of capacitors, these devices can only provide coarse control over the secondary voltages. They cannot control the secondary voltages precisely. Another major issue is a significant voltage drop across service transformers, which is incorrectly assumed to be minimal in the traditional system studies. Traditional voltage regulation practices use primary side measurements and controls to keep the secondary-side customer voltages within ANSI C84.1 limits. In doing so, the voltage drop across the service transformers is ignored assuming it to be trivial. However, the advanced metering infrastructure (AMI) data from the field indicates that the voltage drop across the service transformers is significant, and can vary from 2 to 13 V on a 240 V base which is more than half of the ANSI band (Padullaparti et al. 2016). The decoupling of primary feeder circuit from the secondary, with the lack of voltage control on the secondary-side, forces the voltages on the primary to be maintained at higher values even if few customer load points experience undervoltage. This defeats the objectives of conservation voltage reduction (CVR) which promotes maintaining load voltages close to the lower ANSI limit to save energy.

The above reasons have resulted in the emergence of a new set of advanced voltage regulation devices connected at the grid edge. The term, grid-edge, refers to the secondary-side of the distribution circuit. The low-voltage advanced voltage regulation devices are electronically controlled and are designed to be installed on the secondary-side of the distribution circuit to have more direct control of load voltages for effective voltage regulation. Some of these advanced voltage regulation devices are deployed in the field and have shown effective performance as on date. In-line power regulator-50 (IPR-50), static var compensator-20 (SVC-20) developed by Gridco Systems (Gridco 2016), and the edge of network grid optimization (ENGO) developed by Varentec (Varentec 2016) are examples of such devices commercially available today.

Given the availability of various advanced secondary-side voltage controls lately, utilities today face the challenge of determining the effective locations to deploy them in the distribution circuits. Although, several methods have been reported in the literature to deploy traditional voltage regulation equipment on the primary distribution circuit, techniques for effective deployment of the secondary-side low-voltage control devices are lacking (Padullaparti et al. 2017a). Additionally, precise evaluation of the voltage regulation performance of these devices requires employing the device models in the system studies that reasonably emulate the characteristics of the devices available in the market. In this chapter, characteristics and modeling of the advanced secondary-side voltage controls are discussed. Two approaches to effectively deploy these devices in a utility distribution circuit are proposed and the resulting voltage regulation performance is compared.

The universal power flow controller (UPFC) model (Montenegro 2016) available in OpenDSS (Dugan 2012) replicates the functionality of IPR-50. The SVC-20 can be modeled as a multi-stage single-phase capacitor bank. For emulating the functionality of ENGO, a proprietary model is used in OpenDSS for the results presented in this chapter. The functionality of SVC-20 and ENGO is similar. They both provide var compensation. Generic names such as UPFC, SVC A, and SVC B are used to refer to IPR-50, SVC-20 and ENGO, respectively in this chapter.

This chapter is organized as follows. A brief review of traditional voltage regulation devices is provided in Section 2. In Section 3, the operation, modeling, and characteristics of the advanced voltage control technologies providing direct control over the secondary voltages are presented. The placement methods, details of the distribution circuit used in this study and the base case are presented in Section 4. The simulation scenarios and results are discussed in detail in Section 5. Conclusions are provided in Section 6.

2. Voltage Control using Traditional Voltage Regulation Techniques

The real and reactive power flows in the distribution circuit create voltage drops across lines and service transformers. In a radial distribution circuit with no voltage control, this results in the reduced voltage at the customer loads compared to the voltage at the substation. In general, the loads located farther from the substation tend to experience undervoltages. There are several ways to improve the voltage regulation. A brief overview of the most popular traditional voltage regulation techniques is provided here.

2.1 Substation Load Tap Changer (LTC)

Many distribution transformers installed at the substation are equipped with a load tap changer (LTC). The LTC is installed on one side of the transformer and changes its effective transformation ratio to regulate the secondary-side voltage of the transformer at a desired level. There are three basic settings associated with the LTC control namely voltage level, voltage bandwidth, and time delay. The voltage level (also known as a set point or bandcenter) specifies the desired output voltage of the LTC transformer to keep the load voltages in the feeder within permissible limits. The voltage bandwidth functions as a dead band around the voltage level to change the LTC tap position. For example, if the voltage level is 120 V and the bandwidth is 2 V, then the tap position does not change as long as the LTC transformer secondary voltage is between 119 V and 121 V. The most common settings for the voltage bandwidth on 120 V base are 2.0 and 2.5 volts. Voltage bandwidth settings of 3.0 V and above are used where tight regulation is not required (Harlow 1996). To avoid the tap changer operations for voltage excursions of short duration, an intentional time delay is included which is typically in the range of 30 to 60 s.

2.2 Line Drop Compensator (LDC)

Line drop compensator (LDC) is used to regulate the voltage at a point along the feeder remote from the LTC transformer location. The LDC uses an internal model of the impedance of the distribution line and the line current measurement to calculate the voltage drop from the LTC transformer secondary to the regulation point. It then adjusts the LTC transformer tap position to compensate the voltage drop so that the voltage at the regulation point is maintained at the desired level. The LDC can also be associated with the step voltage regulators.

2.3 Step-Voltage Regulator

Step-voltage regulator is basically an autotransformer that has a shunt winding on one side and a series winding on the other side. The series winding is connected either to aid or oppose the voltage of the shunt winding as shown in Fig. 1. Most regulators can regulate the voltages from −10 to +10 percent of their nominal voltage rating in 32 steps of approximately 5/8 percent voltage change in each step (Short 2004).

The step-voltage regulator can be of single-phase or three-phase. A three-phase regulator is gang operated, i.e., the taps on all windings change simultaneously to the same tap position. Alternatively, a three-phase regulator can be formed by connecting three single-phase voltage regulators. In such a connection, the tap position on each individual phase is controlled separately. Commonly, utilities use single-phase regulators even on three-phase circuits (Short 2004).

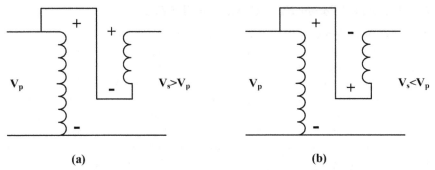

Fig. 1: Single-phase step-voltage regulator (a) step-up connection (b) step-down connection.

2.4 *Fixed and Switched Capacitors*

The capacitor banks installed on the primary wire help improve voltage regulation and reduce circuit losses by supplying the required leading reactive power. In addition, the capacitors help to free up the system capacity by improving the power factor. Proper sizing and siting of capacitor banks along with appropriate controls are important in obtaining the aforementioned benefits. Otherwise, the capacitors can create excessive voltages and additional circuit losses in case of overcompensation. The capacitors can be fixed or switched. The fixed capacitor banks remain in service always irrespective of the loading or other circuit conditions. The switched capacitor banks, on the other hand, can be equipped with local controls to take them out of service when not needed. The typical control options available with the switched capacitor banks are time, temperature, voltage, vars, power factor, and current. The switched capacitor bank controllers generally offer more than one of these control options. Some controllers offer one primary control with voltage override. For these controllers, a voltage band is given beyond which the voltage control overrides the primary control which can be any one of the control options available other than the voltage.

3. Advanced Voltage Regulation Technologies

The advanced voltage regulation devices address many shortcomings of the traditional voltage controls described in the previous section. These electronically controlled devices are designed to install on the secondary networks to offer direct control over the load voltages. A set of these devices can improve voltage regulation in a distribution circuit. In this section, the operation and characteristics of three advanced voltage regulation devices namely UPFC, SVC A, and SVC B are discussed.

3.1 Universal Power Flow Controller (UPFC)

The universal power flow controller (UPFC) is a single-phase power electronics-based multi-function device that can provide voltage regulation and power factor correction. The device specifications of a commercially available 50-kvar unit are given in Table 1. It can be installed on a service transformer secondary and configured to operate in any of the three functional modes (Montenegro et al. 2016). The device performs voltage regulation, power factor correction, and dual control in modes 1, 2, and 3, respectively. For voltage regulation in mode 1, the device can boost or lower the voltage at its output up to 10% of its voltage rating, i.e., 24 V with respect to its input node. As for the power factor correction in mode 2, the device can inject reactive power up to 10% of its rating, i.e., 5 kvar. The reactive power injection can be leading or lagging depending upon the requirement. In mode 3, the device tries to accomplish both voltage regulation and power factor correction simultaneously.

UPFC Device Modeling. A commercially available UPFC device can be emulated using the steady-state UPFC model available. The mathematical model of the UPFC is shown in Fig. 2. The Thevenin equivalent of the series voltage source of the UPFC is represented by the series impedance X_s and shunt current source I_s, where X_s is the impedance of the series transformer of the UPFC. The value of the current source I_s is given by:

$$I_s = \frac{V_{diff} - (V_{out} - V_{in})}{jX_s} + I_s[z-1] \tag{1}$$

$$V_{diff} = (V_{ref} - |V_{in}|)e^{j\theta_{vin}} \tag{2}$$

where $I_s[z-1]$ is a shift register containing the value of the current source I_s calculated in the previous power flow solution iteration. Equations (1) and (2) are used to calculate the value of the current source iteratively until power flow convergence is reached. When the convergence is reached, the value

Table 1: Specifications of UPFC.

Specification	Details
Rating	Single-phase, 60 Hz, 50 kVA
Load voltage regulation range	± 10% with 0.5% accuracy
var compensation range	10% of rating (leading or lagging), i.e., up to 5 kvar
Response time	Less than 1 cycle
Operation modes	Mode 1: Voltage regulation Mode 2: Power factor correction Mode 3: Both voltage regulation and power factor correction

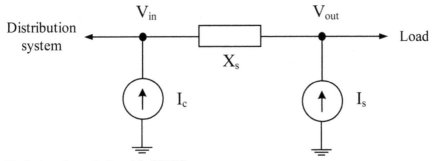

Fig. 2: A mathematical model of UPFC.

of I_s will be equal to $I_s[z-1]$. The value of the current source I_c is calculated as:

$$I_c = \frac{V_{out}}{I_s \times V_{in}} \qquad (3)$$

Equation (3) allows balancing the power at input and output sides of the UPFC. The UPFC converts active power into reactive power to maintain the output voltage at reference voltage setting. Thus, (3) can be reformulated as:

$$I_c = -(\text{real}(I_c) \times \text{Losses} + \text{imag}(I_s)) \qquad (4)$$

where the 'Losses' corresponds to the losses of the device defined by a curve.

UPFC Device Characteristics. To study the characteristics of UPFC, a simple test circuit as shown in Fig. 3 is simulated in OpenDSS (Dugan 2012). The circuit has an ideal voltage source of 12.47 kV and a single-phase 7.2 kV/0.24 kV, 50 kVA service transformer. A load rated 15 kVA with 0.9 power factor lagging is connected to the service transformer secondary through a 100-ft service wire. The parameters of this simple circuit are taken from an actual distribution circuit model. A UPFC device is connected to the service transformer secondary (240 V node). Figure 4 shows the UPFC device power loss characteristic as a function of input voltage defined in the UPFC model. The power losses are minimum at 0.8% of the device rating when the input voltage is 1 pu and increases linearly to 1.43% when the input voltage varies by 0.1 pu in either direction.

The operational characteristics of UPFC in voltage regulation mode are studied by configuring the UPFC in mode 1 to regulate the output voltage (V_{out}) at 1 pu. The input voltage of UPFC (V_{in}) is varied by varying the source voltage in the simple circuit model. Then the input UPFC voltage (V_{in}) and the load voltage (V_{load}) without and with the UPFC device in service are plotted as shown in Fig. 5. The black dotted line is the

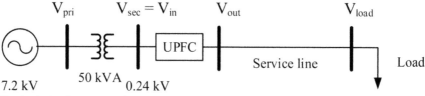

V_{pri} $V_{sec} = V_{in}$ V_{out} V_{load}

UPFC

Service line   Load

7.2 kV   50 kVA   0.24 kV

Fig. 3: A simple circuit with UPFC.

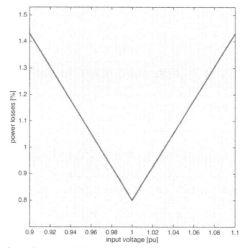

Fig. 4: UPFC power loss characteristic.

load voltage without voltage regulation (no UPFC) neglecting the minor voltage drop across the service line, i.e., the load voltage is equal to the transformer secondary voltage ($V_{load} = V_{sec} = V_{in}$). It is observed that when the UPFC is in service, the load voltage is regulated close to 1 pu in the full UPFC input voltage range between 0.9 pu and 1.1 pu. Furthermore, the UPFC output voltage, indicated by the thick dashed line, is equal to 1 throughout this range. The UPFC injects voltage in series up to ±24 V to regulate the output voltage (V_{out}) to the specified value (V_{sp}). Note that the load voltage is slightly less than 1 pu due to the voltage drop in the service line. The series voltage injection by the UPFC ($V_{inj} = V_{out} - V_{in}$) as the input voltage is deviating from the specified voltage is depicted in Fig. 5(b). Note that as the V_{in} becomes lower than the specified voltage V_{sp}, the UPFC injects series voltage to regulate the load voltage to V_{sp}. The UPFC output voltage is greater than its input voltage ($V_{out} > V_{in}$) in this case. Alternatively, if V_{in} is higher than V_{sp}, the UPFC injects negative voltage to regulate the load voltage to V_{sp}, thus $V_{out} < V_{in}$.

The UPFC is then configured in mode 2, i.e., power factor correction mode with the target power factor specified as 1 (PF = 1) to study the characteristics in this mode. The service transformer primary voltage is

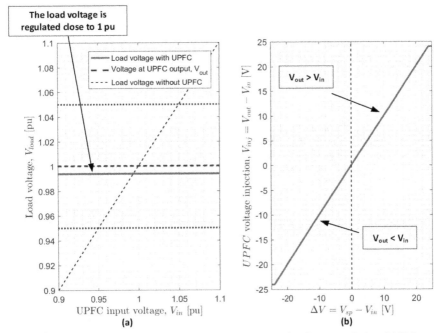

Fig. 5: UPFC operational characteristics in mode 1 (a) Load voltage regulation (b) Voltage injection by UPFC.

varied and the power factors at load and transformer primary are recorded along with reactive powers of load, service transformer, and UPFC. The results are shown in Fig. 6(a). It is observed that the load power factor changes in response to the load voltage changes due to the load model characteristics. Furthermore, although the load power factor is about 0.8 lagging, the power factor at the service transformer primary is close to unity because the reactive power required by the load is partly supplied by the UPFC. The reactive power injected by the UPFC is limited at 5 kvar leading which is the maximum limit of the device (10% of 50 kvar). The load voltage without and with the UPFC in service as the transformer secondary voltage is varied is shown in Fig. 6(b). The load experiences an additional voltage drop across UPFC series impedance because of the presence of UPFC.

The UPFC mode 3 is dual control mode. In this mode, the UPFC attempts to accomplish both voltage regulation and power factor correction simultaneously. To study this behavior, the UPFC is configured in mode 3 and the service transformer secondary voltage (V_{sec}) is varied by varying the source voltage. The load voltage regulation and UPFC voltage injection characteristics in this mode, shown in Fig. 7, indicate that the UPFC performs voltage regulation quite well similar to mode 1. With the

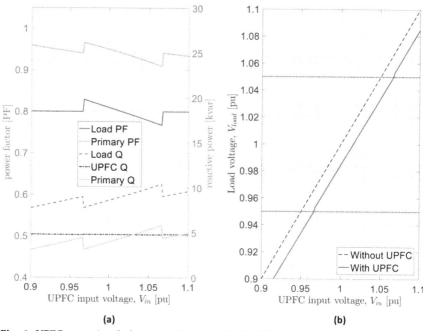

Fig. 6: UPFC operational characteristics in mode 2 (a) Power factor correction (b) Load voltage drop.

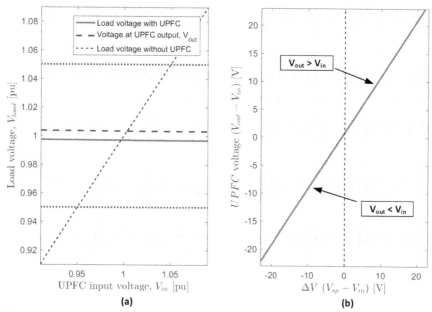

Fig. 7: UPFC operational characteristics in mode 3 (a) Load voltage regulation (b) Voltage injection by UPFC.

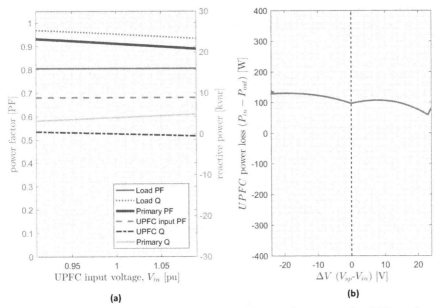

Fig. 8: UPFC operational characteristics in mode 3 (a) Power factor correction (b) Power loss in UPFC.

UPFC in mode 3, the UPFC output voltage is regulated to a value slightly above $V_{sp} = 1$ pu. Furthermore, the UPFC also accomplishes power factor correction well. This is evident from the Fig. 7 as despite the load power factor is close to 0.6, the power factor at the UPFC input is 1 in Fig. 8(a). Note that due to the regulated voltage, the load power consumption seen in Fig. 8(a) is different compared to that shown in Fig. 6(a). The power loss in the UPFC is shown in Fig. 8(b).

3.2 Static Var Compensator A (SVC A)

The static var compensator A (SVC A) is a single-phase shunt-connected device. This device is designed to be installed on the secondary side of a distribution service transformer. A 20-kvar SVC A device, for example, consists of 4 capacitor units each supplying 4.375 kvar leading reactive power at 240 V. The SVCs provide required voltage boost at the installed location by bringing the required number of capacitor units into service. The specifications of a commercially available 20-kvar SVC A are given in Table 2.

The SVC A voltage regulation characteristics are studied from the simple circuit simulations with SVC A model as shown in Fig. 9. The SVC A voltage regulation characteristics as the service transformer secondary voltage V_{sec} (SVC terminal voltage) increases is shown in Fig. 10. In this

Table 2: Specifications of SVC A.

Specification	Details
Rating	Single-Phase, 60 Hz, 50 kvar
Capacitor stages	4 capacitor stages each rated for 4.375 kvar at 240 V
Voltage boost characteristic	Low threshold to switch in the capacitor stages; Start switching out the capacitor stages when terminal voltage is above 3 V of target voltage
Extreme thresholds	> 252 V switch all stages out < 216 V all switch in

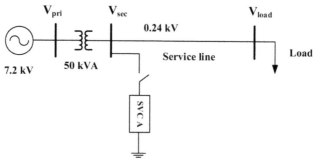

Fig. 9: A simple circuit with SVC A.

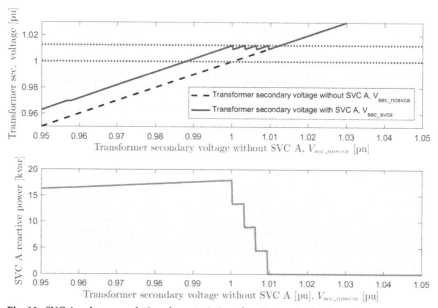

Fig. 10: SVC A voltage regulation characteristics when its terminal voltage increases.

figure, V_{sec} without SVC A (V_{sec_nosvca}) and with SVC A (V_{sec_svca}) are plotted against V_{sec_nosvca}. These characteristics are obtained as V_{sec} is varied by increasing the source voltage from 0.9 pu to 1.1 pu. Note that the SVC A specified voltage and the upper threshold voltages are configured as V_{sp} = 1 pu (240 V) and 1.0125 pu (243 V) by setting ONsetting = 240 V and OFFsetting = 243 V, respectively. From Fig. 10, it is observed that as the transformer secondary voltage without SVC A is initially well below the specified voltage of 1 pu (240 V), all the capacitor stages are in service. Thus, V_{sec} with SVC A is higher than that without SVC A. As V_{sec_svca} increases beyond the upper threshold of 1.0125 pu (243 V), the capacitor stages are switched out sequentially till the voltage becomes less than 243 V. After the last capacitor stage switched out, V_{sec} without and with the SVC A in service are same because the SVC A does not bring any capacitor stages into service. The voltage boost provided by the SVC A is also evident from this figure.

The SVC A voltage regulation characteristics when its terminal voltage decreases are shown in Fig. 11. The characteristics are obtained by decreasing the source voltage from 1.1 pu to 0.9 pu. It is observed that as V_{sec_svca} is above the upper threshold of 1.0125 pu (243 V), none of the capacitor stages are in service. As a result, both V_{sec_svca} and V_{sec_nosvca} align with each other. As V_{sec_svca} voltage decreases below the specified voltage V_{sp} = 1 pu (240 V), the capacitor stages are switched in sequentially to maintain the voltage at V_{sp}.

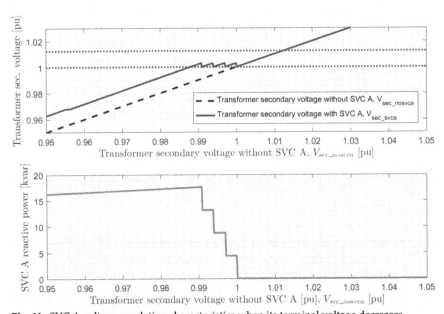

Fig. 11: SVC A voltage regulation characteristics when its terminal voltage decreases.

3.3 Static Var Compensator B (SVC B)

An SVC B is a voltage regulating device that can inject capacitive reactive power in step increments. The rating of an SVC B device is 10 kvar with 10 steps, i.e., an SVC B device can produce reactive power from zero to 10 kvar in the increments of 1 kvar in each step. The important differences between the traditional switching capacitor banks and the SVC B devices are that the latter are electronically controlled and installed on the secondary side of distribution service transformers at 240 V nodes as single-phase units. A single SVC B device connected at a 240 V node maintains the node voltage within a chosen voltage band (V_{band}) around a chosen voltage set point (V_{sp}). For example, if the SVC B voltage regulation settings are V_{sp} = 240 V (1 pu) and V_{band} = 2 V, the node voltage is regulated within +/−1 V around 240 V, i.e., between 239 V (0.996 pu) and 241 V (1.004 pu). To accomplish the voltage regulation, the SVC B device uses proprietary control algorithms to inject reactive power into the secondary circuit as a function of voltage difference between the observed node voltage and SVC B voltage set point, and voltage band settings, i.e., $Q_{SVCB} = f(V_{sp}, V_{sec}, V_{band})$.

The functionality of SVC B is similar to SVC A, both of them regulate the terminal voltage to a specified value by leading var injection. However, there are two major differences between the two devices. First, the SVC B has ten capacitor stages of 1 kvar each, thus, rated for 10 kvar. The SVC A is rated for 17.5 kvar and employs four capacitor stages of 4.375 kvar each. With smaller size of capacitor stages coupled with V_{band} setting, the SVC B can provide fine voltage regulation compared to SVC A. Second, the configuration options are different. The SVC B provides an option to select voltage tolerance level around V_{sp} using V_{band} whereas the SVC A has a fixed band of 3 V.

SVC B Voltage Regulation Characteristics. The voltage regulation characteristics of an SVC B device are obtained by simulating a simple test circuit shown in Fig. 12 in OpenDSS with and without the SVC B device. The parameters of the simple circuit are same as those used for studying the other devices. An SVC B is connected to the service transformer secondary (240 V node) with settings V_{sp} = 240 V and V_{band} = 2 V and the source voltage V_{source} is varied to vary V_{sec}. Then V_{sec} is noted without and with the SVC B connected in the circuit.

The voltage regulation characteristics of the SVC B as its terminal voltage increases are shown in Fig. 13. In this figure, V_{sec} without SVC B (V_{sec_nosvcb}) and with SVC B (V_{sec_svcb}) are plotted against V_{sec_nosvcb}. These characteristics are obtained as the V_{sec} is varied by increasing the source voltage from 0.9 pu to 1.1 pu. Note that the SVC B specified voltage and the voltage band are configured as V_{sp} = 1 pu (240 V) and 2 V, respectively.

It is expected that the SVC B regulates the terminal voltage to be within 0.996 pu (239 V) to 1.004 pu (241 V) band with these settings. From Fig. 13, it is observed that as the transformer secondary voltage without the SVC B is initially well below the specified voltage 1 pu (240 V), all the capacitor stages are in service to boost up the voltage. Thus, V_{sec} with SVC B is higher than that without the SVC B. As V_{sec_svcb} increases beyond the upper limit of the voltage band 1.004 pu (241 V), the capacitor stages are switched out sequentially to regulate the voltage at the upper limit of the voltage band 1.004 pu. After the last capacitor stage switched out, V_{sec} without and with the SVC B in service are the same because none of the capacitor stages of

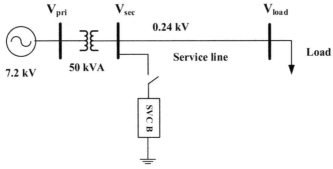

Fig. 12: A simple circuit with SVC B.

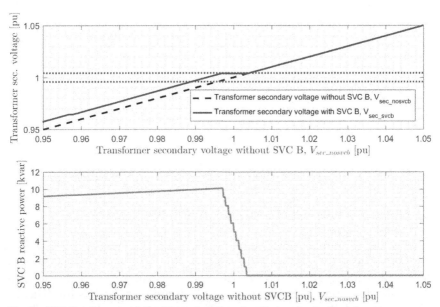

Fig. 13: SVC B voltage regulation characteristics when its terminal voltage increases.

the SVC B are in the circuit. The voltage boost provided by the SVC B is also evident from the Fig. 13.

The voltage regulation characteristics of the SVC B as its terminal voltage decreases are shown in Fig. 14. These characteristics are obtained as the V_{sec} is varied by decreasing the source voltage from 1.1 pu to 0.9 pu. From Fig. 14, it is observed that as the transformer secondary voltage without the SVC B is initially well above the specified voltage of 1 pu, none of the capacitor stages are in service, thus the SVC B terminal voltage is same as without the SVC B in service V_{sec_nosvcb}. As V_{sec_nosvcb} decreases below the lower limit of the voltage band 0.996 pu (239 V), the capacitor stages are switched in sequentially to regulate the voltage at the lower limit of the voltage band 0.996 pu. After the last capacitor stage is switched in, V_{sec_svcb} starts decreasing as there are no additional capacitor stages available to boost up the voltage to maintain at 0.996 pu.

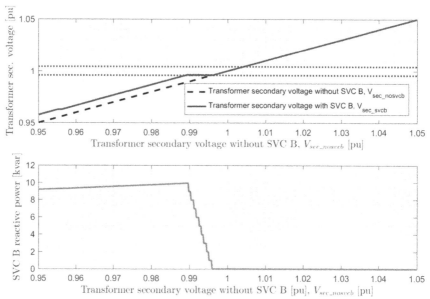

Fig. 14: SVC B voltage regulation characteristics when its terminal voltage decreases.

4. Approaches to Deploy Advanced Voltage Control Technologies

In this section, the methods for placing the advanced voltage control devices at grid-edge to eliminate undervoltage violations in the distribution circuit are described. Two placement methods are considered: simultaneous placement and iterative placement. The placement methods install the devices at the locations with the largest undervoltage area to maximize the performance of the devices. The analysis of the simulation results has shown that the iterative method of placing the devices is effective in eliminating the undervoltage violations compared to the simultaneous placement method. This method can eliminate undervoltage violations in the test circuit.

4.1 Placement Criteria

Advanced voltage control devices are installed in distribution circuits to eliminate undervoltage violations. Three potential criteria for selecting the device locations are described in this section: lowest minimum voltage magnitude, longest undervoltage duration, and the largest undervoltage area formed by voltage magnitude and duration as shown in Fig. 15. To maximize the performance of the voltage control devices, the devices are placed at the locations with the most severe undervoltage according to the quasi-static time-series (QSTS) simulation results performed for a 24-hour period with a 1-minute time resolution on the peak day.

An undervoltage condition at a specific location occurs when the voltage magnitude (RMS) at that location is less than 0.95 pu (NEMA 2016). The characteristic of the undervoltage condition can be described by two basic quantities, i.e., minimum voltage magnitude and undervoltage duration (Dugan et al. 2016). The minimum voltage magnitude at a particular location is the lowest voltage magnitude in RMS at that location during the entire simulation period of 24 hours. The undervoltage duration at a particular location is the total duration of the undervoltage conditions at that location during the 24-hour simulation period. The most severe undervoltage violation location usually has either the lowest minimum voltage magnitude or the longest undervoltage duration. However, it is possible that some locations may have a very low voltage magnitude in a short period (Fig. 16) or a very long undervoltage violation with a voltage slightly lower than 0.95 pu (Fig. 17). These locations may not be desired for placing the devices because the load demand at each node is simply a rough approximation result obtained from the load allocation process and therefore these scenarios might not occur in the real world. Additionally, the minimum voltage criterion is not statistically representative, i.e., this criterion may select a location based on an outlier, one-off situation. Also, it is not effective to place a device at a location with a long undervoltage duration, but the voltage magnitude is close to the threshold voltage. The combination of the two variables represents the undervoltage condition better. Therefore, the undervoltage area criterion, proposed in (Chirapongsananurak et al. 2016), is used for the effective device placement. The undervoltage area at a specific location is defined as the total area where the 24-hour voltage profile curve at that location is under the voltage threshold of 0.95 pu. In other words, the undervoltage area is the area bounded by the voltage profile curve and the voltage threshold line.

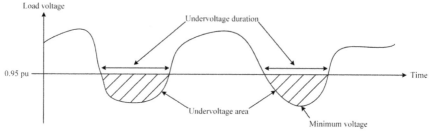

Fig. 15: Voltage profile with minimum voltage magnitude, undervoltage duration, and undervoltage area.

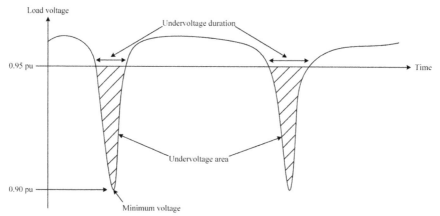

Fig. 16: Voltage profile with a very low voltage magnitude in a short period.

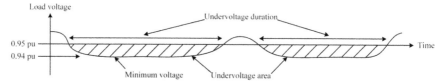

Fig. 17: Voltage profile with a voltage slightly lower than 0.95 pu in a very long duration.

Figure 18 shows a sample voltage profile at a load bus. When the 24-hour voltage profile curves of each load are obtained, the undervoltage areas of each load are calculated. The voltage control devices are placed at the transformer secondaries corresponding to the loads with the largest undervoltage areas.

4.2 Placement Methods

The locations of the advanced voltage control devices are selected to maximize the performance of the devices in eliminating the undervoltage violations. In this section, two placement methods for selecting the locations

Fig. 18: Voltage profile at a load bus.

of advanced voltage control devices are considered, i.e., simultaneous placement and iterative placement. Simultaneous placement places all the available devices simultaneously at the locations with the largest undervoltage areas. However, iterative placement places only one device at the location with the largest undervoltage area at a time. The iterative placement process is repeated until all the available devices are placed. The steps involved in these two methods are described here.

Simultaneous Placement Method. The simultaneous placement procedure involves the following steps:

1. Perform the peak day QSTS simulation without any advanced devices installed in the circuit. Then, determine the load voltage having the highest area of undervoltage violation. Check the rating of the corresponding transformer. If the transformer rating is equal to the size of a UPFC, install a UPFC at the transformer secondary. Otherwise, install an SVC A.

2. Install next UPFC and/or SVC A at the transformer secondaries corresponding to the loads having the next highest areas of undervoltage violation until the planned number of UPFCs and SVC A devices are deployed.

Iterative Placement Method. The iterative placement procedure involves the following steps:

1. Perform the peak day QSTS simulation without any advanced devices installed in the circuit. Then, determine the load with the highest area of undervoltage violation. Check the rating of the corresponding transformer. If the transformer rating is equal to the size of a UPFC, install a UPFC at the transformer secondary. Otherwise, install an SVC A.

2. Perform the peak day QSTS simulation with the device installed in the previous step in service. Then, determine the load with the highest area of undervoltage violation. Taking the limit on the number of devices per location into consideration, check the number of devices already installed at the transformer secondary corresponding to the load with the highest area of undervoltage violation. If the number of previously installed devices is at the limit, install a device at the transformer secondary closest to the limit exceeding location.

3. Repeat the process with the UPFCs and/or SVC A devices installed in the previous steps in service to find out the transformer secondary to install the next device until undervoltage violations are completely removed or the planned number of UPFCs and SVC A devices are deployed.

4. If the number of devices required to eliminate undervoltage violations is less than the planned number of devices, increase the threshold voltage used to calculate the undervoltage violation area, e.g., from 0.95 to 0.96 pu, and repeat the entire process.

4.3 Details of Test Circuit and Base Case

The one-line diagram of the test circuit used for this study, plotted using GridPV tool (Reno 2013), is shown in Fig. 19 (Chirapongsananurak et al. 2016). In this figure, the black lines represent the three-phase circuits and the gray lines represent single-phase circuits. This circuit has a 69/12.47-kV 10.5-MVA substation transformer with a load tap changer (LTC) control, three single-phase mid-line step voltage regulators installed in each phase, and seven switched capacitors (five three-phase capacitors and two single-phase capacitors).

The time-series simulation of the test circuit is performed for a 24-hour period on the peak day with a 1-minute time resolution. The primary-side traditional voltage regulation equipment along with their controllers is in service, but the advanced secondary-side voltage control devices are not in service while performing the peak day QSTS simulation. This circuit condition is considered as the base case in this study. The detailed base case QSTS simulation results are shown in Fig. 20. Figure 20(a) contains the statistical summary of the load voltages. The contour of the probability density function is plotted with the lines corresponding to the minimum and maximum voltages. The color of this contour is associated with a number of loads (in percent) that exhibit a particular voltage magnitude for each time step (Hernandez et al. 2017a,b). The minimum and maximum voltages are the lowest and highest load voltages in the entire circuit at each time step. Figure 20(b) presents the summation of the reactive power produced by all the seven capacitors in the circuit. On the peak day, the

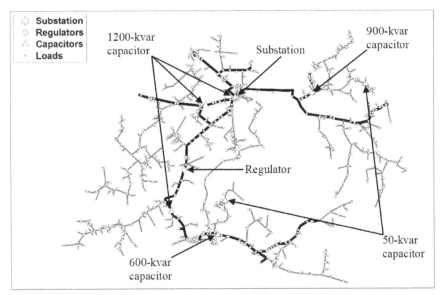

Fig. 19: One-line diagram of the test circuit.

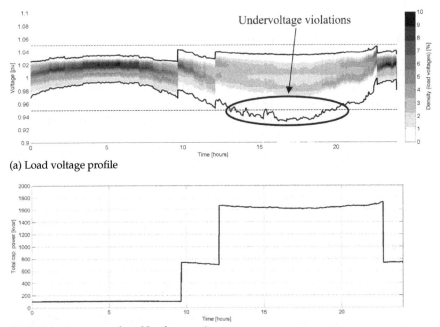

(a) Load voltage profile

(b) Reactive power produced by the capacitors

Fig. 20: Base case time-series simulation results.

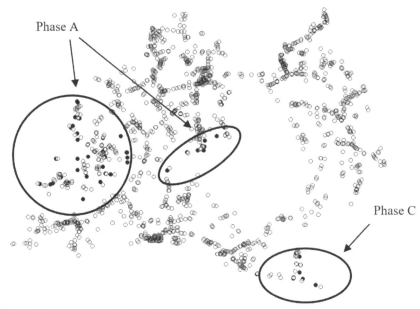

Fig. 21: Base case map of undervoltages.

lowest voltage of 0.9320 pu occurs at 4:46 pm, the undervoltage violations (V < 0.95 pu) occur for 377 minutes between 1 pm and 8 pm, the maximum number of loads with voltages below 0.95 pu is 12.2% (occur at 5:27 pm), and the maximum number of loads with voltages between 0.95 and 0.96 pu is 10.2% (occur at 3:18 pm).

The map of undervoltages is presented in Fig. 21. In this map, each load is represented by a dot. The black dots represent the loads with their minimum voltage below 0.95 pu (observed for 24 hours), while the white dots represent the loads with their minimum voltage above 0.95 pu. The map shows that the locations of undervoltages are clustered in three areas. The two areas on the left and in the middle of the figures are on Phase A, while the area on the bottom right of the figures is on Phase C. These areas are the candidate locations for installing advanced voltage control devices.

5. Simulation Scenarios and Analysis of Results

Two placement scenarios were developed to evaluate the effectiveness of both placement methods described in the previous section. In both scenarios, one UPFC and 15 SVC A devices are available to be placed. The placement results are presented and discussed in the following subsections.

5.1 Simultaneous Placement Scenario

In the simultaneous placement scenario, a UPFC and 15 SVC A devices are placed by using the simultaneous placement method. The number of UPFCs and SVC A devices installed on each service transformer is limited to 1, and the voltage threshold used to calculate the undervoltage area is set to 0.95 pu (ANSI limit).

The QSTS simulation results and the locations of UPFCs and SVC A devices are shown in Fig. 22 and Fig. 23, respectively. Figure 22(a) shows that the circuit is not able to maintain voltage regulation (the undervoltage violations still exist) with the devices deployed by the simultaneous placement approach although all SVC A devices produce rated reactive power during the peak hours as presented in Fig. 22(b). The lowest voltage magnitude in this scenario is 0.9349 pu (occur at 5:27 pm). The maximum number of loads with voltages below 0.95 pu is 2.1% (occur at 5:26 pm), and the maximum number of loads with voltages between 0.95 and 0.96 pu is 2.1%. Note that the lowest voltage magnitude in this scenario is higher than that in the base case, and the maximum number of loads experiencing undervoltages when one UPFC and 15 SVC A devices are in service is less than that in the base case. In other words, the voltage regulation is improved. According to the base case analysis, the locations

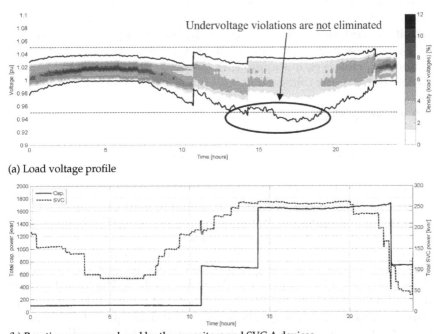

(a) Load voltage profile

(b) Reactive power produced by the capacitors and SVC A devices

Fig. 22: Time-series simulation results for simultaneous placement scenario.

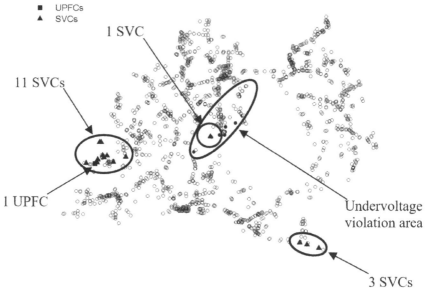

Fig. 23: Locations of UPFCs and SVC A devices for simultaneous placement scenario.

of the undervoltage violations are clustered in three areas, i.e., the areas on the left, in the middle, and on the bottom right of the one-line diagram. The simultaneous approach places one UPFC and 11 SVC A devices in the left cluster and three SVC A devices in the bottom right cluster as shown in Fig. 23. Therefore, the undervoltage violations in both clusters are eliminated. However, this approach places only one SVC A in the middle cluster, so the undervoltage violations in this cluster still exist.

5.2 *Iterative Placement Scenario*

In this scenario, a UPFC and 15 SVC A devices are placed by using the iterative placement method. The number of UPFCs and SVC A devices installed on each service transformer is limited to 1, and the voltage threshold used to calculate the undervoltage area is set to 0.96 pu so that all available UPFCs and SVC A devices can be placed.

The QSTS simulation results and the locations of UPFCs and SVC A devices are shown in Fig. 24 and Fig. 25, respectively. Figure 24(a) shows that the undervoltage violations are eliminated. In other words, the circuit can maintain voltage regulation with one UPFC and 15 SVC A devices deployed by the iterative placement method. The lowest voltage magnitude in this scenario is 0.9558 pu (occur at 5:27 pm). The maximum number of loads with voltages between 0.95 and 0.96 pu is 1.2% (occur at 5:27 pm). Other load voltages are between 0.96 and 1.05 pu. The reactive power produced by the SVC A devices during the peak hour presented

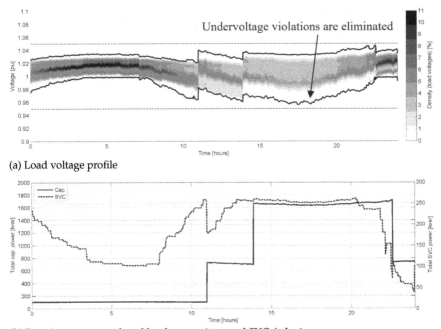

(a) Load voltage profile

(b) Reactive power produced by the capacitors and SVC A devices

Fig. 24: Time-series simulation results for iterative placement scenario.

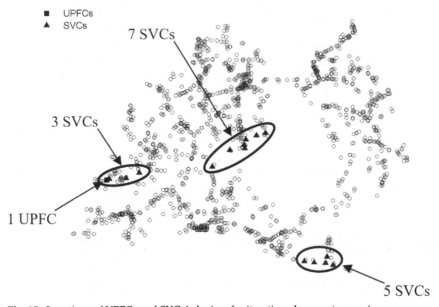

Fig. 25: Locations of UPFCs and SVC A devices for iterative placement scenario.

Fig. 24(b) is similar to the SVC A reactive power in the simultaneous placement scenario. As shown in Fig. 25, the iterative approach places one UPFC and three SVC A devices in the left cluster, seven SVC A devices in the middle cluster, and five SVC A devices in the bottom right cluster. Note that the simultaneous placement approach places only one SVC A in the middle cluster.

6. Conclusion

The results of simultaneous and iterative placement scenarios presented in Section 5 indicate that the device locations change significantly in both the scenarios. This is because the SVC A devices, being shunt capacitors, boost up the voltages in the surrounding locations also with their capacitive reactive power injection. Thus, the locations having high undervoltage areas in the base case scenario may not experience the same undervoltage area when some SVC A devices are already placed in their cluster. Therefore, it is not necessary to place SVC A devices at all those locations having high undervoltage areas in the base case. The iterative placement method considers the influence of the installed SVC A devices in each iteration by performing QSTS simulation with the devices installed in the preceding iterations for the subsequent device placement. Thus, the iterative placement is more effective placement method compared to the simultaneous placement. Furthermore, the iterative placement method checks for the necessity of placing an additional device in each iteration by computing the undervoltage areas. This ensures that a minimum number of devices as necessary to mitigate voltage violations are deployed. With each device costing several hundred dollars and offering a service life of 15 years (Gridco 2016), it is recommended to deploy as many devices as required to completely remove the voltage violations from the circuit to improve service quality.

The traditional voltage regulation equipment is necessary for the distribution circuits for providing voltage regulation. However, with their large size and presence on the primary, they cannot provide precise control of the secondary voltages. For example, switching off a primary capacitor based on its local control can lead to undervoltage violations at few customer loads on the secondary. For effective voltage regulation, a combination of traditional voltage regulation equipment on the primary and a set of advanced voltage control devices on the secondary need to be deployed. For device deployment at the grid edge, techniques such as simultaneous or iterative placement that are targeted to eliminate undervoltage violations can be used for obtaining effective voltage regulation performance. When a set of traditional and advanced voltage regulation devices are deployed, the traditional equipment provides basic voltage regulation and the advanced devices eliminate undervoltage violations where the traditional equipment could not precisely regulate the voltages.

References

Chirapongsananurak, P., N. Ganta, H.V. Padullaparti, S. Santoso, J. Taylor, M. Simms and A. Vukojevic. 2016. Strategic placement of low voltage connected voltage regulation devices. *In*: Proceedings of 2016 CIGRE U.S. National Committee, Grid of the Future Symposium, Philadelphia, PA, USA.

Dugan, R.C. 2012. Reference guide: The open distribution system simulator (OpenDSS). Electric Power Research Institute Inc. Palo Alto, CA.

Dugan, R.C., M.F. McGranaghan, S. Santoso and H.W. Beaty. 2012. Electric Power Systems Quality, 3rd ed. New York: McGraw-Hill.

Gridco. 2016. SVC20: Low Voltage Pole-Mounted 20 kVA static VAR compensator. Gridco Systems. http://gridcosystems.com/products/svc/.

Harlow, J.H. 1996. Load tap changing control. In National Rural Electric Cooperative Association (NRECA), Houston, TX, USA.

Hernandez, M., G. Ramos, H.V. Padullaparti and S. Santoso. 2017a. Statistical inference for visualization of large utility power distribution systems. In Inventions 2: 11.

Hernandez, M., G. Ramos, H.V. Padullaparti and S. Santoso. 2017b. Visualization of time-sequential simulation for large power distribution systems. pp. 1–6. *In*: IEEE Manchester PowerTech, Manchester.

Montenegro, D. 2016. UPFC Model Documentation. http://smartgrid.epri.com/SimulationTool.aspx.

Montenegro, D., J. Taylor and R. Dugan. 2016. Open source unified power flow controller model for quasi-static time series simulation. *In*: CIGRE International Colloquium, Paris, France.

National Electrical Manufacturers Association (NEMA). 2016. American National Standard for Electric Power Systems and Equipment-Voltage Ratings (60 Hertz). Arlington, VA.

Padullaparti, H.V., P. Chirapongsananurak, S. Santoso and J. Taylor. 2017a. Edge-of-grid voltage control: device modeling, strategic placement, and application considerations. In IEEE Power and Energy Technology Systems Journal 4(4): 106–114.

Padullaparti, H.V., M. Lwin and S. Santoso. 2017b. Optimal placement of edge-of-grid low-voltage SVCs in real-world distribution circuits. pp. 1–6. *In*: IEEE Workshop on Power Electronics and Power Quality Applications (PEPQA), Bogota.

Padullaparti, H.V., Q. Nguyen and S. Santoso. 2016. Advances in volt-var control approaches in utility distribution systems. pp. 1–5. *In*: IEEE Power and Energy Society General Meeting (PESGM), Boston, MA.

Reno, M.J. and Kyle Coogan. 2013. Grid integrated distributed PV (GridPV). Sandia National Laboratories. Albuquerque, New Mexico 87185 and Livermore, California 94550.

Short, T.A. 2004. Electric Power Distribution Handbook. CRC Press. Boca Raton, FL.

Varentec. 2016. Edge of Network Grid Optimization (ENGO). Varentec. http://varentec.com/products/engo-v10/.

CHAPTER 13

Large Data Analysis for Advanced Grids

Dumitru-Iulian Nastac,* George Anescu and
Anatoli Paul Ulmeanu

1. Introduction

Over the last decade the concept of an advanced grid has been in fact identical with smart grid technology, which has a central role to integrate the increasing renewable generation, electric vehicle and the internet of things devices. According to a study sponsored by the US Department of Energy (NETL 2008), Smart Grid (SG) can be considered as an ensemble of several applications such as demand response, demand forecast, emergency management, anomaly detection and adaptive pricing, built upon an Advanced Metering Infrastructure (AMI)—a system that measures, collects and analyzes data about energy usage. As it is remarked in Stimmel 2014, smart grid technologies provide universal and clean electrification, alleviate climate change by enabling a variety of efficiencies and renewable generation, and get us closer to a guarantee of affordable, safe and reliable electricity. To fully realize this mandate, utilities have no other course but to transform themselves into data-driven businesses. The data sets collected from the power systems are intrinsically large, heterogeneous and distributed, and therefore pose efficiency difficulties to the traditional approach wherein the data sets

[1] University Politehnica of Bucharest, Romania.
Emails: george.anescu@gmail.com; paul.ulmeanu@upb.ro
* Corresponding author: nastac@ieee.org

are first moved to a centralized location and afterwards are analyzed. In order to efficiently analyze such large data sets, there is a need of new distributed algorithms which are moved closer to the location where the data is collected. The utilities of Big Data analytics technologies refer to the set of technologies within the digital energy ecosystem which can help capture, organize and analyze massive quantity of information as it flows and provides meaningful insight that helps people explain, predict, and expose hidden opportunities in order to assess specific real situations and improve operational and business efficiency.

In this part of the book we will address only essential aspects of the complex and multidisciplinary subject concerning large data analysis for advanced grids. The chapter is organized as follows: Section 2 is a general description of the smart grid in the Big Data analytics field. Then, general information about the Hadoop Stack software technology from Apache Software Foundation is presented in the third section. The fourth Section analyses the advances in Big Data architectural design for smart grids, followed by the fifth Section, which concern data analytics applications for equipment monitoring and fault diagnostics. The nonstationary behavior of the electricity consumption and adaptive forecasting models are analyzed in the sixth section. Then, in the seventh Section are presented the advances in electric vehicles and renewable energy integration. New related applications of data analytics in Hidden Markov Models are briefly described in the eighth Section, before the chapter conclusions.

2. General Description of the Field

McKinsey Global Institute (2012) takes the narrative route to describing Big Data: "[It] refers to datasets whose size is beyond the ability of typical database software tools to capture, store, manage, and analyze." And further, "[The] definition can vary by sector, depending on what kinds of software tools are commonly available and what size of datasets is common in a particular industry". Therefore, in essence, when an organization's data gets so voluminous that it starts to cause problems, it becomes "Big Data".

According to the organization DigitalEurope document from 2015, the key characteristics of Big Data in SG are: **volume**: the immense amounts of data, tending to sizes measurable in PB; **velocity**: data is made available as streaming data, tending to real time or on a daily basis (where it used to be collected a few times a year); **variety**: the various amounts of new data from multiple data sources, presenting both structured and unstructured data, from usual databases, network devices measurements, to web site navigation logs or even images, audio and videos; **veracity**: there is an uncertainty association with data quality and with the increasing volume of information made available, data becoming less reliable and possibly hindering effective analytics.

As it is emphasized in Yu et al. (2015), the large scale data sets specific for SG are posing two important types of challenges: data science challenges and mathematical challenges. Data collected from power systems suffer from three fundamental problems: (1) They are intrinsically incomplete; (2) They are heterogeneous and therefore difficult to join; (3) Systems update or make their data available at different rates. The traditional mathematical methods are not suitable for handling locally very large and high-dimensional distributed data sets. In order to be able to address the high-dimensional aspect, the algorithms coming from the fields of statistics, machine learning and optimization (such as sampling, clustering, classification and linear/nonlinear optimization) need to be readily scalable. Dimension reduction techniques (also scalable) are needed to extract relevant subsets and underlying features while they are controlling the degree of reduction and the estimation's accuracy as the user deems the fittest. Near real-time analysis and control applications are needed in order to cope with spatial distribution system data and high velocity of streaming (such as emergency control and anomaly detection). Furthermore, it may not be feasible or economical to store all raw data and process them later, therefore being more appropriate to devise advanced algorithms able to analyze the data within a single run.

The analysis itself is just a piece of the whole smart grid data analytics puzzle. Before the complex analysis techniques (such as network analysis, cluster analysis, time-series analysis, machine learning, or statistical modeling), the underlying data must be collected and organized, operations which are traditionally performed using ETL software tools (as in *Extract, Transform,* and *Load*). Data collection is in itself a challenge, given the wide variety of available data across the utility. Organizing data is the process of ensuring data coherence by cleaning the raw data (fixing bad values, smoothing and filling in gaps), joining various data sets and storing all the data in a data warehouse. Analysis can then begin, but even advanced analysis is useless without the interpretation and real application of its results. Once analyzed, the processed data must be presented to users in a functional and user-friendly manner so that it improves actions and outcomes. Even very clean data and advanced analytical processes amount to nothing if the resulting information cannot be understood by the users, if conclusions cannot be drawn, and if no action can be taken. The results of the analytical processes generally have four functions: descriptive (explaining what happened), diagnostic (explaining why it did happen), predictive (describing what will happen under various conditions) and prescriptive (describing what to do in order to obtain the most optimal or high-value result).

As it is emphasized in Xydas et al. (2015), data mining is an interdisciplinary process integrating different techniques from statistics, pattern recognition and machine learning, in order to extract information from large datasets. As results of processing associations, hidden patterns, anomalies and significant structures are discovered in the large amounts of data. Data mining is a step in the process known as Knowledge-Discovery-from-Databases (KDD). The KDD process consists of distinct stages of data pre-processing, data formatting and data mining. The Data Pre-processing stage consists of actions like data selection and data clearance. After all data is collected from a database, a preliminary analysis is performed in order to comprehend and select the useful data. The selection action is critical for the extraction of information, as the irrelevant data, if not eliminated, could create noise and lead to incorrect conclusions. Additionally, the data selection reduces the size of the dataset, resulting in lower computational and storage requirements, also reducing the processing time. The selected data is then forwarded to a succession of clearing actions, where missing values are either removed or estimated when it is possible. In addition, outliers are detected and eliminated so that the extracted conclusions are not distorted. The Data Formatting stage placed between the Data Pre-processing and Data Mining processes has the purpose to transform and format the data in preparation for the Data Mining technique of the next stage. Attributes are defined to express the different features in the dataset and then the data are organized in attribute groups that express the same type of information. These processes are important since any potential error during the Data Formatting can greatly impact the outcome. The most important stage of the KDD procedure is the final Data Mining stage, which includes data processing with one or more algorithms, selected according to the goal of the analysis.

The Data Mining algorithms are divided in two main categories: unsupervised learning and supervised learning. The unsupervised learning algorithms include clustering algorithms, useful for an initial understanding of a dataset, and (depending on the application) data partitioning and pattern recognition algorithms. In supervised learning algorithms each data entry is a pair of an attribute vector and a target (desired) value. The initial dataset is divided in the training dataset and the testing dataset. The training dataset is used for learning the correlation between the attributes and the target values. In this way we obtain a trained system that can be employed for similar kinds of data. The testing dataset is then used by the trained model to evaluate its performance. In case the trained model fails to provide the desired output (within a confidence interval), a reconfiguration of the data mining algorithm is performed and the training-testing cycle is repeated. This iterative process is terminated once the desired output is attained, and thereafter the trained model is considered ready to be applied to unclassified data.

The study of Yu et al. (2015), classifies the Big Data applications in SG in two main categories: for short-term operations and for long-term planning studies. Concerning the first category, the paper briefly describes some applications for short-term system operations (such as detection of energy theft, detection of electric vehicles integration, development of more granular load forecasting/renewable generation forecasting, visualization of power distribution systems, state estimation and equipment diagnosis). The second category of applications implies the modeling of customer consumption behavior under various incentives and pricing structures and the transformation of distribution system planning process.

A key and genuine concern preventing the extensive adoption and customer acceptance of the smart grid data technologies is related to data security and customer privacy. Many industries, such as the banking industry and the telecom industry, have been already exposed to large 'data explosions' and are successfully operating after implementing appropriate sets of best practices and regulations to ensure high levels of customer privacy protection, including, where appropriate, customer consent and opt-in/out choices. Such security technologies based on strong principles of authentication, authorization, encryption, detection of policy violations, logging of events and auditing are possible to extend for securing the advanced power grids.

3. Big Data Software Technologies

The data generated by smart grids (characterized by the 4V features of volume, variety, velocity, and veracity) is beyond the ability of commonly used software tools to capture, manage, and process within a tolerable elapsed time, and therefore there is a need in the field for more specific Big Data technologies.

We have to state that the large data analysis involves, especially nowadays, the concept of Big Data and related domains. This includes, of course, data mining, machine learning and many other concepts. There are some related books (for example Bengfort and Kim 2016, Dean 2014) that describe these issues of treating large sets of data and a reader can discover that several concepts which are quite fluid and depend on the author's vision.

In a Big Data problem any significant parameter could be treated as a new dimension of the data model. We have to state that in the vast structure of the Big Data representation, the main idea is to implement the computing part inside the utility (grid) system. Practically, in the case of advanced power grids, the aim is to maximize the performance of a power grid by handling the large number of dimensions and also by working with thousands of variables. In few words, the whole effort in a Big Data application involves three main tasks: making the infrastructure, computing the data (data mining using machine learning for instance), and finally, the analytics part, which is useful to extract the main ideas and provide a clear representation (or visualization) of decisive parameters.

It is also very important to have a proper approach in the field of networks. The graph theory plays a significant role and could provide some of the necessary instruments to tackle the problem of complex electrical grids. A real large power grid is not a graph with uniformly distributed edges. Depending on its structure, geographical positions, various types of consumers and others constraints, there are groups of points highly interconnected, which are somehow dissipated inside the whole grid. The tackling of such grids is not a simple task and could involve particular approaches for an efficient interconnection (see Barabasi 2002).

In order to analyze the data in acceptable time, one of the main ideas is to split the great amount of work into simplest tasks that can be parallelized. The parallel processing capability is becoming important since more than 90% of world's data was created in the last two years (SINTEF 2013), and this trend will probably continue in the future with a similar impact as the Moore's law has in microprocessor's hardware (exponential growth of transistor count in microprocessor's hardware, with a doubling every two years, Moore 1965).

An important source of the increasing trend, concerning the large amount of unstructured data, could be the internet working of physical devices, also named the Internet of things (or IoT devices). And in the power grids field, such devices are already present and will have a tremendous impact in the future.

Traditional relational database technologies, like SQL, have been proven inadequate in terms of response times when applied to very large datasets such as those generated by the advanced power grids (Shrestha 2016). To address this shortcoming, some new technologies are required, most commonly including the use of NoSQL (Not only SQL) data stores and the deployment of Hadoop, a framework from Apache Software Foundation adequate for processing massive data stores in a distributed manner.

Apache Hadoop is an open source software platform for distributed storage and large scale processing of very large data-sets on clusters of low-cost commodity hardware. The first version of Hadoop was created by Doug Cutting and Mike Cafarella in 2005, according to Bappalige (2014). The success of Hadoop is due to its ability to store, manage and analyze large amounts of structured and unstructured data (at petabyte scale) in a scalable and flexible manner, quickly, reliably and at low-cost.

Hadoop is based on distributed processing (storage, management and analysis) of data locally at nodes in large computing clusters so that when a node fails the processing is re-directed to the remaining nodes in the cluster and data is always automatically pre-replicated in anticipation of future node failures. Unlike traditional relational database management systems, Hadoop does not need to apply structured schemas before storing data, it can store data in any format, including semi-structured or unstructured formats and later parse and apply schema to the data when reading it. For a short introductory presentation of the Hadoop technology we recommend article of Bappalige 2014.

Typically a Big Data ecosystem (Demchenko 2013) can be organized as follows:

At the first level, the infrastructure layer, the following three characteristics have to be implemented: reliability, scalability and distributed computing. Apache Hadoop is implementing these three characteristics (see White 2012). In order to provide the reliability characteristic the data is distributed on a large number of local disks on the nodes within the Big Data environment (low cost commodity hardware). Since in a real problem the dynamic management is required, it means that adding or retiring the nodes should be made in a natural way without interrupting the whole system activity. The novelty of a Big Data system is that computing is done on each distributed nodes instead of classic approach where data comes to a single computational node through a queue data structure. Hadoop HDFS is implementing such a management of the file system.

The second level is represented by a programming model for large scale data processing. Such a model can be implemented using the MapReduce algorithm and Spark engine (see Karau et al. 2015, Ryza et al. 2015).

The third level introduces the Big Analytics by aggregating several heterogeneous technologies like NoSQL, R and Python data science programing languages (Ryza et al. 2015). A various number of industries can benefit now of the advantages of such environment by getting new business insights using services like: fraud detection, media pricing, news curation, stock exchange trends and power load forecasting.

4. Advances in Big Data Architecture Design for Smart Grids

An important research topic to be addressed in the SG technology is the design of suitable architectures in order to support large scale data storage and processing required by large scale SG applications. There are several architecture models proposed based on the hierarchical physical topology of the power distribution networks, such as centralized, decentralized, distributed and hybrid. A secure decentralized data-centric information infrastructure for SG describing the challenges in low latency communication protocols, security and publish subscribe data mechanisms for SG is proposed in Kim et al. (2010). A cloud-based SG information management model is proposed in Rusitschka et al. (2010) and Fang et al. (2013), along with a discussion on key challenges. Another cloud-based architecture for demand response (CDR) is proposed in Kim et al. (2011).

In the study of Akshay Uttama Nambi et al. (2014), different data processing architectures for hierarchical power distribution networks are investigated considering realistic deployments in both dense (urban) and sparse (rural) environments and as an aspect of novelty, a comprehensive performance analysis of the proposed architectures is performed based on several key cost indicators, such as energy consumption, processing power, storage requirements and communication bandwidth. The key elements of the used architectural model are: (1) Home Area Nodes (HANs)—a node at the consumer premises that receives energy consumption information from all appliances in the household; (2) Neighborhood Area Nodes (NANs)—acts as an intermediate node between consumers and utility providers, and it serves a small geographical area (i.e., a neighborhood consisting of several houses). NAN receives information from the HANs within the neighborhood and multiple NANs are deployed to cover the utility's territory; (3) Utility Control Unit (UCU)—represents the central control entity of a utility company. This node is responsible for billing, maintaining data, determining electricity price and carrying out demand response management. UCU acts as the root node of the architectural model.

The architecture models considered in the above mentioned study are: (1) Centralized Architecture: in centralized architectures, only the UCU has data processing and storage capability. HANs periodically sense and transmit the energy consumption values to UCU via NANs making information flow unidirectional. No data aggregation is applied in centralized architectures; (2) Decentralized Architecture: in this architecture, only NANs have data processing and storage capabilities. HANs transmit data periodically to the respective NAN, but instead of forwarding the data, the NAN stores and processes this data locally. In decentralized architectures, since complete data is available at the NANs, data aggregation is possible. For instance, NANs can aggregate hourly energy consumption while reporting to the UCU. UCU generates queries to retrieve information from NANs only when required. Thus, NANs act as central entities in this architecture; (3) Distributed Architecture: in distributed architectures, all HANs have data processing and storage capabilities. HANs periodically sense and store the energy consumption values locally. UCU initiates a query to fetch the data, which is forwarded to the NAN and in turn to the HANs. HANs process the query and send the reply to UCU via NAN and are assumed to have adequate processing and data storage capabilities and communicate only upon the reception of a query; (4) Hybrid Architecture: in hybrid architectures, HANs and NANs have both data storage and data processing capabilities. Hybrid architectures are extensions of distributed architectures, where HANs not only sense and store but also transmits aggregated energy values to NANs. For instance, HANs can sense and store energy values periodically and at the end of the day send an aggregate energy consumption reading to the NAN. The data aggregation granularity may vary depending upon the application considered. The proposed architectures have been evaluated in urban and rural environments, based on energy consumption, processing power, storage requirements and communication bandwidth.

As a conclusion, the mentioned study showed that centralized architectures are the cheapest to deploy, but on the other hand, distributed architectures have a higher deployment cost but are more energy efficient.

The system architecture for power distribution systems analytics proposed in Yu et al. (2015) is able to tackle four major classes of data: (1) customer data measured by smart meters, (2) grid data measured by a Supervisory Control and Data Acquisition (SCADA) system deployed on the distribution and transmission network, (3) market data such as prices, and (4) other data such as macro-economic or publicly collected census and text, weather, tweets data. The heterogeneous and complex data sets are transmitted through various types of communication networks and stored in traditional data warehouses, relational databases, file servers, web servers and application servers. The data sets are then loaded into Hadoop clusters and/or distributed in-memory databases, depending on the size, usefulness of the data, and relevance of downstream applications. The in-memory database or High-Performance Computing Cluster (HPCC, open source, data-intensive computing platform developed by LexisNexis Risk Solutions) is a powerful proprietary technology typically more expensive than the Hadoop technology. To perform predictive analytics, optimization and control, code developed in analytical tools such as SAS (software suite for advanced analytics developed by SAS Institute), Mahout (scalable machine learning software solution from Apache Software Foundation), or Revolution R (Microsoft R Server) is moved to Hadoop or HPCC clusters to run Big Data applications.

As it is remarked in Yu et al. (2015), the challenges in developing Big Data applications in power distribution system are two-fold: first, to design a flexible system architecture that accommodates and optimizes Big Data analytic workloads, and second, to develop scalable mathematical tools capable of processing distributed data. The study He et al. (2015), proposes for the first time a Big Data architecture design for smart grids based on a solid mathematical foundation, namely the use of Random Matrix Theory (RMT), as a mathematical modeling tool. According to the proposed methodology, first a standard RMT is systematically formed to map the system. Then, a high-dimensional analysis is conducted (i.e., Empirical Spectrum Density, and Kernel Density Estimation) and the results are compared with the RMT theoretical predictions (i.e., Marchenko-Pastur Law, and Ring Law) to discriminate signals from white noises. Within the mathematical procedure, a high-dimensional statistic, mean spectral radius (MSR), is proposed to indicate the data correlations. More than that, MSR also clarifies that the parameter interchanged among the utilities under the group-work mode for distributed calculation. For a specific zone-dividing interconnected system, each site is able to form a small matrix only with its own data. In this way, the integrated matrix can be divided into blocks for distributed calculation and vice versa, for the overall system, it is possible to conduct high-dimensional analysis by integrating the regional matrices or even by processing a few regional high-dimensional parameters. The algorithm of the proposed RMT based architecture is a fixed objective procedure with simple logic and fast processing speed. The non-asymptotic framework of RMT enables to conduct high-dimensional analysis for real systems, even with relatively moderate datasets. Five case studies were conducted by applying the proposed architecture and the summarized results of the study are as follows: (1) The comparisons between the experimental findings and the RMT predictions, as well as the proposed indicator MSR, are sensitive to event detections. In addition, data in different dimensions are correlative under high-dimensional perspective; (2) The MSR is somehow qualitatively correlated with the quantitative parameters of system performance; (3) The architecture, besides for event detections, can also be used as a new method to find the critical active power point at any bus node, taking account of probable grid fluctuations; (4) The architecture is compatible with the block calculation only using the regional small database and practical for real large scale distributed systems. In addition, the high-dimensional comparative analysis is sensitive to situation awareness for grids operation, even with imperceptible different measured data; (5) The architecture is suitable not only for the power flow analysis, but also for the fault detection.

5. Data Analytics in Equipment Monitoring and Fault Diagnosis

Through the acquisition of rich and highly reliable data collections, with multiple dimensional parameters, the smart grid allows a better condition monitoring of electric power equipment. In this way it is necessary to provide an early fault detection and diagnosis mechanism and also reduce the downtimes, increase of equipment's reliability and optimize of the plant maintenance plan.

Unlike current solutions, which focus mostly on detecting a particular anomaly on a single component or a subsystem (component level), in the study of Li et al. (2015), the main concern was to implement a Condition Monitoring System (CMS) for detecting anomalies of Wind Power Plants (WPPs) at system level, in this way optimizing the maintenance plan of whole wind power plant. The modeling of WPPs on system level is a very challenging problem due to the high complexity of WPPs and its harsh working environment. Through continuous monitoring of WPPs, a huge amount of data are collected, which represents the statuses of WPPs against the time and can be analyzed using classic techniques or modern artificial intelligence (AI) techniques. The study of Marquez et al. (2012) has reviewed the classic techniques and methods used for condition monitoring of WPPs, such as vibration analysis, oil analysis, signal processing methods, etc., while modern algorithms used to detect the system anomalies based on AI, such as Support Vector Machines (SVM), k-nearest neighbor (k-NN), etc., were reviewed in Wang and Sharma (2014). In the previously mentioned study (Li et al. 2015), by using cluster analysis, the system normal behaviors are learned automatically from system observations without understanding the complex relations between anomalies and data, the learned model showing the conditions of the whole WPPs. In addition, the influence of the harsh environment was also modeled through analysis of the environment data in relation with the WPPs data. Furthermore, the proposed approach did not depend on any particular structure of WPPs, thus being universally applicable for condition monitoring of WPPs. The data used in the evaluation was collected over a period of 4 years from 11 real WPPs in Germany with 10 minutes resolution. The dataset consisted of 12 variables which described the work environment (e.g., wind speed, air temperature) and the status of WPPs (e.g., power capacity, rotation speed of generator, voltage of the transformer). The main idea of the proposed solution was to automatically learn a model of normal system behaviors from observing normal behavior data using cluster analysis. Clustering is primarily an unsupervised machine learning method, but in this case it was applied in a semi-supervised manner. In the first step the observed data set of the system normal behaviors was preprocessed using Principal Component Analysis (PCA) (Wold et al. 1987) with the purpose to reduce the data dimensionality. After that, clustering algorithms were applied on the preprocessed data set to generate a model on system level. Two clustering algorithms, DBSCAN (Density-Based Spatial Clustering of Applications with Noise, Ester et al. 1996) and Spectral Clustering (von Luxburg 2007), were explored for system level, modeling the normal behaviors of WPPs. In the final step, the learned system model was used for automatic anomaly detection, based on the assumption (Chandola et al. 2009) that "normal data instances lie close to their closest cluster centroid, while anomalies are far away from their closest cluster centroid." An anomaly detection method was developed and evaluated with real WPPs data, which used Mahalanobis distance and Marr wavelet as metric of the similarities between the new observations and the learned normal behavior model. The accuracy of spectral clustering based anomaly detection was much better than the DBSCAN method, which is based on modeling the WPPs with a F1-Measure of 84.71% and a balanced accuracy of 87.40%. Therefore, the spectral clustering was finally chosen as more appropriate to generate the normal behavior model of WPPs in the proposed solution.

In many cases the reliability of collected data can be negatively affected by anomalous data which can be caused by sensor faults, communication failures, collecting device failures and so on. Therefore, the effective filtering of the anomalous data becomes an important research problem in the smart grid. The traditional methods for filtering out the anomalous data are mainly based on threshold and the statistic of measurement of outliers, such as three sigma criteria (Pauta Rule) (Zhang et al. 2015). In the real application, these methods can filter out some part of the normal data or even real failure data of equipment, which leads to the accuracy degradation of acquisition data and reduces the performance of fault diagnosis.

In Zhang et al. (2015), the OPTICS (Ordering Points To Identify the Clustering Structure, Ankerst et al. 1999) clustering-based condition monitoring anomalous data filtering algorithm was proposed. First, the OPTICS density clustering algorithm was used for mining the distribution characteristics of acquisition data. Then, an anomalous data filtering strategy was designed, combined with electric power equipment state standards and threshold decision rules, etc. The effectiveness of detecting anomalous data was verified through the experiments on one 110 kV substation equipment transformer oil chromatography and the GIS (Gas Insulated Substation) SF6 density micro water. The proposed method was compared with other three traditional methods (Pauta Rule, Grubbs and Dixon), and the OPTICS Clustering-based algorithm showed significant performance improvement in identifying the features of anomalous data as well as filtering condition monitoring anomalous data. Noises were reduced effectively and the overall reliability of condition monitoring data was also improved. The recognition accuracy rate was improved by 30% and the reported average accuracy of anomalous recognition was about 87%. The algorithm proposed in Zhang et al. (2015) showed better results than traditional methods and it was concluded to be more suitable for condition monitoring through multidimensional data collection and analysis.

6. Advances in Prediction of Electricity Consumption

The forecasting of electricity demand is of great economical interest for players on the global electricity market, since an accurate prediction of the consumption is required in order to obtain the best prices on the day-ahead market and to avoid purchasing on the more expensive real-time spot market. As it is remarked in a literature review on the problem of predicting the energy consumption in buildings by Zhao and Magoulès (2012), it is a complex and difficult challenge, due to the many factors that influence the energy performance in buildings, such as ambient weather conditions, building structure and characteristics, the operation of sub-level components like lighting and Heating, Ventilation and Air-Conditioning (HVAC) systems, occupancy and their behavior. The problem can be simply posed as the prediction of a univariate continuous variable (hourly total energy consumption in a building) by means of a model that may contain multiple exogenous variables (independent variables that affect a model without being affected by it), and endogenous variables (variables that affect a model but are also affected by it). Three approaches have been explored for predicting electricity consumption in buildings: the engineering methods, the statistical methods and the machine learning (artificial intelligence) methods.

The engineering methods use physical models that are based on building characteristics, such as the wall materials or the characteristics of each room and external parameters, like weather, to predict electricity consumption. These methods range from simple manual estimations to detailed computational simulations (see Al-Homoud 2001). A detailed computational simulation would estimate the electricity consumption by means of existing software tools based on a finely grained picture of all the building components and characteristics. A directory of such building energy software tools is maintained by the US department of energy.

In the literature review section of Dagnely et al. (2015), the results of some statistical approaches, machine learning approaches and comparative studies of the three types of approaches are presented:

The statistical approaches are based on classical statistical techniques. Typically in a statistical approach, historical data that represents the past behavior of a building—past consumption, weather or sensor data such as occupancy—is used to fit a model, for example, a regression model. However, predicting electricity consumption can be posed as a time series problem, where data points at time t depend on data points at a time $t - i$ and therefore, methods such as Auto Regressive Integrated Moving

Average (ARIMA) are widely used, as to account for the interdependency of data points. In Newsham and Brit (2010), an ARIMAX model is evaluated (i.e., an ARIMA model with exogenous inputs that influence the noise terms) for predicting the electricity consumption of a three-story building of 5800 m², comprising laboratories and 81 individual work space. Due to the presence of laboratories, this building does not represent a conventional office building. As predictors, they used past consumption data, weather data and occupancy data (through sensor monitoring logins but not logoffs). By adding the login predictor, the accuracy of the model improved slightly. Nevertheless, they conclude that measuring logoffs or better sensors, such as a camera, may have a positive impact on the accuracy in a more conventional office building. Regular regression analyses have also shown good accuracy, despite the potential violation of the assumption of independence among data points. In Ansari et al. (2005), the cooling consumption of a building is computed by means of a linear regression function with the temperature difference between inside and outside as a predictor. Cho et al. (2004), studied the impact of the training dataset size when applying linear regression to predict the energy consumption of a building. They used the average outdoor temperature as a predictor for the energy consumption of the heating system. With one day of hourly measurement data, they obtained a very inaccurate system with an error range from 8 to 117%. A training dataset of one week of daily data gave an error range of 3 to 32%, while for three weeks of training data the errors ranged from 9 to 26%. They concluded that in a training set shorter than one month the outdoor temperature variability is a more important cause of error than the length of the training set. In contrast, with a training set of more than one month of daily data, the length of the measurement period strongly influenced the accuracy. As an example, with three months of daily data, their model was able to make electricity consumption predictions with errors ranging from only 1 to 6%.

Machine learning approaches are also data-driven, but they use techniques coming from the field of Artificial Intelligence. Two of the most used methods in energy consumption prediction are Artificial Neural Networks (ANN), (Jain et al. 1996) and Support Vector Machines (SVMs), (Cristianini and Shawe-Taylor 2000). ANNs create a network of input nodes, in-between nodes, and output nodes, all connected by weighted links. The output nodes are thus a function of the input nodes, but their relationship can be obfuscated by hidden in-between nodes and the relationships created by the links. SVMs are typically used for classification problems. To resolve non-linear classification problems, SVMs transpose the data points to a higher dimensional space by means of a kernel. In this higher dimensional space, the data points may have a linear separation. The linear separation is then found by fitting a hyperplane between the classes. The hyperplane has to maximize the margins, i.e., the largest distance to the nearest training data point from the classes. As a related approach to SVM, an extension called Support Vector Regression (SVR) (Smola and Schölkopf 2004), was adapted to regression problems. In SVR the hyperplane has to minimize the error cost of the data points outside of a given margin, while the error cost of the data points inside the margin are considered as null. The method prioritizes the minimization of the upper bound of the training error rather than the (global) training error. In Gonzalez and Zamarreño (2005), an Auto-associative feed-forward Neural Network was used to predict the electricity consumption of two buildings, with input variables such as the temperature difference, the hour of the day, the day of the week and the previous consumption. The study showed that the previous consumption reflects other parameters such as the occupancy level, the models performing relatively accurately during recurring holidays, although this information was not directly encoded as predictor. In Ekonomou (2010), an ANN is used to predict the (yearly) Greek long-term energy consumption (2005–2015) using the past thirteen years as training data (from 1992 to 2004). The inputs for fitting the model were the yearly ambient temperature, the installed power capacity, the yearly per resident electricity consumption and the gross domestic production. The study tested various models, using the combination of 5 back-propagation learning algorithms, 5 transfer functions and 1 to 5 hidden layers with 2 to 100 nodes in each hidden layer. The best model had a compact structure and a fast training procedure with 2 hidden layers of 20 and 17 nodes. For the past ten years, SVM methods are on the rise and many studies have been performed. One of the earliest applications of a SVM to the (monthly) forecast of building energy consumption is Dong et al. (2005), where the method was successfully applied to four commercial buildings with a better accuracy than other machine learning methods. The study emphasized on the fact that the SVM methods require the tuning of fewer parameters when compared to ANN methods. In a more recent study of Jain et al. (2014), the impact of various spatial and temporal variables to predict the energy consumption of a multi-family residential building was examined. They applied a SVM with the following predictors: the past consumption, temperature, solar flux, weekend/holidays and hour of the day. The study tested the impact of spatial (i.e., by apartment, by floor or for the whole building) and temporal (i.e., every 10 min, hourly or daily) granularity of the data. The model that was based on data per floor and per hour was found to be significantly more accurate than the other possible combinations.

In some studies the comparison of the three types of approaches was performed. In Neto and Fiorelli (2008) the EnergyPlus tool (Crawley et al. 2004) was compared to a machine learning approach based on Feed-forward neural networks. The result of this comparison revealed that both approaches are similar in term of prediction accuracy. However, the Feed-forward neural network approach turned out to be more straightforward, as it only relied on 17 months of past consumption and weather data. In contrast, the engineering model was cumbersome to construct, as it depended on the availability of domain experts and a precise knowledge of all the building characteristics. In Azadeh et al. (2007) ANN models using a supervised multilayer perceptron network were compared to a conventional regression technique for predicting the monthly electricity consumption in Iran. The ANN model outperformed significantly more accurate than conventional regression. The literature review of Zhao and Magoulès (2012) mentions that some specific cases of ARIMA models are usually easy to estimate, but they lack in accuracy in comparison with machine learning methods. However, in Amjady (2001), a modified ARIMA taking into account the domain knowledge of experts (e.g., to define manually a starting point for the parameter tuning) was tested and ANN and ARIMA where similar in term of accuracy, but the modified ARIMA succeeded to outperform slightly both models.

According to Dong et al. (2005), from a mathematical perspective, and in terms of optimization, SVM present some advantage in comparison with ANN. SVM methods always lead to a unique and globally optimal solution, whereas (back propagation) ANN methods may lead to a locally optimal solution. From an application perspective, both ANN and SVM produce accurate results, but need a sufficient amount of representative historical data to train the model. Usually, however, SVM models are slightly more accurate, as shown in Ahmad et al. (2014).

Many other machine learning approaches have been compared. Tso and Yau (2007), applied decision trees, which have the advantage of being easily interpretable (in contrast to ANN and SVM). Huo et al. (2007), experimented with genetic programming, which mimic the natural evolution to make evolutionary programs by recombination, mutation and selection to find good solutions. Yang et al. (2010) studied the use of fuzzy rules, through a combination of the Wang-Mendel method and a modified Particle Swarm Optimization (PSO) algorithm. However, a promising approach could be the use of a hybrid method, i.e., a combination of different methods. Hybrid methods may be able to go beyond the accuracy of a SVM or ANN, as shown in Kaytez et al. (2015), where a SVM was combined with Least Squares as a loss function to construct the optimization problem. Nie et al. (2012), investigated the combination of ARIMA, to predict the linear part of the consumption and SVM to predict its non-linear part. Li et al. (2010), tested a General Regression Neural Network, i.e., an ANN where the hidden layer consists of Radial Basis Functions (RBF).

In the comparative study of Dagnely et al. (2015), the focus was on predicting as precisely as possible the (hourly) electricity consumption of a medium-scale office building. Three regression models were investigated as candidate methods: the more advanced Ordinary Least Squares (OLS) regression and SVR methods and a naive autoregressive baseline. The study found that for a given realistic dataset it was impossible to significantly improve on that autoregressive baseline with the two more advanced regression models, despite their increased computational cost. Given the computational cost and the deployment and maintenance complexity of the considered two more advanced regression models, the study concluded that from the industrial applicability perspective an autoregressive model is the most effective methodology at present to do prediction of energy consumption in buildings. The evolution of the electric load in a vast electric grid has usually a non-stationary behavior and depends on significant number of parameters.

In a national power system a correct planning of power generation is required by transmission system operators in order to ensure an efficient distribution of electric load. The management of electrical distribution should operate in safe conditions when it has a reliable prediction of the electric load evolution. This implies a short term forecasting, which frequently involves a window of time no longer than couple of hours (that is enough even in a semi-automated distribution management). Conventional power generating units (coal, fuel, gas fired, nuclear, hydro, gas turbines, diesel power plants, etc.) is operated in conjunction with Renewable Energy Sources (RES). In many countries, the growing amount of fluctuating production from RES, especially wind/photovoltaic power, creates new conditions in system operation and control. Therefore, for power systems which include RES, a good forecasting is crucial. This usually involves a large amount of data that could be efficiently processed especially using Artificial Neural Networks (ANNs). A good forecasting model should include, at the input, various histories of relevant parameters and also a history of the output (using delay vectors) in a recurrent manner (see Nastac and Ulmeanu 2013, Nastac et al. 2013). In these studies, the proposed systems are adaptive in order to fine tune the predictions with the last acquired data, especially when working within non-stationary environments. For a fast recalibration of the ANN, the reference network weights were reduced with a scaling factor γ ($0 < \gamma < 1$) and further used as the initial weights of a new training sequence, with the expectation of a better accuracy. This retraining technique is successively applied for discrete values of γ (such as $\gamma = 0.1, 0.2, ..., 0.9$) and then the whole process is repeated several times (by using the parameter N_{rep} – number of repetitions). In order to enhance the procedure a shaking parameter (Q_{shake}), of the initial weights, could be involved. There will be a double *"for"* loop that will control this process:

for i = 1:N$_{rep}$

    for j = 0.1:0.1:0.9

$$Net_{new\_weights} = Net_{previous\_weights} \cdot (j + Q_{shake} \cdot (rand(1) - 0.5)); \qquad (1)$$

where Q_{shake} has a small value (Q_{shake} << j, like 0.0001) with the aim not to disturb significantly the values of the weights after theirs rescaling.

Moreover, a series of predictors (like the one described above) could be arranged in a cascade (see Nastac et al. 2013) for improving the accuracy of the forecasting in a specified window. In the previously mentioned paper, six different predictors were used in order to have the best possible view of the evolution of the electric load, in every moment, for the next six hours.

Most research on forecasting electrical energy usage has been done with good results on large networks (Cavallo et al. 2015), but it is becoming increasingly important to do forecasting in very small-scale power networks (micro-grids), as these systems should be able to operate autonomously.

A study by Subbayya et al. (2013), was focused on applying some classical times-series forecast methods, such as Autoregressive Integrated Moving Average (ARIMA), Autoregressive Moving-Average (ARMA), Multi-Regression (MR), Double Exponential Smoothing and Random Forest on 6 micro-grids. The study showed that from these methods ARIMA was the most accurate. In another study by Kivipõld and Valtin 2013, it was found that below a certain number of houses, energy usage patterns tend to become more stochastic and, therefore, the classical times-series forecast methods are expected to not be very accurate. In a study (Cavallo et al. 2015) the forecasting is done on a day-ahead basis, with different statistical methods evaluated in a very small network (50 synthetic groups of 30 houses). The demand of these groups were based on smart-meter data recorded from a set of 900 Irish houses through a trial operated by the Irish Commission of Energy Regulation in 2011 (Cavallo et al. 2015). Five methods were evaluated: ANN, Wavelet Neural Networks (WNN), ARMA, MR and Autoregressive Multi-Regression (ARMR). The study concluded that for very small-scale networks, the ANN method is the most efficient technique for predicting the following day's demand, while ARMA models are the least efficient in such scales.

7. Advances in Electric Vehicles and Renewable Energy Integration

Due to the current global concerns about greenhouse gas emissions determined by the extensive use of fossil fuels, the EVs are seen in recent years as a clean alternative to traditional gas fueled vehicles. However, considering that the power system is a major source of greenhouse gas, the extensive adoption of EVs can induce an increased demand in the power system and consequently can have contrary effects to the goal of reducing greenhouse gas emissions. The solution to this problem seems to be the extensive use of renewable energy sources to charge PEVs, but it is raising another challenging problem where big data techniques are required: how to efficiently and simultaneously integrate renewable energy sources and EVs in SG. The challenge of EVs and renewable energy integration is an inherently multidisciplinary problem that depends heavily on robust and scalable computing techniques and the domains of data analytics, pattern recognition and machine learning have a lot to offer in this field.

Some companies involved in the development of Home Energy Management (HEM) systems are estimating the potential of adoption of their products (and implicitly their revenue) by linking it with some favorable elements, including electric mobility and local renewable power generation. In a study developed for the territories of Germany, Vidal de Oliveira and Kandler (2015), it was considered that the installation of a HEM system is facilitated if a household already has one or more favorable elements such as a garage with an electric car, a photovoltaic system (PV system) or a heat pump. During the study a large amount of relevant data related to the local generation of renewable energy and electric mobility were gathered and analyzed. The built data model allowed to enter dependencies between elements and use ranges as inputs. The study found that 6.3% of the houses with a charging system could have potential for installing a HEM system in 2015 and predicted an increase up to 25% in 2020. The model also predicted that the number of one-family households with at least two favorable elements to HEM systems installation could reach almost 4 million in 2050. The implemented data model can be easily expanded for using additional favorable elements (such as battery banks or wind turbines) and can be applied for prediction studies in other territories in the world where accurate data can be collected.

According to Xydas et al. (2015), one of the most restraining factors in adoption of electric vehicles on a large scale is the concern that the EVs drivers have related to the lack of a mature charging infrastructure. In order to manage the charging of EVs in distribution networks, the Distribution System Operators (DSOs) will have to upgrade their infrastructure or implement smart control techniques in parallel with the development of regulative measures to serve these new customers. In order to reduce the impacts of this additional electricity demand on the electric power system, a low cost solution would be to implement effective methods for controlling and coordinating EV charging. Due to the fact that more than 90% of the cars are parked during the day, there is an opportunity to shift the electricity consumption caused by EV charging to times with lower electricity demand (Papadopoulos et al. 2011). Smart charging control algorithms can take advantage of this opportunity and reduce peak loads, realize a valley filling effect on the demand curve or charge EVs preferentially from renewable energy sources. These algorithms determine the charging schedules of EVs according to their objective (e.g., peak reducing, valley filling, frequency regulation, etc.). The charging events that occur at various times and locations create a large volume of data, which is recorded and stored by back up offices or individual charging stations. According to Eichinger et al. (2013), the collection and management of scattered data in a central point is unrealistic, and therefore, distributed data collection centers are introduced for the management of data from a group of charging stations. The main role of such centers (also called "aggregators") is to aggregate and reduce the data collected from many charging stations. The databases aggregate information related to (i) the time and place of charging events, (ii) the amount of requested energy and optionally (iii) the ID of the EV and/or the charging station and the information is further analyzed in order to extract patterns for the charging and travel habits of the EVs and patterns for the activity at charging stations.

The study of Xydas et al. (2015), proposed a decentralized control algorithm for managing both responsive and unresponsive EVs. The employed model applied a performant SVM data mining algorithm based on a Gaussian Radial Basis Function (RBF) kernel for EV short term load forecasting. The performance of the proposed algorithm was tested in two case studies for a generic distribution network deployed on a geographical area in UK with 3,072 customers. The first case study considered an EV fleet consisting of both responsive and unresponsive EVs charging at 11 kW charging stations and it concluded that when the EV forecasting module is activated the EVs are adapting their charging schedule in order to reduce the impact of the unresponsive EVs on the demand curve. The second case study investigated the effect of the charging station's power rate on the performance of the decentralized control model and it concluded that when the forecasting module is activated there is a demand peak reduction for every combination of responsive/unresponsive EVs considering charging rates of 3, 11 and 22 kW. The final conclusion of the study was that a smart management strategy of EVs charging is based on aggregation and that EV load forecasting could be beneficial for both DNOs and vehicle owners.

8. Data Analytics in Hidden Markov Models

In the scientific literature there is a vast range of applications based on Hidden Markov Models (HMMs). So far the studies related to smart energy applications acknowledged the use of HMMs in the following fields:

— Forecasting energy consumption. An important scientific literature is dedicated to forecasting the energy consumption relying on Markov/ Monte Carlo models. Ardakanian et al. (2011), adopts the Markov models for predicting the domestic electricity consumption. Bondu and Dachraoui (2015), propose the Markov chains combined with co-clustering models to simulate individual electricity consumption. The state space models are proposed in Kang et al. (2014) and Wang et al. (2010), in order to predict short-term power consumption. An important number of researchers (see, Ardakanian et al. 2014, Ardakanian et al. 2011, Bondu and Dachraoui 2015, Zhou and Xie 2011, Wang and Meng 2008, Zhou et al. 2008, Dongxiao et al. 2010) rely on the Grey Markov models to forecast also the energy consumption.

— Load profiles. In the scientific literature the domestic energy load demand is strongly related to user activities. In Collin et al. (2014), the authors combined the behavior characteristics using the Monte Carlo method with aggregated load for developing a methodology for low voltage load models. The electric load profiling analysis was also highlighted by Richardson et al. (2008), as a combination of occupancy patterns and the generated electricity demand of all major appliances found in the domestic environment, with a resolution of one minute. The approach was based on constructing Markov-chain occupancy profile and appliance-activity mapping. The study was validated on 22 dwellings by showing good daily correlation between the approach and collected data, Richardson et al. (2010). In the EU strategy on heating and cooling (European Commission 2016), the authors were able to build a thermal profile of energy consumption using Hidden Markov Models for demand-response (DR) programs that focus on the temperature-sensitive part of residential electricity demand. A Hidden Markov Model is also used in Zia et al. (2011), in order to identify the electric loads. This problem of energy disaggregation, i.e., the decomposing of the household's energy consumption into individual appliances in order to improve energy utility efficiency, has been addressed by many researchers (cf., e.g., Zoha et al. 2013, Wang et al. 2013, Kolter and Jaakkola 2012, Chen et al. 2013, Heracleous et al. 2014, Kelly and Knottenbelt 2014, Zoha et al. 2012). Different mathematical techniques have been used for this purpose: Factorial hidden Markov models (see Zoha et al. 2013, Wang et al. 2013, Kolter and Jaakkola 2012), clustered regression (Chen et al. 2013) and conditional random fields (Heracleous et al. 2014). Note also the work of Kelly and Knottenbelt (2014), on a metadata schema for energy disaggregation and the nice survey of Zoha et al. (2012), on non-intrusive load monitoring methods and techniques for disaggregated energy. An interesting work is the one of Ardakanian et al. (2014), that also addressed the problem of electricity consumption profiles based on a tractable time-series autoregressive model (Periodic Auto Regression with eXogenous variables—PARX) that is able to isolate the effect of external temperature on electricity consumption;

— Efficient strategies and heating/cooling solutions in the residential buildings. Here an important research direction is related to determine the occupancy profile in residential buildings and to estimate the probability of the presence of the occupants in the building. The importance of the user occupancy is related to both energy consumption and user behavior. Richardson et al. (2008), presented a detailed analysis regarding occupancy in UK households. The authors used Markov models to generate synthetic data at a ten-minute resolution, using a different scenario for weekdays and weekends and validated by using 2000 Time Use Survey (TUS) data. The approach indicated that the model performs very well in terms of producing data with statistical characteristics similar to the original TUS data. Andersen et al. (2014), used also the Markov processes for dynamic modeling of the presence of occupants towards achieving reliable simulation of energy consumption in buildings.

9. Conclusions

This chapter describes the interdisciplinary field of large data analysis for advanced power grids. The Big Data techniques, already in general use in other industries, have a great potential to address the data management and analysis issues faced by energy market participants in a rapidly evolving regulatory environment. Unfortunately, the support and acceptance for Big Data technologies in the energy sector from both customers' and utilities' sides are still at an early stage due to specific concerns. From the customers' perspective, they have security and privacy concerns and they do not really understand the benefits of the smart grid and resent fees or charges that support its commissioning. From the utilities perspective, a first obstacle to adoption of Big Data analytics is represented by the deficiency of practical implementation examples proving the advantages of converting Big Data into valuable operational intelligence. The second obstacle to adoption of Big Data analytics is insufficient research results on system architecture designs and advanced mathematics for large collections of data. The interrelated fields (like machine learning, data mining, but also the hardware and architectural solutions) are very active research fields with rapid developments and there is also the anxiety manifested from the utilities' side to invest in solutions that can rapidly become obsolete. The last obstacle to adoption of Big Data analytics is the fear of failing to adhere to the data protection and data privacy standards. There is hope that the obstacles from both sides will be soon surpassed, once the technology becomes mature and standardized and both sides start to recognize undeniable benefits.

References

Ahmad, A.S., M.Y. Hassan, M.P. Abdullah, H.A. Rahman, F. Hussin, H. Abdullah and R. Saidur. 2014. A review on applications of ANN and SVM for building electrical energy consumption forecasting. Renew. Sustain. Energy Rev. 33: 102–109.

Akshay Uttama Nambi, S.N., M. Vasirani, R. Venkatesha Prasad and K. Aberer. 2014. Performance analysis of data processing architectures for the smart grid. In: Innovative Smart Grid Technologies Conference Europe (ISGT-Europe), IEEE PES.

Al-Homoud, M.S. 2001. Computer-aided building energy analysis techniques. Build. Environ. 36: 421–433.

Amjady, N. 2001. Short-term hourly load forecasting using time-series modeling with peak load estimation capability. IEEE Trans. Power Syst. 16: 498–505.

Andersen, P.D., A. Iversen, H. Madsen and C. Rode. 2014. Dynamic modeling of presence of occupants using inhomogeneous Markov chains. Energy and Buildings 69: 213–223.

Ankerst, M., M.M. Breunig, H.P. Kriegel and J. Sander. 1999. OPTICS: ordering points to identify the clustering structure. pp. 49–60. In: SIGMOD '99 Proceedings of the 1999 ACM SIGMOD International Conference on Management of Data. ACM Press, New York.

Ansari, F.A., A.S. Mokhtar, K.A. Abbas and N.M. Adam. 2005. A simple approach for building cooling load estimation. Am. J. Environ. Sci. 1: 209–212.

Ardakanian, O., S. Keshav and C. Rosenberg. 2011. Markovian models for home electricity consumption. pp. 31–36. In: Proceedings of the 2nd ACM SIGCOMM workshop on Green Networking (GreenNets '11), ACM, New York, NY, USA.

Ardakanian, O., L. Golab, N. Koochakzadeh, S. Keshav and R.P. Singh. 2014. Computing electricity consumption profiles from household smart meter data. Proceedings of the Workshop of the EDBT/ICDT 2014 Joint Conference, Athens, Greece.

Azadeh, A., S.F. Ghaderi and S. Sohrabkhani. 2007. Forecasting electrical consumption by integration of neural network, time series and ANOVA. Appl. Math. Comput. 186: 1753–1761.

Bappalige, S.P. 2014. An introduction to Apache Hadoop for big data. https://opensource.com/life/14/8/intro-apache-hadoop-big-data, las accessed in March, 2017.

Barabasi, A.L. 2002. The New Science of Networks. Perseus Books Group.

Bengfort, B. and J. Kim. 2016. Data Analytics with Hadoop—An Introduction for Data. O'Reilly Media.

Bondu, A. and A. Dachraoui. 2015. Realistic and Very Fast Simulation of Individual Electricity Consumptions, 2015 International Joint Conference on Neural Networks (IJCNN), Killarney, pp. 1–8.

Cavallo, J., A. Marinescu, I. Dusparic and S. Clarke. 2015. Evaluation of forecasting methods for very small-scale networks. pp. 56–75. In: Woon, W.L., Z. Aung and S. Madnick (eds.). Data Analytics for Renewable Energy Integration. Springer, Switzerland.

Chandola, V., A. Banerjee and V. Kumar. 2009. Anomaly detection: A survey. ACM Comput. Surv. 41: 15:1–15:58.

Chen, H.-H., P.-F. Wang, C.-T. Sung, Y.-R. Yeh and Y.-J. Lee. 2013. Energy disaggregation via clustered regression models: a case study in the convenience store. pp. 37–42. In: Proceedings of the 2013 Conference on Technologies and Applications of Artificial Intelligence (TAAI '13). IEEE Computer Society, Washington, DC, USA.

Cho, S.-H., W.-T. Kim, C.-S. Tae and M. Zaheeruddin. 2004. Effect of length of measurement period on accuracy of predicted annual heating energy consumption of buildings. Energy Convers. Manage. 45: 2867–2878.

Collin, A.J., G. Tsagarakis, A.E. Kiprakis and S. McLaughlin. 2014. Development of low-voltage load models for the residential load sector. IEEE Transactions on Power Systems 29(5): 2180–2188.

Crawley, D.B., L.K. Lawrie, C.O. Pedersen, F.C. Winkelmann, M.J. Witte, R.K. Strand, R.J. Liesen, W.F. Buhl, Y. Joe Huang and R.H. Henninger. 2004. EnergyPlus: new, capable, and linked. J. Architectural Plann. Res. 21: 292–302.

Cristianini, N. and J. Shawe-Taylor. 2000. An Introduction to Support Vector Machines and Other Kernel-based Learning Methods. Cambridge University Press, New York.

Dagnely, P., T. Ruette, T. Tourwé, E. Tsiporkova and C. Verhelst. 2015. Predicting hourly energy consumption. Can regression modeling improve on an autoregressive baseline? pp. 105–122. In: Woon, W.L., Z. Aung and S. Madnick (eds.). Data Analytics for Renewable Energy Integration. Springer, Switzerland.

Dean, J. 2014. Big Data, Data Mining, and Machine Learning: Calue Creation for Business Leaders and Practitioners. John Wiley & Sons, Hoboken, New Jersey.

Demchenko, Y., C. Ngo and P. Membrey. 2013. Architecture Framework and Components for the Big Data Ecosystem. SNE technical report SNE-UVA-2013-02. pp. 1–31.

DigitalEurope. 2015. Energy Big Data Analytics "Unlocking the benefits of Smart metering and Smart Grid Technologies", www.digitaleurope.org.

Dong, B., C. Cao and S.E. Lee. 2005. Applying support vector machines to predict building energy consumption in tropical region. Energy Build. 37: 545–553.

Dongxiao, N., W. Yanan, L. Jianqing, X. Cong and W. Junfang. 2010. Analysis of electricity demand forecasting in inner Mongolia based on gray Markov model, 2010 International Conference on E-Business and E-Government, Guangzhou, pp. 5082–5085.

Eichinger, F., D. Pathmaperuma, H. Vogt and E. Muller. 2013. Data analysis challenges in the future energy domain. pp. 182–233. In: Yu T., N. Chawla and S. Simoff (eds.). Computational Intelligent Data Analysis for Sustainable Development. Chapman and Hall/CRC, Boca Raton.

Ekonomou, L. 2010. Greek long-term energy consumption prediction using artificial neural networks. Energy 35: 12–517.

Ester, M., H.P. Kriegel, J. Sander and X. Xu. 1996. A density-based algorithm for discovering clusters in large spatial databases with noise. pp. 226–231. In: Proceedings of the 2nd International Conference on Knowledge Discovery and Data Mining (KDD96). AAAI Press.

European Commission. 2016. An EU Strategy on Heating and Cooling.

Fang, X., D. Yang and G. Xue. 2013. Evolving smart grid information management cloudward: A cloud optimization perspective. IEEE Transactions on Smart Grid 4(1): 111–119.

González, P.A. and J.M. Zamarreño. 2005. Prediction of hourly energy consumption in buildings based on a feedback artificial neural network. Energy Build. 37: 595–601.

He, X., Q. Ai, C. Qiu, W. Huang, L. Piao and H. Liu. 2015. A Big Data Architecture Design for Smart Grids Based on Random Matrix Theory. arXiv preprint arXiv:1501.07329.

Heracleous, P., P. Angkititraku, N. Kitaoka and K. Takeda. 2014. Unsupervised energy disaggregation using conditional random fields. IEEE PES Innovative Smart Grid Technologies, Europe, Istanbul, pp. 1–5.

Huo, L., X. Fan, Y. Xie and J. Yin. 2007. Short-term load forecasting based on the method of genetic programming. pp. 839–843. In: International Conference on Mechatronics and Automation, ICMA 2007, IEEE.

Jain, A.K., J. Mao and K. Mohiuddin. 1996. Artificial neural networks: a tutorial. IEEE Computer. 29: 31–44.

Jain, R.K., K.M. Smith, P.J. Culligan and J.E. Taylor. 2014. Forecasting energy consumption of multi-family residential buildings using support vector regression: Investigating the impact of temporal and spatial monitoring granularity on performance accuracy. Appl. Energy 123: 168178.

Kang, Z., M. Jin and C.J. Spanos. 2014. Modeling of end-use energy profile: An appliance-data-driven stochastic approach, IECON 2014 - 40th Annual Conference of the IEEE Industrial Electronics Society, Dallas, TX, pp. 5382–5388.

Karau, H., A. Konwinski, P. Wendell and M. Zaharia. 2015. Learning Spark: Lightning-Fast Big Data Analysis 1st Edition, O'Reilly, U.S.A.

Kaytez, F., M. Cengiz Taplamacioglu, E. Cam and F. Hardalac. 2015. Forecasting electricity consumption: a comparison of regression analysis, neural networks and least squares support vector machines. Int. J. Electr. Power Energy Syst. 67: 431–438.

Kelly, J. and W. Knottenbelt. 2014. Metadata for energy disaggregation. 2014. IEEE 38th Annual International Computers, Software and Applications Conference Workshops, pp. 578–583.

Kim, H., Y.-J. Kim, K. Yang and M. Thottan. 2011. Cloud-based demand response for smart grid: Architecture and distributed algorithms. In IEEE SmartGridComm.

Kim, Y.-J., M. Thottan, V. Kolesnikov and W. Lee. 2010. A secure decentralized data-centric information infrastructure for smart grid. Communications Magazine, IEEE.

Kivipõld, T. and J. Valtin. 2013. Regression analysis of time series for forecasting the electricity consumption of small consumers in case of an hourly pricing system. pp. 127–132. *In*: 15th WSEAS International Conference on Advances in Automatic Control, Modelling & Simulation (ACMOS '13) Brasov, Romania, June 1–3, 2013.

Kolter, J.Z. and T. Jaakkola. 2012. Approximate inference in additive factorial HMMs with application to energy disaggregation. Proceedings of the Fifteenth International Conference on Artificial Intelligence and Statistics (AISTATS-12) pp. 1472–1482.

Li, P., J. Eickmeyer and O. Niggemnann. 2015. Data driven condition monitoring of wind power plants using cluster analysis. pp. 131–136. *In*: Proceedings of 2015 International Conference on Cyber-Enabled Distributed Computing and Knowledge Discovery (CyberC). IEEE, NJ.

Li, Q., P. Ren and Q. Meng. 2010. Prediction model of annual energy consumption of residential buildings. pp. 223–226. *In*: 2010 International Conference on Advances in Energy Engineering (ICAEE), IEEE.

Marquez, F.P.G., A.M. Tobias, J.M.P. Perez and M. Papaelias. 2012. Condition monitoring of wind turbines: Techniques and methods. Renewable Energy 46: 169–178.

McKinsey Global Institute. 2012. Big Data: The next frontier for innovation, competition, and productivity. Report, June, 2012, https://www.mckinsey.com/mgi.

Moore, G. 1965. Cramming more components onto integrated circuits. Electronics 38(8): 114–117.

Nastac, D.I. and A.P. Ulmeanu. 2013. An advanced model for electric load forecasting. IEEE Workshop on Integration of Stochastic Energy in Power Systems (ISEPS), Bucharest, November 7 2013, pp. 14–19.

Nastac, D.I., I.B. Pavaloiu, R. Tuduce and P.D. Cristea. 2013. Adaptive retraining algorithm with shaken initialization. Revue Roumaine des Sciences Techniques – Série Électrotechnique et Énergétique 58(1): 101–111.

National Energy Technology Laboratory (NETL). 2008. Advanced Metering Infrastructure. https://www.smartgrid.gov/files/advanced_metering_infrastructure_02-2008.pdf.

Neto, A.H. and F.A.S. Fiorelli. 2008. Comparison between detailed model simulation and artificial neural network for forecasting building energy consumption. Energy Build. 40: 2169–2176.

Newsham, G.R. and B.J. Birt. 2010. Building-level occupancy data to improve ARIMA-based electricity use forecasts. pp. 13–18. *In*: Proceedings of the 2nd ACM Workshop on Embedded Sensing Systems for Energy-Efficiency in Building, ACM Press.

Nie, H., G. Liu, X. Liu and Y. Wang. 2012. Hybrid of ARIMA and SVMs for short-term load forecasting. Energy Procedia 16: 1455–1460.

Papadopoulos, P., O. Akizu, L.M. Cipcigan, N. Jenkins and E. Zabala. 2011. Electricity demand with electric cars in 2030: comparing Great Britain and Spain. Proc. Inst. Mech. Eng. Part A: J. Power Energy (PIA) 225(A): 551–566.

Richardson, I., M. Thomson and D. Infield. 2008. A high-resolution domestic building occupancy model for energy demand simulations. Energy and Buildings 40(8): 1560–1566.

Richardson, I., M. Thomson, D. Infield and C. Clifford. 2010. Domestic electricity use: A high-resolution energy demand model. Energy and Buildings 42(10): 1878–1887.

Rusitschka, S., K. Eger and C. Gerdes. 2010. Smart grid data cloud: A model for utilizing cloud computing in the smart grid domain. In IEEE SmartGridComm.

Ryza, S., U. Laserson, S. Owen and J. Wills. 2015. Advanced Analytics with Spark: Patterns for Learning from Data at Scale 1st Edition, O'Reilly, U.S.A.

SINTEF. 2013. Big Data, for better or worse: 90% of world's data generated over last two years. ScienceDaily, 22 May 2013.

Shrestha, R.B. 2016. Connecting big data to big insights. Appl. Radiol. 45(3): 38–41.

Smola, A.J. and B. Schölkopf. 2004. A tutorial on support vector regression. Stat. Comput. 14: 199–222.

Stimmel, C.L. 2014. Big Data Analytics Strategies for the Smart Grid. Auerbach Publications Boston, MA, USA.

Subbayya, S., J. Jetcheva and W.-P. Chen. 2013. Model selection criteria for short-term microgrid-scale electricity load forecasts. pp. 1–6. In: Innovative Smart Grid Technologies (ISGT), 2013 IEEE PES.

Tso, G.K.F. and K.K.W. Yau. 2007. Predicting electricity energy consumption: a comparison of regression analysis, decision tree and neural networks. Energy 32: 1761–1768.

Vidal de Oliveira, A.K. and C. Kandler. 2015. Correlation analysis for determining the potential of home energy management systems in Germany. pp. 94–104. In: Woon, W.L., Z. Aung and S. Madnick (eds.). Data Analytics for Renewable Energy Integration. Springer, Switzerland.

Von Luxburg, U. 2007. A tutorial on spectral clustering. Statistics and Computing 17: 395–416.

Wang, J., J. Kang, Y. Sun and D. Liu. 2010. Load forecasting based on GM - Markov chain model, 2010 Second Pacific-Asia Conference on Circuits, Communications and System, Beijing, pp. 156–158.

Wang, K.S. and V.S. Sharma. 2014. SCADA data based condition monitoring of wind turbines. Advances in Manufacturing 2: 61–69.

Wang, L. and X. Luo and W. Zhang. 2013. Unsupervised energy disaggregation with factorial hidden Markov models based on generalized backfitting algorithm, 2013 IEEE International Conference of IEEE Region 10 (TENCON 2013), Xi'an, pp. 1–4.

Wang, X.-P. and M. Meng. 2008. Forecasting electricity demand using grey-Markov model. 2008 International Conference on Machine Learning and Cybernetics, Kunming, 1244–1248.

White, T. 2012. Hadoop: The Definitive Guide 4th edition, O'Reilly, U.S.A.

Wold, S., K. Esbensen and P. Geladi. 1987. Principal component analysis. Chemometrics and Intelligent Laboratory Systems 2: 37–52. Elsevier, Amsterdam.

Xydas, E., C. Marmaras, L.M. Cipcigan and N. Jenkins. 2015. Smart management of PEV charging enhanced by PEV load forecasting. pp. 139–168. In: Rajakaruna, S., F. Shahnia and A. Ghosh (eds.). Plug in Electric Vehicles in Smart Grids, Power Systems. Charging Strategies. Springer.

Yang, X., J. Yuan, J. Yuan and H. Mao. 2010. An improved WM method based on PSO for electric load forecasting. Expert Syst. Appl. 37: 8036–8041.

Yu, N., S. Shah, R. Johnson, R. Sherick, M. Hong and K. Loparo. 2015. Big data analytics in power distribution systems. In: Innovative Smart Grid Technologies Conference (ISGT), 2015 IEEE Power & Energy Society.

Zhang, Q., X. Wang and X. Wang. 2015. An OPTICS clustering-based anomalous data filtering algorithm for condition monitoring of power equipment. pp. 123–134. In: Woon, W.L.,

Z. Aung and S. Madnick (eds.). Data Analytics for Renewable Energy Integration. Springer, Switzerland.

Zhao, H. and F. Magoulès. 2012. A review on the prediction of building energy consumption. Renew. Sustain. Energy Rev. 16: 3586–3592.

Zhou, H., W. Wang, W. Ni and X. Xie. 2008. Forecast of residential energy consumption market based on grey Markov chain. IEEE International Conference on Systems, Man and Cybernetics, 2008, Singapore, pp. 1748–1753.

Zhou, L. and D. Xie. 2011. Prediction of rural electricity consumption based on grey-Markov model. 2011 Second International Conference on Mechanic Automation and Control Engineering, Hohhot, pp. 525–528.

Zia, T., D. Bruckner and A. Zaidi. 2011. A hidden Markov model based procedure for identifying household electric loads. pp. 3101–3106. *In*: IECON 2011-37th Annual Conference on IEEE Industrial Electronics Society.

Zoha, A., A. Gluhak, M.A. Imran and S. Rajasegarar. 2012. Non-intrusive load monitoring approaches for disaggregated energy sensing: a survey. Sensors 12(12): 16838–16866.

Zoha, A., A. Gluhak, M. Nati and M.A. Imran. 2013. Low-power appliance monitoring using Factorial Hidden Markov Models. 2013 IEEE 8th International Conference on Intelligent Sensors, Sensor Networks and Information Processing 12: 527–532.

CHAPTER 14

Power Electronic Application for Power Quality

Igor Papič, Boštjan Blažič, Leopold Herman* and Ambrož Božiček

1. Introduction

Until recently, the main requirement of electricity consumers was security, especially in terms of non-interrupted supply. With the increasing presence of loads, powered using power electronics, the quality of electricity is becoming increasingly important. Namely, inadequate power quality in industry can interrupt the industrial processes, which may lead to very high financial losses. High power quality is important also in commercial buildings, where the share of equipment, that is sensitive to low quality, is increasing. The contribution of modern electronically controlled devices to the current situation is twofold: on one hand, these devices are the main culprit for the deterioration of the quality level, but on the other hand, these devices are the most sensitive to poor power quality.

A traditional solution for the problems, mentioned above, would involve reinforcing the grid with additional power lines and/or transformers. This is practically an unacceptable solution, especially due to high costs. Thus, in order to limit the adverse impact of the connected loads and to improve the level of quality of electricity, passive *LC* filters are still the most common solution. Disadvantages of these filters are mainly the following: fixed compensation characteristics (limited number

Faculty of Electrical Engineering, University of Ljubljana, Slovenia.
Emails: igor.papic@fe.uni-lj.si; bostjan.blazic@fe.uni-lj.si; ambroz.bozicek@fe.uni-lj.si
* Corresponding author: leopold.herman@fe.uni-lj.si

of harmonics), possibility of unwanted resonance interaction with other elements of the system and their physical size, which all greatly limits their usefulness and effectiveness.

An alternative approach to these problems involves use of semiconductor switching devices, controlled reactive elements and active compensation devices. All these devices are based on the power electronics. Controlled reactive elements and active compensation devices allow dynamic compensation, whereas the compensators are also capable of dynamically responding to the changing conditions in the network.

Compensation devices that are nowadays most commonly used in electricity networks are presented in detail in the Section 2. First, main semiconductor switching elements will be described, which are the basic building block of all devices. This will be followed by a description of three types of harmonic filters—passive, hybrid and active harmonic filters. Next, Static Var Compensator (SVC), Unified Power Quality Conditioner (UPQC) and at the end Dynamic Voltage Restorer (DVR), will be presented.

2. Power Converters

Semiconductor switching elements are the basic building block of active compensators (Ghosh and Ledvich 2002, Hingorani and Gyugyi 2000). Power converters can nowadays cover a wide voltage and power range— tens of kV and hundreds of MVA.

Each switching element is composed of one or more semiconductor power switches, the damping circuit and the circuit for generating the control *on* and *off* signals. The typical nominal values of the majority of the semiconductor switches amount to about 1–5 kA and 5–10 kV, with the useful area somewhere between 25% and 50% of the nominal current value and about 50% of the nominal voltage value. In addition to voltage and current limits, the following characteristics can also be of significant importance:

- Losses in the conductive state, which causes heating and thus the need for cooling.
- Switching speed, which causes the time that elapses between the complete on and off state and vice versa, and dictates the design of the damping circuit to damp voltage and current spikes.
- Switching losses—when switched in, the current increases before voltage drops to zero and when switched off, the voltage begins to rise, even before the current drops to zero. The simultaneous presence of voltage and current presents the loss that dictates the maximum switching frequency with which the switch can operate.

Roughly, power semiconductor switches can be divided into three groups: diodes, transistors and thyristors. Characteristics of individual families of elements are described below.

2.1 Semiconductor Components

2.1.1 Diode

The diode is a two-layer element (the *p-n* junction), which can conduct current only in the conducting direction, e.g., from the anode to the cathode, when the anode has a sufficiently high positive potential to the cathode. The diode blocks current in the blocking direction, when the cathode has a positive potential with respect to the anode.

In power compensation devices, fast diodes are used, together with asymmetric GTOs and the IGBTs. Their current and voltage nominal values must match the nominal values of components, which they are used with.

2.1.2 Transistor

Transistors are a family of three-layer elements. Transistor conducts when one of its electrodes (i.e., the collector) is polarized positively against the other (i.e., the emitter) and when there is a control signal present at the base electrode. In high-power semiconductor devices, mainly IGBT transistors are used.

IGBT (Insulated Gate Bipolar Transistor)

IGBT acts as a transistor with high voltage and current ratings. An *on-off* step is carried out through the MOSFET transistors. The advantage of the IGBT's is particularly fast *on* and *off* transition, and low switching losses, which enable its use at higher firing rates and hence the use of pulse-width modulation. The main weakness of the IGBT's is a higher voltage drop in the conductive state, compared to the thyristors.

2.1.3 Thyristor

Thyristor is a four-layer semiconductor element. In comparison with the transistor, it has a poor switching characteristic—i.e., longer switching times and higher switching losses. On the other hand, the thyristor offers lower losses in the conductive state and it is built for higher power ratings.

GTO

GTO has the structure and properties that are similar to the ordinary thyristor. The control current, required to turn off the GTOs, is quite large in comparison to the current needed to activate it. GTOs require large and long switch-off pulses, thus they can only be operated with a relatively low switching frequency (a few hundred Hz). This is a major drawback of the GTO compared to the IGBT. On the other hand, GTO may reach much higher voltage and current ratings, compared to the IGBT, and is therefore used in power applications of hundreds of MVA.

2.2 Basic Converter Topologies

Power converter, which will be presented in this chapter, form a basic building component of a whole family of modern compensation devices (Ghosh and Ledvich 2002, Hingorani and Gyugyi 2000). Devices based on the voltage or current power converters can be described as a controlled voltage or current sources.

Due to better operational characteristics and lower prices, voltage source converters are used more often. Below, we present two types of converters, but the focus will be on the voltage-type. This will be followed by a more detailed description of the three-phase voltage source converter.

2.2.1 Voltage and Current Source Converters

A basic three-phase current source converter is composed of six semiconductor switches, which can block positive as well as negative voltage between the anode and the cathode (Fig. 1).

On the dc side of the converter, a reactor is used as a dc-current source. With an appropriate switching of semiconductor switches, converter generates a set of three phase currents of variable frequency, amplitude and phase angle. In doing so, the output current is proportional to the dc current through the reactor. If we assume that the generated currents do not contain harmonics and that converter exchanges only reactive current with the network, than the voltage drop on the reactor is close to zero. The size of the reactor in this case does not play a significant role. Current source converter should have capacitors connected in parallel to the output terminals, which will limit the size of the induced voltage, caused by the rapid changes in current.

The current source converter is rarely used in practice. Mainly because of the requirement to use symmetric switches, that are more expensive than asymmetric ones.

Figure 2 shows basic structure of three-phase voltage source converter. It is built with six semiconductor switches with antiparallel diodes. On dc side of the converter a capacitor is connected. By properly switching of semiconductor converter switches, a series of three-phase voltage of variable frequency, amplitude and phase angle are generated at the output terminals. The inverter must be connected to the network via coupling reactance, which limits the current inrush between the inverter and network.

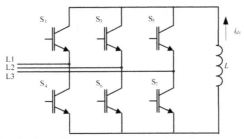

Fig. 1: Basic structure of a three-phase current source converter.

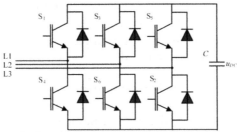

Fig. 2: Basic structure of a three-phase voltage source converter.

The source voltage converter is conducting the current in both directions, thus it needs switching elements (combination of semiconductor switches and diodes) that allow this. On the other hand, the polarity of the voltage at the dc-capacitor does not change; therefore, the switching elements do not need the ability to block the voltage in the blocking direction. Consequently, the voltage source converter is constructed of asymmetric semiconductor switches, with the possibility of current interruption (e.g., GTO and IGBT) with diodes connected in antiparallel. Some semiconductor switches have such a diode integrated into a uniform structure, which is suitable for use in a voltage source converter.

The capacitance of the capacitor on the dc side must be high enough to withstand current inrushes without significant fluctuations in the dc voltage. Assuming, that the converter generates only a basic voltage component and that only reactive current is exchanged with the network, then the size of the dc-capacitor has no significant role, since DC current is close to zero.

2.3 *Passive Filters*

Passive harmonic filters represent a conventional solution to the compensation of reactive power and harmonic filtering. Before the installation of such devices in the electricity power system, thorough analysis of their impact on the network, are required. Although, passive filters do not generate harmonics, they can significantly affect the level of harmonic distortion in the system. Namely, each network containing capacitances and inductances has one or more natural resonance frequencies. When one of these frequencies matches the harmonic in the network, this harmonic component can be amplified significantly. The main disadvantage of these compensators is therefore a risk of resonance between the compensator and the network.

This is more or less successfully prevented by proper selection of components of the compensator (with its tuning), nevertheless we can only speak about the reduced probability of a resonance. Positive features of these devices include robust design, high-power designs (few MVAr) and relatively low cost (low maintenance costs) (Peng et al. 1990, Das 2004, Shwedi and Sultan 2000, Pires et al. 2006).

In Fig. 3 the basic implementations of passive filters are shown. These topologies form the basis for the derivation of more complex filtering structures.

Any combination of passive components (R, L and C), when used in the frequency domain is called a passive filter. In power engineering, the filters are mainly used for filtering harmonic current. They are divided according to their behaviour in the frequency domain:

- low-pass filter,
- high-pass filter and
- band-pass filter.

According to the operation principle of passive filters in the field of power engineering, they can be divided into tuned and detuned versions. Tuned filters are tuned to represent low impedance for specific harmonics. They are usually composed of several units, each of them tuned to a specific harmonic. Normally, they are installed near current harmonic source, in order to prevent harmonic currents from entering the network and the associated adverse effects.

The purpose of detuned filters is not preventing flow of harmonic currents into the network, but primarily to compensate reactive power. If properly tuned, the probability of resonance is substantially reduced. Due to the lower flow of harmonic currents through the detuned filter compared

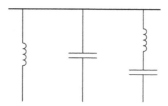

Fig. 3: Basic passive harmonic filters.

to the tuned one, the electrical dimensions of the filter components are much smaller, which results in a lower price (Pires et al. 2006).

2.3.1 Configuration of Passive Filters

Compensation devices can be divided into series and parallel (depending on the method of connection). While the first are mainly used for filtering the voltage harmonics, the latter are used for filtering load current harmonics. Below, both are briefly presented; however, hereafter we will focus primarily on the parallel connection (see Fig. 4) (Senini and Wolfs 1999).

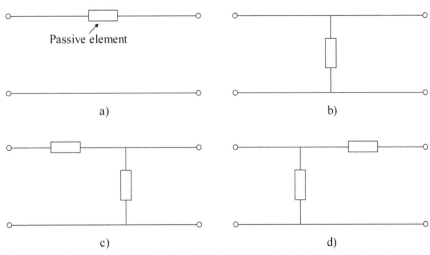

Fig. 4: Basic possibilities of connecting compensator to the network; (a) series connection, (b) parallel connection, (c) and (d) a combination of series and parallel connection.

2.3.2 Series and Parallel Filters

- **Series filters**

    In Fig. 5(a), the topology of the series filter is shown. The filter represents a variable impedance between two adjacent network nodes, enabling control of the power flow between the nodes. It also allows filtering of harmonics, by imposing high impedance to current harmonics from the load side.

    It should be noted that this structure is not suitable for filtering the current harmonics caused by the loads that operate as a rigid harmonic current sources. When we are dealing with this type of load, a large harmonic voltage drop may occur across the filter impedance, when loads are connected in series. For filtering this type of loads, parallel harmonic filters presented below, are more effective.

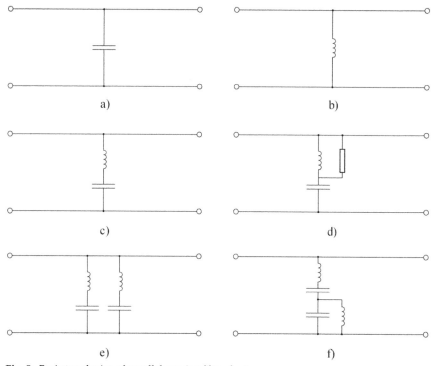

Fig. 5: Basic topologies of parallel passive filter devices.

By using this filter topology, the voltage on the load side is distorted in accordance to the operation of the load and the voltage on the power supply side is filtered.

- **Parallel filters**

  The parallel topology of a passive filter is schematically shown in Fig. 5(b). Parallel compensators represent low-impedance current path to the filtered current harmonics, caused by the operation of the non-linear loads. These filters are mainly used for reactive power compensation and harmonic current filtering.

  A parallel compensator is analogous to the series one and is used with non-linear loads that are producing current harmonics. It is not suitable for use with loads that operate as rigid voltage harmonic sources. To filter these type of loads, a series element that limits the harmonic current through filter is compulsory. Structures that allow such operation are shown in Fig. 5(c) and(d).

2.3.3 Operation

Parallel passive filters are often used to improve the parameters of power quality. In Fig. 5, basic topologies of parallel passive filter are shown. The simplest filter is a parallel-connected capacitor, as shown in Fig. 5(a). It is primarily used for reactive power compensation; however, it may slightly reduce the current harmonic distortion as well, because its impedance decreases with increasing frequency. When using this type of filter, there is always a risk of resonance between the filter and the network. In Fig. 5(b), another simple element is shown, which is rarely used as a filter. Sometimes they are used to compensate unsymmetrical loads.

For filtering the harmonics of the load current, topologies in Fig. 5(c) and (d) are used very often. Filter in Fig. 5(c) is a single-tuned parallel filter. It is tuned so that the series-resonance between the reactor and the capacitor corresponds to the filtered harmonic (Pires et al. 2006). Because of the series resonance, filter represents low impedance at the tuned harmonic, and thus diverts harmonic currents into the filter. Figure 5(d) shows a high-pass filter, which represents low impedance for all harmonics higher than the tuned frequency.

The structure in Fig. 5(e) is one of the options for filtering multiple current harmonics. This topology consists of number of tuned filters, each being designed for a specific harmonic. A high-pass filter can be added (Fig. 5(d)), to enable filtering of higher harmonics. An alternative topology is shown in Fig. 5(f), which shows a double-tuned passive filter. This filter is equivalent to two single-tuned ones, but has a few advantages. As seen in the figure, only one capacitor is exposed to the full rated voltage and only one reactor is exposed to the full rated current. This allows savings in terms of reduced dimensions of the elements of the filter. This topology is often used in high-voltage applications, for example in HVDC systems.

2.4 Active Filters

With the development of semiconductor switches, active filters are growing in importance. Switching losses of the inverters are becoming smaller and they are available with high rated power (Senini and Wolfs 1999). A very simplified scheme of the active filter is shown in Fig. 6. With a properly designed control algorithm, active filters are able to tackle a number of tasks in the field of power quality, for example:

- reactive power compensation,
- harmonics filtering,
- asymmetry compensation,
- flicker compensation,
- voltage regulation at the PCC...

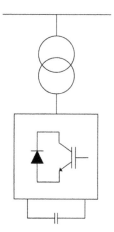

Fig. 6: Active filter.

Of course, better performance comes at a higher price and this is a key reason that in practice there is still a dilemma, whether to use passive or active filters.

The main part of these devices is a power converter (voltage or current), which can be controlled so that the device improves the selected power quality parameters. Today, the most problematic parameters of power quality are primarily interruption of power supply, voltage fluctuations, flicker, harmonic distortion, voltage asymmetry. ... All of these can be effectively limited by the installation of active filters.

Active filters can be used for power factor correction, without the risk of causing resonance amplification of the harmonic components. They also have the ability to operate independently from the impedance of the system, which allows them to be used in demanding situations where, due to the resonance problem, use of passive filters is not possible. Active filters are tracking changes in the filtered harmonic current in real-time.

Some disadvantages of active filters are:

1. construction of large and fast power inverters is very demanding,

2. investment and operating costs of active filters are high.

The main configurations of the active filters will be presented below.

2.4.1 Configuration of Active Filters

Parallel active filters (Fig. 7) are used to improve power quality parameters. They can be controlled in a way to operate as one or more passive compensators, without negative effects between filter and network. These compensators can be used for reactive power compensation. The active element generates a capacitive current of fundamental frequency, so that the mains current power factor gets close to one.

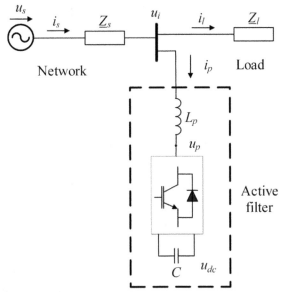

Fig. 7: Basic scheme of a parallel active filter.

Parallel active compensator can be used for harmonic filtering. Often the two functions, compensation of the reactive current (base component) and harmonic filtering are combined in one device. Such compensators are used quite often; there are also several commercial implementations (Chen et al. 2008, Shui et al. 2009).

Since usually power dimensions of these compensators are quite large, topologies consist of several converter units. With series or parallel connection of several units together, it is possible to achieve the necessary high rated values and a wide range of filtering.

Figure 8 shows a basic scheme of a series connected active filter, where u_s represent the system voltage, i_s system current, Z_s system impedance, u_i voltage at the point of common coupling, i_l load current, Z_l load impedance, i_p converter current, L_p converter (coupling) inductance, u_p converter output voltage and u_{dc} dc-side converter voltage. The voltage source converter is connected in series with network through series transformer.

2.4.2 *Operation*

The basic scheme of a parallel-connected active filter is shown in Fig. 7. Most often, a voltage source converter with a capacitor or a power source (e.g., battery) on the dc side is connected in parallel to the network. The main function of the parallel active filter is to compensate the load current i_L (to filter harmonics and compensate reactive power). As a regulated source of reactive power, it can also participate in voltage regulation, for

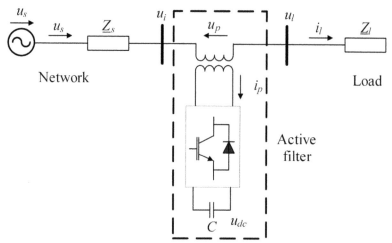

Fig. 8: Basic scheme of a series active filter.

example, as a voltage support during large motor starting. Compensation of flicker is also possible with a proper voltage control. If the converter is implemented in a four-wire configuration, it is able to generate a zero current component (neutral wire current) and can compensate it.

Source current i_s can be written as the sum of the load current i_l and the converter current i_p:

$$i_s = i_l + i_p \tag{1}$$

Load current will be fully compensated, when the converter will generate such a current that the sum of $i_l + i_p$ will form a sinusoidal, symmetrical currents of fundamental network frequency, which will be in phase with the mains voltage. Nominal current of the inverter is:

$$i_p = i_{ref} - i_l \tag{2}$$

where i_{ref} is a reference current.

Active filter can also exchange active power with the network, if dc side is connected to an energy source. The energy source may be an appropriate sized capacitor, or for example, batteries, together with the controller of dc-voltage.

2.5 Hybrid Filters

In order to combine the positive properties of passive filters (robust design, easy to use, simplicity) and active filters (flexibility), hybrid filters have been widely proposed in the literature in recent years. A number of different hybrid filter configurations have been proposed and they can be categorized in several ways. The structure of the filter or its topology, defines the basic operation mode of the device and the nature of the power quality problem that can be compensated by such a topology. Control algorithm and method for setting the reference values of the regulator will determine the effectiveness of the active part of the filter in performing the required function.

2.5.1 Configuration of Hybrid Filters

The most commonly discussed hybrid active filter topologies are series connected active filters with passive filter added in parallel (Lascu et al. 2007), and a parallel connected passive filter with an active filter in series (Teodorescu et al. 2006, Bojoi et al. 2005). Some other possible topologies are summarized in (Fukuda and Endoh 1995). It is clear that not all topologies are effective in performing certain specific tasks, so when planning a hybrid filter, it is first necessary to precisely define the requirements for its operation.

Figure 9(a) shows parallel combination of passive and active part (Fukuda and Endoh 1995). The passive filter is tuned to several harmonics and thus it takes the main load in filtering harmonics. The active part normally performs resonance damping between the compensator and the network, or some other function. The required size of the active part is in this way smaller than with the stand-alone active filter, but it is still relatively high, since the inverter must be designed for the full nominal voltage.

Figure 9(b) shows a hybrid filter as a series combination of the passive and active part. In this topology, the voltage on the active part is reduced due to the voltage drop on the passive part. In this way, the necessary voltage dimensions of the filter are greatly reduced, but there is still full

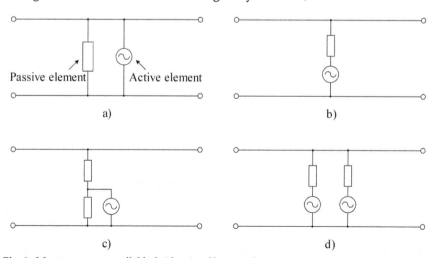

Fig. 9: Most common parallel hybrid active filter topologies.

rated fundamental current that active part needs to withstand, when the filter is used for reactive power compensation.

This does not apply to the topology in Fig. 9(c), which can achieve both, low voltage and current dimensions (Hafner et al. 1997). Figure 9(d) shows a topology with multiple hybrid filters in parallel (Bhattacharya et al. 1997, Cheng et al. 1998, 1999), where each filter, filters out only one harmonic, so the size of the individual branches in comparison to the load, may be very small.

2.5.2 Operation

Configuration and a basic idea of a hybrid filter operation is shown in Fig. 10. The control algorithm of the power inverter measures the current into the

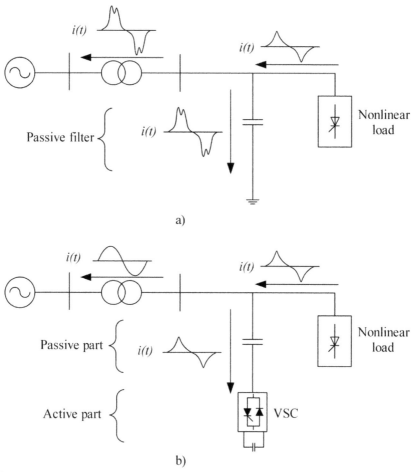

Fig. 10: Basic HAPF operation; (a) Stand-alone passive filter operation, (b) VSC added in series to the passive part.

network and actuates the switch of the inverter to filter the current of the network to a sinusoidal waveform. Of course, there are also other possible operation modes of hybrid filters.

As it is evident from the figure, the upgrade of the passive compensator to the hybrid filter does not require anything other than adding the voltage inverter in series to the existing passive part.

2.6 *Static Var Compensator (SVC)*

Static Var Compensator (SVC) would generally be described as a parallel connected passive compensator with the possibility of adjusting the reactive power output in both, the inductive as well as the capacitive range. The naming static comes from the fact that it does not have rotating parts when compared to the synchronous compensator (synchronous machine).

2.6.1 VSC (FC-TCR) Configuration

In the most common configuration (Fig. 11), the VSC consists of a 6-pulse thyristor-controlled reactor and several fixed capacitors (FC), which are all connected to the busbars through breakers. Fixed capacitors are usually implemented as filters, which are tuned to particular frequencies. The configuration with thyristor-switched capacitors (TSC) is also possible.

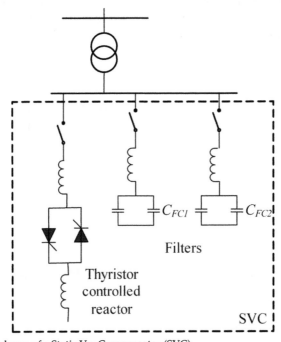

Fig. 11: Basic scheme of a Static Var Compensator (SVC).

Capacitors provide a constant source of capacitive reactive power and, in combination with the controlled reactor, that can generate the inductive reactive power, the compensator can achieve the desired exchange of reactive power between the compensator and the network at any given time. Compensator can be connected to the network with a transformer—for adjusting the voltage levels—or directly, as far as electrical dimensions of thyristor switches permit it.

The overall reactive power regulation range of the SVC depends on the power rating of the TCR and ratings of fixed capacitors. If the nominal power of capacitors is Q_{FC} and the power of the TCR is Q_{TCR}, than the regulating range of the reactive power is: from $Q_{SVC} = Q_{FC} - Q_{TCR}$ to $Q_{SVC} = Q_{FC}$.

2.6.2 *Operation*

Figure 12 shows a general *U-I* characteristic of the SVC.

The characteristic depends on the structure of the compensator, network properties and mains voltage. Thickened line indicates the characteristic at the rated mains voltage. Having thyristor-controlled reactor completely switched off, compensator generates maximal capacitive current that is determined by the fixed capacitors. By decreasing the thyristors firing angles, the reactor current is growing until at the full conduction angle, it reaches the maximum (inductive) value. If the mains voltage is still rising with the fully open thyristors, the compensator behaves like a linear impedance to the point, where semiconductor switches reach the maximum allowable thermal load.

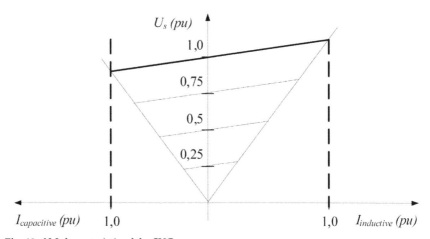

Fig. 12: *U-I* characteristic of the SVC.

2.7 Unified Power Quality Conditioner (UPQC)

Use of the synchronous compensator in configuration, which is characteristic for STATCOM, together with the series synchronous source, represents the most complete way to compensate power quality. This approach is known as the Unified Power Quality Conditioner (UPQC). The concept was first introduced by Laszlo Gyügye (1992).

2.7.1 Configuration of the UPQC

The Unified Power Quality Conditioner consists of a combination of series and parallel filters (Fig. 13), with series and parallel voltage source converters having common dc-circuit. UPQC generally allows the implementation of all the functions that series and parallel devices do (series and parallel active filters)—such as voltage dips and voltage harmonic compensation and simultaneous compensation of the load reactive current. The device can be operated in such a way that with the proper regulation, the series and parallel part of the UPQC ensures minimal exchange of active power with the network. According to the literature, operating modes, where one of the converters (usually parallel) provides all of the necessary active power for the operation of the second inverter and the dc-voltage regulation, are most common.

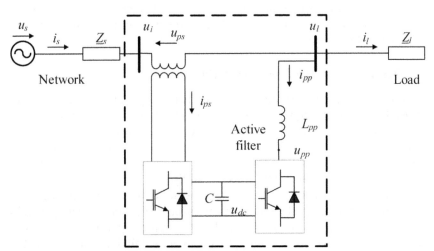

Fig. 13: A connection scheme of UPQC.

2.7.2 Operation

Let us assume that a series injected voltage u_{ps} is controlled without restrictions. Irrespective of the current through the system line, the phase shift of the injected voltage can take a value between 0° and 360°, with amplitude between zero and specified maximum value. Than it follows, that the UPQC can generate or absorb both active as well as reactive power, where reactive power is generated or absorbed within the converter itself, whereas the exchange of active power is provided by the dc circuit.

By adding the injected voltage to the mains voltage u_i, UPQC provides all the basic functions of power quality improvement. Injected voltage can be divided into the component that modifies the amplitude of the mains voltage, into the component, which represents the compensation of the series impedance and the component that provides the phase shift of the network voltage.

The task of the parallel converter is usually to supply or receive active power dictated by the series branch. The converter is connected to the network via coupling impedance. The parallel arm can generate or consume reactive power, which enables independent reactive power compensation.

We can conclude that the UPQC is an effective and versatile concept for power quality improvement. With the possibility of simultaneous changing of all the power parameters and with fast response of the device it is suitable for many uses, where there is need for effective regulation of voltages and currents.

2.8 Dynamic Voltage Restorer (DVR)

A static synchronous voltage source with an appropriate converter can be used for effective dynamic series compensation. Static voltage source, or Dynamic Voltage Restorer (DVR), produces a corresponding voltage of arbitrary frequency, which is injected into the series power line. This voltage can compensate the voltage drop across the inductive impedance of the line, which is caused by the current flow.

Nominal values of the equipment determine the maximum amplitude of the injected voltage. This maximum value of the series voltage can be used for the capacitive as well as inductive area of operation. The size of the injected voltage can be changed with a very small delay and regardless of the size of the current flowing through the line.

Losses are mainly dependent on the size of current through the line and less on the size of injected voltage. The presence of harmonics is determined by the size of the injected voltage and will greatly depend on the structure of the converter.

2.8.1 Configuration of the DVR

Figure 14 shows a basic scheme of DVR. Voltage source converter is connected in series with the line through a series transformer. It is used to protect sensitive loads against voltage sags and swells, as well as for filtering harmonic distorted line voltage. The first series active filter was installed in 1996 in the USA (Woodley et al. 2009). Installing the device is certainly sensible in these cases, where costs of downtime are large, compared to the price of the device. The series filter also enables control of power flow in the line.

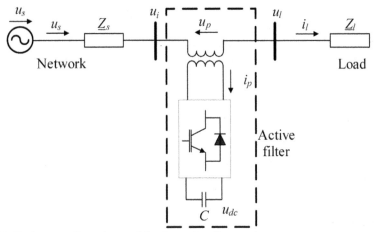

Fig. 14: Basic connection scheme of the dynamic voltage restorer.

2.8.2 Operation

The voltage at the connection point of the load u_l can be written as a sum of the voltage on the busbars u_i and the generated converter voltage u_p:

$$u_l = u_i + u_p \tag{3}$$

The voltage on the load busbars will be fully compensated when the inverter voltage is generated so that the sum $(u_i + u_p)$ will form a sinusoidal, symmetrical voltage of fundamental mains frequency. The reference inverter voltage is:

$$u_p = u_{ref} - u_i \tag{4}$$

where u_{ref} is the reference voltage, i.e., sinusoidal and the symmetrical voltage of the fundamental mains frequency.

Similarly as with the parallel active filter, the design of the dc-side capacitor is of crucial importance. For an effective operation of the device, a relatively constant dc-voltage is a key factor. The control algorithm enables minimal exchange of active power between the inverter and network. In this case, mainly reactive power is exchanged with the network, which reduces the required size of the dc-capacitor.

3. Conclusion

This chapter provides an overview of the modern compensation devices, which are intended for use primarily in distribution or industrial networks. These devices generally provide the means of achieving an appropriate level of power quality and an uninterrupted power supply.

This chapter introduced, the concepts of modern compensation devices. Semiconductor switching devices were also presented in detail and were followed by the presentation of three families of compensation devices:

- passive compensation devices,
- active compensation devices and
- hybrid compensation devices.

The structures of these devices, their functionality and basic operation principles were described.

References

Akagi, H. 1994. Trends in active power line conditioners. IEEE Transactions on Power Electronics 9(3).

Akagi, H. 1996. New trends in active filters for power conditioning. IEEE Transactions on Industry App. 32(6): 1312–1322.

Akagi, H. 1998. The state-of-the-art power electronics in Japan. IEEE Transactions on Power Electronics 13(2): 345–356.

Aredes, M., K. Heumann and E.H. Wantabe. 1998. An universal active power line conditioner. IEEE Transactions on Power Del. 13(2): 545–551.

Bhattacharya, S., P.T. Cheng and D.M. Divan. 1997. Hybrid solutions for improving passive filter performance in high power applications. IEEE Transactions on Industry Applications 33(3): 732–747.

Blažič, B. and I. Papič. 2003. Sodobne kompenzacijske naprave v distribucijskih omrežjih. 6. Konferenca Slovenskih elektroenergetikov, Portorož.

Blažič, B. and I. Papič. 2006. Improved D-StatCom control for operation with unbalanced currents and voltages. IEEE Transactions on Power Del. 21(1): 225–233.

Bojoi, R.I., G. Griva, V. Bostan, M. Guerriero, F. Farina and F. Profumo. 2005. Current control strategy for power conditioners using sinusoidal signal integrators in synchronous reference frame. IEEE Transactions on Power Electronics 20(6).

Chen, L., Y. Xie and Z. Zhang. 2008. Comparison of hybrid active power filter topologies and principles. International Conference on Electrical Machines and Systems, Oct. 17–20, 2008.

Cheng, P.T., S. Bhattacharya and D.M. Divan. 1998. Control of square wave inverters in high power hybrid active filter systems. IEEE Transactions on Industry Applications 34(3): 458–472.

Cheng, P.T., S. Bhattacharya and D.M. Divan. 1999. Line harmonics reduction in high power systems using square wave inverters based dominant harmonic active filter. IEEE Transactions on Power Electronics 14(2): 265–272.

Das, J.C. 2004. Passive filters—potentialities and limitations. IEEE Trans. on Industry Applications 40(1).

Fukuda, S. and T. Endoh. 1995. Control method for a combined active filter system employing a current source converter and a high pass filter. IEEE Transactions on Industry Applications 31(3): 590–597.

Genji, T., O. Nakamura, M. Isozaki, M. Yamada, T. Morita and M. Kaneda. 1994. 400V class high-speed current limiting circuit breaker for electric power systems. IEEE Trans. on Power Delivery 9(3).

Ghosh, A. and G. Ledvich. 2002. Power Quality Enhancement Using Custom Power Devices. Kluwer Academic Publishers, USA.

Gole, A.M., A. Keri, C. Kwankpa, E.W. Gunther, H.W. Dommel, I. Hassan, J.R. Marti, J.A. Martinez, K.G. Fehrle, L. Tang, M.F. McGranaghan, O.B. Nayak, P.F. Ribeiro, R. Iravani and R. Lasseter. 1997. Guidelines for modeling power electronics in electric power engineering applications. IEEE Transactions on Power Delivery 12(1): 505–514.

Gyugyi, L. 1992. A unified power flow control concept for flexible ac transmission systems. IEE Proceedings-C 139(4): 323–331.

Gyugyi, L. 1993. Solid-state synchronous voltage sources for dynamic compensation and real-time control of AC transmission lines. Emerging Practices and Technology, IEEE Standards Press, USA.

Hafner, J., M. Aredes and K. Heumann. 1997. A shunt active filter applied to high voltage distribution lines. IEEE Transactions on Power Delivery 12(1): 266–272.

Hingorani, N.G. and L. Gyugyi. 2000. Understanding FACTS: Concepts and technology of flexible AC transmission systems. IEEE Press, New York.

Lascu, C., L. Asiminoaei, I. Boldea and F. Blaabjerg. 2007. High performance current controller for selective harmonic compensation in active power filters. IEEE Transactions on Power Electronics 22(5).

Mokhtari, H., S.B. Dewan and M.R. Tarvani. 2000. Performance evaluation of thyristor based static transfer switch. IEEE Trans. on Power Delivery 15(3).

Nielsen, J.G., M. Newman, H. Nielsen and F. Blaabjerg. 2004. Control and testing of a Dynamic Voltage Restorer (DVR) at medium voltage level. IEEE Transactions on Power El. 19(3): 806–813.

Papič, I. 2000. Mathematical analysis of FACTS devices based on a voltage source converter – Part I. Electric Power System Research 56: 139–148.

Papič, I. and P. Žunko. 2003. UPFC converter-level control system using internally calculated system quantities for decoupling. International Journal of Electrical Power and Energy Systems 25: 667–675.

Peng, F.Z., H. Akagi and A. Nabae. 1990. A new approach to harmonic compensation in power systems – A combined system of shunt passive and series active filters. IEEE Transactions on Industry App. 26(6): 983–990.

Pires, D.F., C.H. Antunes and A.G. Martins. 2006. Passive and active anti-resonance capacitor systems for power factor correction. Presented at the International Power Electronics and Motion Control Conference – EPE-PEMC, 2006.

Senini, S. and P.J. Wolfs. 1999. Analysis and comparison of new and existing hybrid filter topologies for current harmonic removal. Australasian Universities Power Engineering Conference pp. 227–232.

Singh,B., K. Al-Haddad and A. Chandra. 1999. A review of active filters for power quality improvement. IEEE Transactions on Industrial Elect. 46(5): 960–971.

Shuai, Z., A. Luo, W. Zhu, R. Fan and K. Zhou. 2009. Study on a novel hybrid active power filter applied to a high-voltage grid. IEEE Transactions on Power Delivery 24(4).

Shwehdi, M.H. and M.R. Sultan. 2000. Power factor correction capacitors; essentials and cautions. IEEE Power Engineering Society Summer Meeting 3.

Teodorescu, R., F. Blaabjerg, M. Liserre and P.C. Loh. 2006. Proportional-resonant controllers and filters for grid-connected voltage-source converters. IEE Proceedings Electr. Power Applications 153(5).

Woodley, N.H., L. Morgan and A. Sundaram. 1999. Experience with an inverter-based dynamic voltage restorer. IEEE Trans. on Power Delvery 14(3): 1181–1185.

Zyborsk, J., T. Lipski, J. Czucha and S. Hasan. 2000. Hybrid arcless low-voltage AC/DC current limiting interrupting device. IEEE Trans. on Power Delivery 15(4).

CHAPTER 15

Voltage Sag Reporting and Prediction

André Quaresma dos Santos[1] and
Maria Teresa Correia de Barros[2,*]

1. Introduction

Power quality deals with a wide range of voltage disturbances, which may be permanent in time (continuous phenomena) or of short duration (events) (European Committee for Electrotechnical Standardization 2010). Events are unexpected disturbances that happen randomly overtime with reference to any specific location, and randomly in location with reference to any given instant of time.

Voltage sags are events characterized by a sudden reduction of the voltage magnitude followed by voltage recovery after a short period of time (European Committee for Electrotechnical Standardization 2010). A voltage sag example is depicted in Fig. 1.

Voltage quality becomes an important subject to end users and network operators due to the increase of susceptibility of equipment, and also due to the increase emission of disturbances by equipment. Additionally, end-users with critical equipment are becoming more aware of these issues and are demanding a better provided service.

Depending on the voltage magnitude and on the disturbance duration, voltage sags can affect the operation of electric equipment connected to the power system.

[1] Rede Eléctrica Nacional, S.A., Av. EUA, 55, 1749-061 Lisboa, Portugal.
[2] Instituto Superior Técnico, Universidade de Lisboa, Av. Rovisco Pais, 1049-001 Lisboa, Portugal.
* Corresponding author

Voltage sags are mainly caused by power system faults, and their characteristics at a given network site depend on the fault location and other fault characteristics. Furthermore, the voltage sag characteristics depend on the network topology and the protection systems' performance. As regards the network topology, mesh or radial topologies and short-circuit power affects the amount of voltage magnitude reduction observed at a network site during the event. Furthermore, it affects the size of the network region around a site, where faults can produce voltage sags. As regards the protection systems' performance, this affects the sag duration as this characteristics is related to the total fault clearance time.

Most transmission network faults occur in overhead lines. As shown in (REN 2013), these represent more than 90% of the recorded faults. They have different origins, the most common being lightning, birds, fires, and pollution. Although being difficult to set typical percentages of faults per origin, lightning is usually the most relevant cause. Table 1 highlights

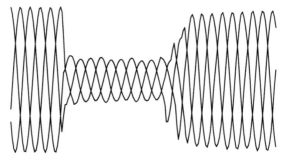

Fig. 1: An example of a voltage sag.

Table 1: Transmission line faults per 10 km from 2004 to 2012 (REN 2013).

Year	Lightning	Birds	Fires	Pollution	External*	Unknown	Total
2004	1.05	0.82	0.43	0.39	0.14	0.17	2.99
2005	0.18	0.41	2.24	0.75	0.26	0.27	4.10
2006	0.81	1.04	0.54	0.06	0.13	0.20	2.78
2007	0.89	0.77	0.04	0.15	0.22	0.18	2.25
2008	0.84	0.52	0.03	0.15	0.08	0.36	1.98
2009	0.96	0.49	0.25	0.00	0.07	0.58	2.35
2010	1.01	0.49	0.20	0.04	0.31	0.54	2.59
2011	1.52	0.84	0.18	0.02	0.10	0.16	2.81
2012	0.50	0.64	0.38	0.01	0.04	0.19	1.75
Average	0.79	0.67	0.59	0.25	0.15	0.29	2.74

* External object that reaches the phase conductors.

these characteristics showing the transmission line faults per 100 km in the Portuguese Transmission Network from 2004 to 2012 by fault cause.

As power system faults occur randomly in time and location, voltage sags have a stochastic nature.

2. Description of Voltage Sags

As defined in both EN 50160 an IEC 61000-4-30 standards, voltage sags are disturbances in the voltage supply, at a point in the electrical system, characterized by a sudden reduction of the voltage magnitude below an established threshold, followed by voltage recovery after a short period of time, from a few cycles to a few seconds. The voltage magnitude is defined by its r.m.s. value, refreshed every cycle or half-cycle according to the monitoring equipment class, S or A respectively. A complete representation of the disturbance requires information on the three phases. The measured voltages may be phase-to-ground or phase-to-phase, depending on the network earthing system. In transmission systems, the earthing system being solidly grounded, phase-to-ground voltages are measured.

The voltage sags are described by two characteristics:

1. A single voltage magnitude value, either residual voltage or sag depth and;
2. A duration value.

The residual voltage is defined as the lowest voltage magnitude recorded during the disturbance, while the depth is defined as the difference between a reference voltage and the residual voltage. A minimum value of the residual voltage is defined, below which the disturbance is no longer classified as a sag. According to EN 50160, this value is 5% of the reference voltage. The reference voltage is defined as the pre-disturbance voltage magnitude averaged over a specified time interval, or simply the declared voltage at the monitored site.

The sag duration is determined by predefined voltage threshold values, which may be different for the sag start and end. The start and end thresholds are conventionally defined as 90% and 92% of the reference voltage, respectively. The minimum duration is 10 ms due to the defined process to compute the voltage r.m.s. value, and the maximum is defined as up to 60 s (International Electrotechnical Commission 2008).

2.1 Polyphase Aggregation

Polyphase aggregation, as described in the EN 50160, consists in defining an equivalent event for polyphase disturbances, characterized by (1) a **single residual voltage** and (2) a **single duration**.

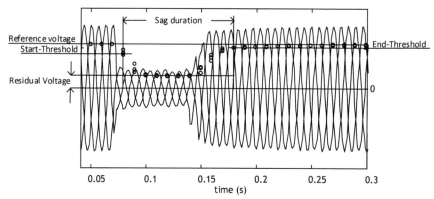

Fig. 2: Description of the residual voltage and duration concepts. Phase to ground voltage waveform and corresponding r.m.s. values refreshed each half-cycle (circle) during a three-phase voltage sag.

The single residual voltage value (or sag magnitude) is the minimum value of the residual voltages observed in the three phases. The single duration is established to comprise the overall duration of the disturbance, in all the three phases. It is defined by the time interval between the voltage start threshold being reached in one of the phases, and the sag end threshold being reached in all the three phases. Therefore, the equivalent sag may start and end in different phases. The sag residual voltage and duration concepts are graphically presented in Fig. 2, that shows an example of a three-phase voltage sag.

A disturbance will be classified as an interruption if the residual voltages, in all three phases, fall below the defined 5% minimum value in the EN 50160 standard.

In the present chapter both terms, residual voltage and sag amplitude are used with exactly the same meaning.

2.2 Time Aggregation

Voltage sags often appear in clusters. This characteristic is normally due to weather conditions, such as lightning or wind, being favorable to produce consecutive faults, and also due to permanent faults combined with reclosure attempts, a common practice in overhead transmission line operation. Experience shows that there is a high probability that faults are not permanent and consequently circuit breaker successful reclosure has a high probability. For instance, in the Portuguese Transmission Network, more than 75% of the reclosure attempts are successful (REN 2013). The remaining 25% will originate voltage sag clusters.

More than one reclosing attempts are possible, although in transmission voltage levels not more than two are allowed as visual inspection becomes necessary for security reasons. In the Portuguese Transmission Network, the first attempt is usually made in less than 1 second after fault clearance, and the time between attempts is not more than 2.5 minutes.

Time aggregation consists in defining an equivalent sag in the case of successive sags over a defined period of time (European Committee for Electrotechnical Standardization). The aggregation period, as well as the set of rules that define the equivalent sag, are regulated in national standards.

2.3 Sag Type

The joint CIGRE/CIRED/UIE working group has proposed to classify voltage sags according to the number of affected phases CIGRE/CIRED/UIE. The classification distinguishes three types of voltages sags:

1. **Type III**—these sags type are characterized by a similar voltage drop in all phases, and are typically caused by three-phase faults. The measured phase voltages contain mainly positive sequence component.

2. **Type II**—these sags type are characterized by a voltage drop, predominantly in two phases, and are typically caused by line-to-line faults or high resistive double line-to-ground faults. The measured phase voltages contain mainly positive and negative sequence components, of almost equal value.

3. **Type I**—these sags type are characterized by a voltage drop predominantly in only one phase and are typically caused by single line-to-ground faults. The measured phase voltages contain positive, negative and zero sequence component.

The phasor diagram of the pre-disturbance voltages $\bar{V}_{pFabc}$ and disturbed voltages $\bar{V}_{abc}$ for the three sag types is presented in Fig. 3.

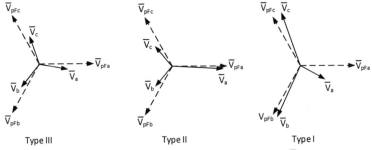

Fig. 3: Phasor diagram of the pre-disturbance measured voltage $\bar{V}_{pFabc}$ and the disturbed measured voltage $\bar{V}_{abc}$ for three types of voltage sags.

3. Voltage Sag Reporting

Voltage sags are reported by the Network Operators using Table 2, which is defined as a site index in EN 50160. The table shows the number of equivalent sags, recorded during a monitoring period, with duration and residual voltage within pre-defined intervals. Voltage sags are typically reported on a trimester or yearly basis.

System indices may also be reported using the same table, showing or the average, or a percentile value (e.g., 75% or 95%), or the maximum number of reported sags in each cell from all measured sites. Further classifications may be included based on the type of network users (distribution network, energy producer, consumer, etc.) or based on the network type (voltage level, rural with a predominance of overhead transmission lines, urban with a predominance of underground cables, network earthing, etc.).

As regards Table 2, it should be noted that it contains much more information than the simple number of sags. In fact, the cell to which a specific sag belongs reveals characteristics of the fault, which originated the sag, namely it's location and clearing process.

The residual voltage value is related to the distance to the fault, sags with residual voltages lower than 40% being caused by faults in the vicinity of the monitored site.

As regards duration, within the [0.01 s, 0.2 s] column, sags are caused by faults correctly cleared by the primary protection systems (International Electrotechnical Commission 2010). Sags located within the [0.2 s, 0.5 s] column are caused by faults cleared by a backup protection, either the remote backup protection, such as the transmission line distance protection Zone 2, or the local backup protection, such as the breaker failure protection.

Several reasons can explain the operation of the backup protection:

1. Communication channel failure: the failure of the communication channel in the transmission line protection system may cause fault clearance with Zone 2 trip delay, typically less than 0.5 s.

Table 2: Reporting of voltage sags according to EN 50160.

Residual voltage [%]	Duration (s)				
	[0.01, 0.2]	[0.2, 0.5]	[0.5, 1]	[1, 5]	[5, 60]
$90 > U \geq 80$	A1	A2	A3	A4	A5
$80 > U \geq 70$	B1	B2	B3	B4	B5
$70 > U \geq 40$	C1	C2	C3	C4	C5
$40 > U \geq 5$	D1	D2	D3	D4	D5
$U < 5$	X1	X2	X3	X4	X5

2. Busbar protection failure: the failure of the busbar protection in clearing a bus fault requires the Zone 2 operation of the remote distance protections, typically with a delay less than 0.5 s.

3. Circuit breaker failure: the failure of a circuit breaker to extinguish a short-circuit current requires the operation of the breaker failure protection, which will trip all surrounding breakers after a short delay, typically 0.15 s to 0.25 s.

Sags located in the third and fourth columns [0.5 s, 1 s] and [1 s, 5 s], correspond to either high resistance faults, cleared by inverse time delay overcurrent protections, or are due to severe protection system failures, i.e., hidden failures, which require the operation of very delayed protection functions. Evidence of hidden failures is very rare but can cause system blackouts (Horowitz and Phadke 2006).

In transmission networks, it is not expected to observe sags in the last column [5 s, 60 s], as most protections are settled with operating times shorter than 3s, due to the equipment thermal rating.

4. Voltage-Tolerance Curves

Most recorded voltage sags do not impact the normal operation of the end-user equipment. The equipment capability to ride-through a voltage sag is considered to depend on the sag type, duration, and residual voltage, although other parameters may also be relevant, such as the magnitude of the pre-disturbance supply voltage, the voltage recovery time, and the equipment control and the protection setting.

In 1977, the Computer and Business Equipment Manufacturers Association's (CBEMA) developed a standard to address computer equipment performance when exposed to magnitude deviations of the supplied voltage. The CBEMA curve[3] has been extensively used since then, growing from a curve describing the performance of mainframe computers, to a curve applied for specification criteria for electronic equipment or integrated into power quality performance contracts between network operators and large industrial customers.

More recently, the SEMI Standard F47-0706 (SEMI 2006) and the IEC 61000-4-11 and IEC 61000-4-34 standards (International Electrotechnical Commission 2004, International Electrotechnical Commission 2005) were published aiming at establishing a common reference for evaluating the equipment susceptibility to voltage sags.

The SEMI Standard F47-0706 defines the voltage sag immunity required for semiconductor processing, metrology, and automated test

---

[3] http://www.itic.org/.

equipment. This includes, amid others, power supplies, computers and communication systems, robots and factory interfaces, AC Contactor coils and AC relay coils, chillers, pumps and adjustable speed drives. The standard requires the equipment survival to the phase-to-neutral and phase-to-phase sags listed in Table 3.

The IEC 61000-4-11 and IEC 61000-4-34 standards are not specific to a given industry, as the SEMI Standard. They refer to electrical and electronic equipment connected to low-voltage power supply networks and define the immunity test methods and range of preferred test levels. These standards characterize equipment voltage sag tolerance as Class 1, 2, 3 and X, although only for Class 2 and 3 the immunity requirements are specified, according to Table 4.

Immunity requirements for Class X have been proposed by CIGRE/CIRED/UIE using a residual voltage versus duration curve, the so-called voltage-tolerance curve. This curve can be draw in a voltage sag plot, with axis the sag duration and the residual voltage, and where sags are represented as dots, Fig. 4.

Table 3: SEMI Standard F47-0706 (SEMI 2006). Required voltage sag immunity.

Residual voltage	Duration at 50 Hz	Duration at 60 Hz
50%	10 cycles	12 cycles
70%	25 cycles	30 cycles
80%	50 cycles	60 cycles

Table 4: IEC 61000-4-11 and IEC 61000-4-34 Standards. Preferred test level and duration for voltage sags.

Class	Residual voltage and durations for voltage sags (50 Hz/60 Hz)		
Class 2	70% during 25/30 cycles		
Class 3	40% during 10/12 cycles	70% during 25/30 cycles	80% during 250/300 cycles

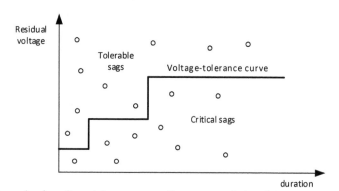

Fig. 4: Example of a voltage-tolerance curve. Sags represented as dots.

It is expected that a Class X equipment is capable to survive to sags located above its declared voltage-tolerance curve, while sags located below may cause equipment outage.

In order to describe the equipment performance, voltage sag immunity classes, A to E, were defined in, with associated voltage-tolerance curves, for different voltage sag types.

The equipment immunity description allows equipment owners to evaluate the adequacy of the equipment, as regards the chosen network connection site, giving the acceptable number of expected outages per year. This can be assessed by using historical voltage quality data, such as the SARFI-curve value (IEEE 2014), which corresponds to the number of voltage sags, recorded at a network site during one year, which are below a specific voltage-tolerance curve. This value is assumed equal to the number of equipment outages. Alternatively to historical data, voltage sag prediction methods can be used.

5. Voltage Sag Responsibility—Sharing Curves

So far, no limits for voltage sags have been introduced in the form of national regulations but, very recently, an important public consultation by European Regulators (European Regulators' Group for Electricity and Gas 2013), has proposed the EN 50160 revision, namely the introduction of the concept of duties and rights for all parties involved: network operators and users. Voltage sag responsibility is being discussed by the Council of European Energy Regulators (CEER) (CEER 2012), including the responsibility sharing concept. Acceptable disturbance levels are being discussed, considering network operators expenditure to improve voltage quality, as well as costs directly supported by customers, due to equipment malfunctioning/outages or to improve equipment immunity.

As a first outcome of this discussion, it was introduced the concept of voltage sag responsibility-sharing curve. The curve sets limits above which users are responsible for equipment misoperation or outages, and below which the network operators are responsible, within a regulatory framework.

A possible choice for the responsibility-sharing curve has been proposed by CEER (CEER 2012) based on the C1 class curve for types I/II sags, as per IEC 61000-4-11/34 Class 3 at 50 Hz, Fig. 5.

6. The Stochastic Nature of Voltage Sags

Power system faults occur randomly in time and location. Consider the monitoring of all power system faults that occur in a network during a

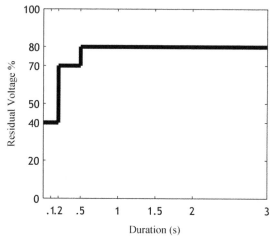

Fig. 5: The first responsibility-sharing curve proposed by CEER (2012). Equivalent to IEC 61000-4-11/34 Class 3 at 50 Hz.

period, for example a year, and the monitoring outcome of the set of all faults, each being characterized by: (1) fault instant; (2) location; (3) type; (4) resistance and; (5) duration. When repeating the monitoring period, for example in the following years, different outcomes will occur, and we have a stochastic process which is a function of two variables (Papoulis 1965): (1) the time and (2) the outcome obtained from monitoring.

Figure 6 presents an example where all power system faults, named F, occurring during four different monitoring periods of equal duration, are indicated along timelines. Fault data is complemented by the faults characteristics given in Table 5. The interpretation of the monitoring outcomes is as follows:

1. A specific outcome is a single time function of the fault occurrences and their characteristics.

2. For a specific time of the monitoring period, the probability that the network is in a fault condition is a random variable depending on the set of outcomes.

3. A specific outcome and a specific time of the monitoring period is a mere network state.

As with power system faults, the resulting sags occur randomly in time and, as a result, different monitored periods will have different recorded sags, as regards time of occurrence and characteristics. This is what is shown in Fig. 7, where the monitored sags, in the same network site during two consecutive years, 2013 and 2014, are presented in a jittered

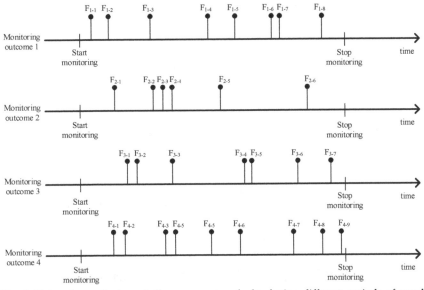

Fig. 6: Network monitoring of all power system faults during different periods of equal duration.

Table 5: Reporting table of the monitoring outcome 1 consisting of monitoring all power system faults that occur in a network during a period.

Fault	Location	Type	Resistance	Duration
F_{1-1}	$l_{F_{1-1}}$	$T_{F_{1-1}}$	$R_{F_{1-1}}$	$d_{F_{1-1}}$
F_{1-2}	$l_{F_{1-2}}$	$T_{F_{1-2}}$	$R_{F_{1-2}}$	$d_{F_{1-2}}$
F_{1-3}	$l_{F_{1-3}}$	$T_{F_{1-3}}$	$R_{F_{1-3}}$	$d_{F_{1-3}}$
F_{1-4}	$l_{F_{1-4}}$	$T_{F_{1-4}}$	$R_{F_{1-4}}$	$d_{F_{1-4}}$
F_{1-5}	$l_{F_{1-5}}$	$T_{F_{1-5}}$	$R_{F_{1-5}}$	$d_{F_{1-5}}$
F_{1-6}	$l_{F_{1-6}}$	$T_{F_{1-6}}$	$R_{F_{1-6}}$	$d_{F_{1-6}}$
F_{1-7}	$l_{F_{1-7}}$	$T_{F_{1-7}}$	$R_{F_{1-7}}$	$d_{F_{1-7}}$
F_{1-8}	$l_{F_{1-8}}$	$T_{F_{1-8}}$	$R_{F_{1-8}}$	$d_{F_{1-8}}$

one-dimensional scatter plot along time (Chambers 1983). This plot is used for perception of the random nature of sag occurrences along time.

In conclusion, voltage sags are stochastic events in power systems and, this important characteristic has to be included in any type of study of this phenomenon.

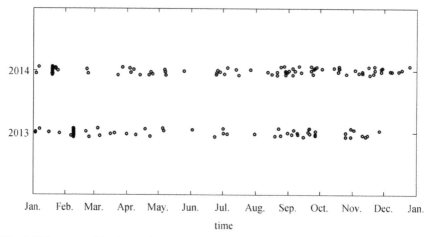

Fig. 7: Voltage sags (dots) recorded during two consecutive monitored years.

7. Examples of Voltage Sag Monitoring and Reporting

One year sag monitoring, in three representative 220 kV sites of an overhead transmission network, is reported in Table 6–8, according to the RQS and similar to the EN 50160 reporting table. The sites are representative of urban, urban/rural and rural locations. The first corresponds to a substation, close to a large city, which feeds an urban 60 kV distribution network, mostly underground. The second site is a substation close to a large city, feeding a mix overhead/underground 60 kV distribution network. The third site is a substation that feeds a rural 60 kV overhead distribution network.

Monitoring results show that most sags are of very short duration (first column). A few sags show longer durations (second to fourth column), corresponding to faults cleared by backup protections. No sags were recorded in the fifth column. This fact is in accordance with the maximum time delay settled in the transmission network protection system (1 to 3 s).

Table 6: Voltage sags monitored in the urban site.

Residual voltage [%]	Duration (s)				
	[0.01, 0.2]	[0.2, 0.5]	[0.5, 1]	[1, 5]	[5, 60]
90 > U ≥ 80	29	1			
80 > U ≥ 70	5				
70 > U ≥ 40	18		1		
40 > U > 0	1				
Total	53	1	1		

As regards depth, results show that sags with less than 40% residual voltage are seldom.

The sag type, as well as the origin, was identified for each recorded sag of the three network sites. As regards sag type, this characteristic was identified for each sag using data collected from voltage quality monitoring devices. As regards the origin, the monitoring data was complemented with information collected by the SCADA and by fault analysis.

Results are shown in voltage sag plots, per sag type, as presented in Fig. 8. A responsibility-sharing curve, based on class C1 is drawn, and the operator's responsibility area is subdivided into partitions.

Results show that sags recorded at the transmission level may have their origin in the 60 kV distribution network. This distinction is made in the presented results by using red and blue for sags originated in the transmission and the distribution networks, respectively.

Table 7 presents an overview of the recorded sags, classified by type and origin, for the three network sites. Results show the predominance of type I/II sags. Notwithstanding, no type I/II sags have been recorded at the urban site with origin in the distribution network. This is in accordance

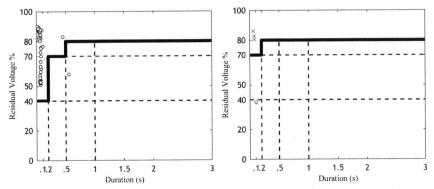

Fig. 8: One-year sag monitoring at a urban site. Sags type I/II (left) and type III (right). Sags with origin in transmission network faults (circle) and distribution network faults (cross).

Table 7: Classification of recorded voltage sags according to their type and origin.

Site	Urban
Number of sags	55
Number of sags type III	4
Number of sags type I/II	51
Number of sags with origin in the transmission network	52
Number of sags with origin in the distribution network	3
Number of sags in the operators' responsibility area	2

to this being underground. Indeed, the number of recorded type I/II sags is related to the existence of overhead line and their overall length.

As regards sags origin, most recorded sags are originated in the transmission network, and the number of sags originated in the distribution network is influenced by this being predominantly overhead or underground.

The recorded sags are mostly above the responsibility sharing curves. The number of sags in the network operators' responsibility area is 2 at the urban site, 3 at the urban/rural site, and 6 at the rural site. One of the three sags recorded at the rural/urban site was found to have its origin in the distribution network. This is an important issue as regards responsibility sharing between system operators.

8. Regulatory Trends and Proposed Voltage Sag Indices

So far, regulation of the maximum number of sags in the operators' responsibility area is not defined. Regulation may consider one unique index value, or alternatively, the responsibility areas could be subdivided, and different index values defined.

Accordingly the indices *SARFI-X-T*, based on the System Average RMS Variation Frequency Index *SARFI* are proposed. The indices, count the number of voltage sags type T in each cell X, on a yearly basis per network site, Table 8 and 9. The meaning of the *SARFI-X-T* indices are exemplified as follows:

4. *SARFI-D1-I/II* corresponds to the number of sags type I and II in cell D1 recorded at a network site during one year. The sags are characterized by a magnitude drop in one or two phase voltages, with duration from 0.01 s to 0.2 s and residual voltage lower than 40%.

5. *SARFI-C2-III* corresponds to the number of sags type III in cell C2 recorded at a network site during one year. The sags are characterized by a magnitude drop in the three phase voltages, with duration from 0.2 s to 0.5 s and residual voltage from 40% to 70%.

A voltage sag plot can also be used to present the corresponding *SARFI-X-T* regions as shown in Fig. 9.

Further to the *SARFI-X-T* indices, an index based on the responsibility-sharing curve can be defined, the *SARFI-RC*. This index is based on the *SARFI-curve*, which corresponds to the number of voltage sags below the responsibility curve. The index may be calculated summing the *SARFI-X-T* values of all cells beneath the responsibility curves.

Table 8: The *SARFI-X-I/II* indices used in network planning considering the responsibility-sharing curve equal to the voltage-tolerance curve class C1. Based on EN 50160 reporting table.

Residual voltage [%]	Duration (s)				
	[0.01, 0.2]	**[0.2, 0.5]**	**[0.5, 1]**	**[1, 5]**	**[5, 60]**
90 > U ≥ 80	*SARFI-A1-I/II*	*SARFI-A2-I/II*	*SARFI-A3-I/II*	*SARFI-A4-I/II*	*SARFI-A5-I/II*
80 > U ≥ 70	*SARFI-B1-I/II*	*SARFI-B2-I/II*	*SARFI-B3-I/II*	*SARFI-B4-I/II*	*SARFI-B5-I/II*
70 > U ≥ 40	*SARFI-C1-I/II*	*SARFI-C2-I/II*	*SARFI-C3-I/II*	*SARFI-C4-I/II*	*SARFI-C5-I/II*
40 > U > 0	*SARFI-D1-I/II*	*SARFI-D2-I/II*	*SARFI-D3-I/II*	*SARFI-D4-I/II*	*SARFI-D5-I/II*

Table 9: The *SARFI-X-III* indices used in network planning considering the responsibility-sharing curve equal to the voltage-tolerance curve class C1. Based on EN 50160 reporting table.

Residual voltage [%]	Duration (s)				
	[0.01, 0.2]	**[0.2, 0.5]**	**[0.5, 1]**	**[1, 5]**	**[5, 60]**
90 > U ≥ 80	*SARFI-A1-III*	*SARFI-A2-III*	*SARFI-A3-III*	*SARFI-A4-III*	*SARFI-A5-III*
80 > U ≥ 70	*SARFI-B1-III*	*SARFI-B2-III*	*SARFI-B3-III*	*SARFI-B4-III*	*SARFI-B5-III*
70 > U ≥ 40	*SARFI-C1-III*	*SARFI-C2-III*	*SARFI-C3-III*	*SARFI-C4-III*	*SARFI-C5-III*
40 > U > 0	*SARFI-D1-III*	*SARFI-D2-III*	*SARFI-D3-III*	*SARFI-D4-III*	*SARFI-D5-III*

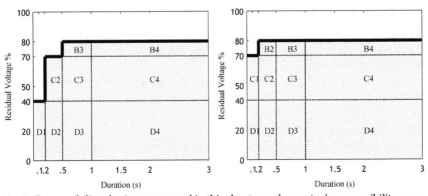

Fig. 9: Responsibility-sharing curve used in this chapter and operator's responsibility areas. Sag type I/II (left) type III (right).

The proposed indices can be adopted to assess the network operation by regulators. Furthermore, thresholds may be defined to guide system operators during network planning and operation.

9. Voltage Sag Prediction

9.1 Overview on Voltage Sags Prediction

The pioneer work by Conrad et al. (1999) gave a first extensive overview of the voltage sag phenomenon, identifying the need to combine different analysis tools to predict sags residual voltage, duration, frequency and economic impact. The residual voltage is predicted by modeling power system faults, the duration by taking typical values of fault clearance time, and the frequency by considering transmission lines and busses fault rates. The economic impact is assessed by the yearly costs due to equipment/ process outage before and after the adoption of mitigation measures.

These authors have proposed the method of Fault Position for predicting the number of voltage sags at a given network site. The method calculates the voltage sags observed in network sites due to many possible faults located in the network. Site characterization, concerning observed sags residual voltage and duration, is made considering all fault positions spread through the network. Each fault representing an equivalent fault in a certain part of the network, i.e., in its vicinity. The failure rate in the representative part of the network is associated with each equivalent fault, and the expected number of sags as a function of residual voltage and duration is calculated. Network sites are characterized by a probability density curve of the residual voltage, resulting from the combination of the observed residual voltages and the frequency of occurrence. Fault Position method has become the most common method used by the subsequent authors.

Conrad et al. 1999 have introduced the graphical representation of the influence of the distance to the fault on the residual voltage at a particular network site, based on contour lines of equal residual voltage superimposed on a geographic map.

These authors have identified some mitigation options to reduce the sag frequency, residual voltage and duration. Concerning frequency, changes in transmission line tower design and ground resistance reduction were proposed. Concerning residual voltage, the authors proposed the introduction of more impedance between the fault and the sensitive equipment by means of operating more efficient network configurations. Finally, concerning duration, the authors proposed the refurbishment of circuit breakers and protection devices, including time setting optimization, to reduce fault clearance time. Furthermore, the authors have identified the possibility to change equipment specification or control settings for fault ride-through capability improvement.

The work carried by Conrad et al. 1999 has inspired Qadar et al. (1999) to apply the methodology to transmission networks and obtain exposed area. The exposed areas are network regions limited by fault positions that cause sags of equal residual voltage at a considered site. The fault positions define contour lines that are overlaid in the network single-line diagram. These authors have introduced the voltage sag table, to characterize network sites, as a different way of presenting results from the calculated residual voltage probability density curves, computed using the Fault Position method.

Fonseca and Alves (2002) developed a methodology for network site characterization using Monte Carlo simulation. The methodology simulates the network operation during a defined time span (several years), and transmission line faults are simulated according to the considered fault rate. The methodology required as input transmission line fault rate, fault type distributions and, typical fault clearance times. The typical fault clearance time considers faults clearance by distance protection Zone 1 or Zone 2 operation. The calculated fault voltages in all network busses and associated fault clearance times are stored for post-simulation analysis. Results are presented as averages values of all sags generated during the Monte Carlo simulation.

The Fault Position and the Monte Carlo simulation methods were compared by Olguin et al. (2005), who have concluded that the last gives a more complete network site characterization as it includes the yearly variability, whereas the first gives only long-term mean values. The yearly variation was included in the Monte Carlo simulation by these authors assuming that the transmission line failure rates over the simulated years follow a normal distribution. The authors apply the methods to predict the sag's residual voltage and results are presented using the SARFI index.

Predicting voltage sags' duration has been a difficult topic to address due to its straight relation with protection system performance. Several authors have avoided the prediction of sag duration and have used typical fault clearance times, although without reasonable technical justification. Aung and Milanovic (2006) made a very interesting work concerning this issue. The authors used the concept of primary and backup protections to characterize the fault clearance time and consequently the voltage sag duration. The primary protection is expected to have priority in fault clearance, while the backup protection is intended to have a delayed operation when the fault is not cleared by the primary protection due to its failure or inability. Voltage sag durations are characterized by faults cleared by primary protection or cleared by backup protection. The authors have modeled the probability of failure of the primary protection system using Fault-Tree analysis, and results were combined with the Fault Position method by adding to the expected residual voltage an expected sag duration.

A different approach was made by Wamundsson and Bollen (2008) who proposed a method for predicting the residual voltage and duration, in which both fault location and fault clearance time are treated as random variables within in a Monte Carlo simulation. The sag duration is modeled by a random variable with a probability density function described by a combination of triangular distributions, associated with the transmission line distance protection operation Zones 1, 2 and 3. The triangular distributions were obtained from disturbance recordings and data from voltage-quality measurements. Results were presented in a voltage sag plot.

Oliveira et al. (2009) have reapplied the Monte Carlo simulation method developed by Olguin to compare monitoring and simulation results aiming to analyze the accuracy of sag characteristics obtained from short monitoring periods.

9.2 Overview on Probabilistic Short-Circuit Studies

Correct modeling of power system faults is a key issue in a diversity of power system studies, such as network planning, equipment design, and protection systems coordination. The currently used methods, underlying international guidelines, and standards are deterministic, based on worst-case scenarios (ANSI/IEEE 1999, International Electrotechnical Commission 2001, Diário da República 2010). These are typically associated with extreme events leading to the highest or lowest short-circuit currents. The deterministic approaches result into over-dimensioning during most of the operating time. Furthermore, as it disregards the likelihood of occurrence, it does not include the uncertainty in the information supporting investment decisions.

Contrary to deterministic methods, a probabilistic approach quantitatively incorporates uncertainty, based on quantifying the likelihood of each event, and its associated uncertainty.

A probabilistic short-circuit analysis of power systems has been developed by El-kady (1985) based on Monte Carlo simulation, aiming to provide probabilistic distributions of fault current in a network region or at a particular bus, considering random changes of generation load, circuit outages and, fault conditions, obtained from network historical data. The author has considered both bus and line faults and results were presented in the form of a histogram. Three types of histograms were proposed by El-kady: (1) the frequency of fault current at a particular network bus; (2) the frequency of fault current in a network region; and (3) the frequency of short-circuit current flowing in certain transmission lines of the network.

The Monte Carlo simulation of short-circuits was also applied by Vaahedi et al. (2000) to probabilistic transient stability assessment on large-scale systems, for which the transient stability limits were calculated. These authors have considered, based on historical statistics, the random nature of the load, network topology, fault location, type and disturbance sequence (fault clearance followed by successful or unsuccessful autoreclose attempts). Results were compared to the ones obtained using traditional deterministic approaches and have concluded the conservative nature of those.

Analytical methods for probabilistic short-circuit studies have also been developed. The early work from Ford and Sengupta (1981), for probabilistic short-circuit design of substation equipment, took advantage of the reduced computational effort by the analytical methods when compared to the Monte Carlo simulation. The developed methodology was applied to radial networks and these authors made some considerations on how the method could be extended to meshed networks.

9.3 Overview on Power System Faults Characterization

The correct fault characterization is one of the key factors for successful application of the probabilistic studies. The use of historical data has been the most efficient way to describe the frequency of occurrences of a certain type of faults and their locations. This has been done by El-Kady (1985) and by Vaahedi et al. (2000). The information related to fault resistance is not commonly available in the system operators historical data, as it is not traditionally recorded in the fault databases, and therefore hardly used in short-circuit calculations.

It is known that several factors may affect the fault resistance such as the network earthing. Field data in medium voltage distribution networks (12 kV to 60 kV) with isolated and compensated neutrals show that, in unearthed networks, there are two distinguished fault resistance ranges (Hanninen and Lehtonen 1998, Aucoin 1985, Lee and Bishop 1985): the first in the range up to 400 Ω and the second in the range higher than 2 kΩ. In compensated networks, the fault resistance ranges from 20 Ω to 10 kΩ with mean value around 400 Ω.

In transmission or sub-transmission networks, where directly or reactance earthed neutrals are applied, the results of an extensive study of power system faults to ground conducted by Edison Electric Institute & Bell System, have shown that faults are not usually of high resistance (Gilkenson et al. 1937).

The use of a fixed value for the fault resistance has been a common practice (deterministic approach). In distribution networks, a 40 Ω fault resistance is typically used in short-circuit analysis for coordination of protective devices on electrical distribution systems, as reported by Dagenhart (2000).

For transmission networks attempts to decompose the fault resistance have been made by Andrade and Sorrentino (2010), by evaluating the so-called effective ground resistance of the towers and adding it to the electric arc resistance. In transmission lines with ground wires, when a single line to ground fault occurs, the fault current flows through the tower grounding system and through the ground wires, and the effective grounding resistance accounts for this effect. The typical expected values of the effective grounding resistance were evaluated for several voltage levels, from 69 kV to 400 kV, and results show that values increase with voltage.

Concerning the arc resistance, and for the short-circuit power typically present in transmission networks, this term is negligible, as shown by Terzija (2004).

As regards the probabilistic approach to fault resistance modeling, different distributions have been used, although with little support from field data. Studies concerning stochastic prediction of voltage sags, conducted by Martinez and Arnedo (2004) and Zhang (2008), have used the normal distribution, with mean value $\mu = [10, 15]\ \Omega$ and standard deviation $\sigma = [1, 4.5]$. However, further to not describing the non-symmetrical fault resistance distribution, shown in experimental results, the normal distribution can generate negative resistance values. In overcurrent protection studies, applied to distribution networks, the Weibull distribution was used by Barnard and Pahwa (1993), with parameters $\delta = 63\ \Omega$ and $\beta = 1.5$. The dependence of the fault resistance probabilistic distribution on the network voltage level was not addressed. However, experimental data reveal that such dependence does exist.

10. Stochastic Modelling of Power System Faults

Fault rate and individual fault characteristics are crucial information used to support power system planning and operation strategies, both technically and economically. Fault rate is a key index to evaluate bulk power system performance, as it reveals very important features such as: correct equipment design, suitable maintenance, equipment aging and abnormal environmental conditions. Additional important information is provided by the individual fault characteristics. For instances, a large number of line to line faults may indicate an inadequate line geometry design as regards environmental conditions, such as wind, and a high fault resistance occurrence may indicate special needs for sensitive earth fault protection.

Individual fault characterization starts by identifying where, in the bulk power system, the fault has occurred. This means identifying which network element was affected and, in the case of transmission lines, its exact location. Most power system faults occur in transmission lines and busbars, and they correspond to the highest short-circuit current values, hence they are the most relevant for equipment specification.

Short-circuits are also characterized by the number of affected phases. Faults may affect one single phase and ground (SLG), two phases (LL), two phases and ground (2LG) and three phases (3PH). Fault type distribution depends on the voltage level, line geometry, terrain relief and weather conditions. Therefore, this characteristic is strongly system dependent.

The resistance that characterizes each fault depends on the short-circuit type and cause. Its value aggregates the arc and the grounding resistance. Additionally, depending on the specific fault, it may include the tower or the touching objects resistance. The fault resistance may be time constant or vary with time, as in case of arc elongation by wind effect. For faults not involving the ground, it is assumed that only the arc contributes to the fault resistance, thus its value being negligible (Terzija 2004).

Some of the identified fault characteristics are not independent. Indeed, the underlying phenomena indicate, and operational data confirm, that fault type as well as fault resistance, depend on the fault location. However, at the planning and design stage, uniform characteristics are considered, and the fault characteristics may be treated as independent of the fault location.

Fault duration is also an important characteristic for system stability, equipment design and voltage dips duration. This characteristic is related to the fault clearance time and is affected by the reliability of circuit breaker, protection relays, measurement transformers, etc.

10.1 Fault Modeling Approach

A stochastic fault modeling, based on Monte Carlo simulation, is developed. The occurrence of faults in a network element, such as a transmission line or a busbar, is simulated during a time span of representative significance. The simulation time span is divided into an equal number of time intervals (trials) of equal duration, such as a second, an hour, or a year. The methodology output is a set of occurrences, each characterized by fault time, location, type, and resistance. Simulation results are analyzed using the same statistical tools as used in power system monitoring and field data reporting, thus allowing a straightforward comparison between simulation results and field data.

The stochastic description of power system faults applied to short-circuit calculation allows obtaining valuable information that enriches the current knowledge of the system planner and designer, which is usually based on maximum and minimum values. The obtained typical values, their ranges, and distributions, broaden the overall knowledge the power system at the planning and design stage, leveraging studies based on risk analysis.

10.2 Fault Rate

Transmission line and busbar fault rates are two indices that characterize a network performance. These rates are key information used to support investment and maintenance decisions. This is done by analyzing trends over several consecutive years. Also important is to pinpoint atypical years, during which the number of faults is abnormally low or high, and identify the corresponding causes. These may be abnormal weather conditions or other unexpected environmental events.

The developed methodology receives as input the network fault rate average value, for transmission lines and busbars. As output, the network fault rates are calculated on a yearly basis, along the chosen simulation time span.

During a Monte Carlo simulation, the transmission lines and busbars elements will fail at a given rate $z(t)$, which individually characterizes each element. During a trial t_0, the probability of fault occurrence P_F in a network element is defined as (Shooman 1968):

$$P_F(t_0) = 1 - \exp\left[-\int_0^{t_0} z(\tau)d\tau\right] \tag{1}$$

For each trial, a random number is generated according to a uniform distribution over the interval [0,1]. The generated number is compared to the corresponding probability of fault occurrence P_F and, if higher, the corresponding network element is assumed to fail during the trial.

Recorded network performance data usually refer to yearly line faults per 100 km, λ_l, and yearly busbar faults per 100 busses, λ_b. Accordingly, a constant failure rate $z_i(t) = \lambda_i$ is assumed for each network element. All busbars will have the same failure rate and transmission lines will have different failures rates, according to their length l_i:

$$\lambda_{b_i} = \lambda_b / 100 \tag{2}$$

$$\lambda_{l_i} = l_i \lambda_l / 100 \tag{3}$$

Simulation results obtained for the network elements are then converted into system performance indices. The analysis can be done by voltage level or for the overall network.

10.3 Line Fault Location

The probability of a fault occurring at a given location along a transmission line is considered uniform when the stochastic model of power system faults is intended to be used during the planning and design stages. Indeed, at this stage, a planning criterion for the transmission line fault rate is considered. This criterion is unique, thus the line fault rate being considered constant in time and along the all line length. The line design, including shielding, the towers, and the grounding resistance, are chosen to meet this planning criterion.

Naturally, the planning criterion is not fulfilled during line operation, and in some line segments, corrective measures are assessed in a cost/benefit analysis. Recently, the Portuguese Transmission System Operator has been working on this issue, analyzing historical fault data of the transmission lines with the highest fault rates. Results were obtained for the fault rate distribution along the line, identifying line segments with high fault rates and others that are practically fault free. Field data do not allow establishing a mathematical model for the fault distribution rate and therefore network planners continue to use a uniform distribution.

Accordingly, to assess the location of each transmission line fault, a random number is generated considering a uniform distribution over the interval [0,1]. The generated value gives the fault location in per unit, this being the line length.

10.4 Fault Type

Fault type probabilistic distributions are defined, on a yearly basis, for transmission lines and busbars.

To assess the fault type that characterizes a particular fault generated by Monte Carlo simulation, the interval [0,1] is divided into four segments, with lengths given by the probability of occurrence of each fault type. A random number is associated with each fault, generated according to a uniform distribution over the defined interval. The segment corresponding to the generated number defines the fault type.

10.5 Fault Resistance

Only transmission line faults involving the ground are assumed to be characterized by a fault resistance. The fault resistance associated to SLG and 2LG faults is described by a probability density function $f(r_f)$ where r_f is the fault resistance value.

The fault resistance value associated with SLG and 2LG faults is assessed using the inverse transform method: a random number v is generated according to a uniform distribution over the interval [0,1], and this value is used in the cumulative function distribution F to find the corresponding fault resistance value:

$$r_f(v) = F^{-1}(v) \qquad (4)$$

where

$$F^{-1}(v) = 1 - e^{-(v/\delta)^\beta} \qquad (5)$$

11. Stochastic Prediction of Voltage Sags

The developed methodology for predicting voltage sags is based on simulating the network operation, during a defined time span, by means of Monte Carlo simulation. The underlying approach combines the developed methodology for stochastic modeling of power system faults, with the reliability analysis of transmission line and busbar protection systems (dos Santos et al. 2015, dos Santos and Correia de Barros). The Monte Carlo simulation outcomes are the indices values *SARFI-curve*, *SARFI-X-T* and *SARFI-RC*, computed per simulated year. The methodology outputs are the results from statistical analysis to the simulation outcomes.

Figure 10 presents the schematic view of the proposed methodology encompassing four tasks grouped in two steps: fault characterization and sag characterization.

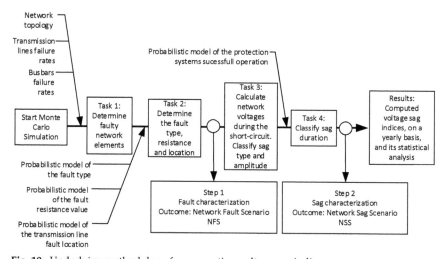

Fig. 10: Underlying methodology for computing voltage sag indices.

11.1 *Fault Characterization*

In the first step, the Network Fault Scenario NFS is settled up, for a transmission network, using the methodology presented in the previous chapter. The network is characterized by its topology, including the generators, the transmission lines, and the transformers.

The network operation is simulated, by means of Monte Carlo simulation, during a defined time span. This time span is divided into an equal number of time intervals (trials) with one-hour duration, during which more than one network element, transmission lines or busbars, can have a fault, as the probability of the network element failures are independent.

The first task consists in identifying the faulty network elements (transmission lines and busses) according to the corresponding failure rates. The second consists in determining the fault characteristics using the probabilistic models presented in dos Santos et al. (2014).

The simulation outcome, which will be referred as Network Faults Scenario (NFS), is the sequence of faults observed in the network during the simulation time span, each fault being fully characterized by:

1. Time—the trial at which the fault occurs;
2. Faulty network element—the transmission line or busbar where the fault occurs;
3. Fault type—3PH, SLG, LL or 2LG;
4. Fault resistance—different from zero in case of a transmission line fault affecting the ground;
5. Fault location—in case of transmission line faults.

The computational flowchart used in creating a NFS is presented in Fig. 11. At this stage, for each identified fault, a random value, between 0 and 1, is also generated, to be used in a second Monte Carlo simulation, for assessing the successful operation of the primary protection system. This parameter is identified as the protection system successful operation index (PSSOI).

The amount of information contained in a NFS obviously depends on the network size and simulation time span, but it also depends on the reliability of the network elements. Considering the flowchart in Fig. 11, the number of faults in a NFS is:

$$N_{NFS} = \sum_{i=1}^{N_T} F_i \tag{6}$$

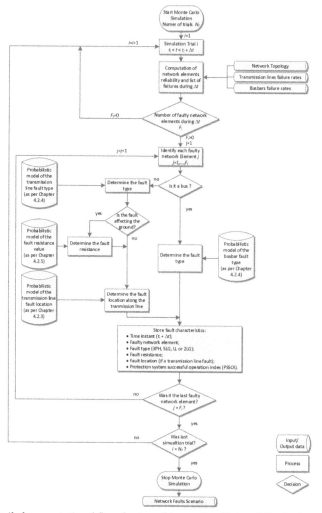

Fig. 11: Detailed computational flowchart used in creating Network Faults Scenarios.

where F_i is the number of faults occurring during the ith trial and N_T the total number of trials in the simulation time span.

11.2 Sag Characterization

Step 2 corresponds to the setting up of the Network Sag Scenario NSS, which contains all sags observed in all network sites of interest, in the form of time series. The sags being characterized by type, amplitude, and duration.

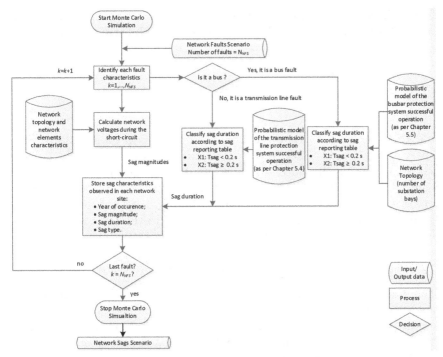

Fig. 12: Computational flowchart used in creating Network Sags Scenarios.

A second Monte Carlo simulation uses the NFS as input and computes, for each network fault of the NFS, the corresponding three phase fault voltages in all network sites of interest. The voltages are then used to determine the corresponding sag magnitude and type. The probabilistic models of the protection systems successful operation, developed in (dos Santos et al. 2015) dos Santos and Correia de Barros are included in the simulation in order to assess the sag duration. Sags are classified as X1 or X2, the first corresponding to faults cleared by the primary protection (duration within [0.01 s, 0.2 s]), and the second by the backup protection (duration within [0.2 s, 0.5 s]). For this purpose, the PSSOI value associated with a fault is compared to the probability of successful operation of the primary protection system. If the PSSOI value is lower than that probability value, the sag is classified as X1, otherwise as X2.

In case of a transmission line fault, the probability of successful operation is affected by the fault location, as well as by the protection scheme. In the case of a busbar fault, the protection architecture and the number of bays in the substation affect the probability of successful operation.

Figure 12 presents the computational flowchart used in predicting voltage sags. The first outcome of this Monte Carlo Simulation is the

Network Sags Scenario NSS, which contains all sags recorded at all network sites of interest, each being characterized by four parameters:

1. Year of occurrence;
2. Sag magnitude;
3. Sag duration;
4. Sag type.

The NSS is used to compute, on a yearly basis of the simulated time span, different voltage sag indices depending on the nature of the voltage sag study. In case the study focus on quantifying equipment outages, the computed index is the *SARFI-curve*, in case it focus on assessing network performance based on responsibility-sharing, part of regulation or planning criteria, the computed indices are the *SARFI-X-T* and *SARFI-RC*. Results are time series of the computed index values.

11.3 Statistical Analysis of the Simulation Outcomes

The index values, contained in the computed time series, are statistically analyzed by means of computing the corresponding density $f(n)$ and cumulative distribution $F(n)$ functions of the number of sags n counted at the network sites on a yearly basis of the simulated time span:

- Number of sags in each cell X and per sag type T

$$f_{SARFI\text{-}X\text{-}T}(n) \qquad (7)$$

$$F_{SARFI\text{-}X\text{-}T}(n) \qquad (8)$$

- Number of sags below the responsibility-sharing curve

$$f_{SARFI\text{-}RC}(n) \qquad (9)$$

$$F_{SARFI\text{-}RC}(n) \qquad (10)$$

- Number of sags below the voltage-tolerance curve

$$f_{SARFI\text{-}curve}(n) \qquad (11)$$

$$F_{SARFI\text{-}curve}(n) \qquad (12)$$

with n the index value.

Site characterization will be made using representative values extracted from the distributions, these being the average or percentile values:

- Number of sags in each cell and per sag type

$$SARFI\text{ - }X\text{- }T_{av} \qquad (13)$$

$$SARFI\text{ - }X\text{- }T_{xx} \qquad (14)$$

- Number of sags below the responsibility-sharing curve

$SARFI\text{-}RC_{av}$ (15)

$SARFI\text{-}RC_{xx}$ (16)

- Number of sags below the voltage-tolerance curve

$SARFI\text{-}curve_{av}$ (17)

$SARFI\text{-}curve_{xx}$ (18)

were *av* means average and *xx* the percentile (ex. 95% or 99.7%).

Choosing the type of representative value depends on the nature of the study. The use of an average value only describes long-run trends and can lead to some optimistic characterization, as the dispersion of results is not measured. Percentile values do not describe a trend but rather a threshold, ensuring that in a certain percentage of the simulated years, the index value was lower than this threshold.

It is recommended to use percentile values instead of average, as the study resultants can be analyzed on the safe side.

12. Application Example

The proposed method was applied to the test network IEEE RTS to calculate the distribution of the maximum interrupted current by the circuit breakers. Furthermore, the application example uses the methodology to predict equipment outages due to voltage sags. The study compares network sites, as regards connecting sensitive equipment, and identifies the most adequate equipment voltage-tolerance curve, considering the minimization of the expected number of outages per year. The study assumes that voltage sags located below the equipment voltage-tolerance curve, cause equipment outages.

The study includes the following assumptions:

1. The network protection system comprehends a redundant distance protection sharing single aided communication channel in all transmission lines and a centralized busbar protection in all network busses.
2. The maximum failure rate of the protection system units is considered.

12.1 Network Topology

The IEEE RTS comprehends: 32 generators corresponding to 3.4 GW installed power; 33 circuits, with two voltage levels, 230 kV and 138 kV, consisting of 1630 km transmission lines and 24 stations, including 5 substations, with one transformer each. In the present application, it is assumed that each station contains 2 busbars per voltage level.

12.2 Fault Rate

Published data (Santos and Correia de Barros 2015) is used to describe the 138 kV and 230 kV transmission lines of the IEEE RTS network. The average value in the time span 2006 to 2012 was used for this purpose: 2.80 faults/(100 km x year) (138 kV) and 1.9 faults/(100 km x year) (230 kV).

As regarding busbar faults, the average value in the time span 2006 to 2012 was considered: 0.29 faults/(100 busses x year) (138 kV) and 0.46 faults/(100 busses x year) (230 kV).

12.3 Fault Type

The fault type distribution data obtained from the published data (Santos and Correia de Barros 2015), and the average values of the 150 kV and 220 kV networks have been considered as input data for the simulation. Concerning busbar faults in both networks, an 80%–0% distribution was considered.

12.4 Fault Resistance

It was assumed that the fault resistance distributions for the 138 kV and 230 kV follow the Weibull distribution with parameters presented in Table 10 (Santos and Correia de Barros 2015).

Table 10: Fault resistance - Weibull distribution parameters (Santos and Correia de Barros 2015).

Voltage	$\delta(\Omega)$	β
138 kV	33.116	1.459
230 kV	38.271	1.842

12.5 Simulation Results—Fault Characterization

Network operation was simulated by Monte Carlo over 80 years and statistical information was extracted from simulation results.

Line fault rate was computed for the 138 kV networks and results are presented in Fig. 13, where the pre-defined yearly average fault rate is highlighted. The calculated average value is 2.83 faults/100 km/year being consistent with the pre-defined values. Simulation results reproduce atypical years as happen in real transmission systems.

Fault type distribution was computed for the 138 kV network, the time series of results is shown in Fig. 14, including corresponding average and standard deviation. Results show the random nature of fault type distribution, respecting the predefined average value. Fault type

5
4
3
2
1
0
Faults/year/100km
1
11
21
31
41
51
61
71
81
Simulated Years

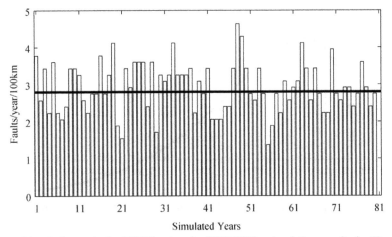

Fig. 13: Line fault rate in the 138 kV network of IEEE RTS – simulation results for 80 years of network operation.

1
Fault type 138kV
40
av
std
SLG
76.2
57.1
69.4
11.8
LL
9.5
0.0
3.9
4.5
2LG
4.8
21.4
12.3
9.3
3PH
9.5
21.4
14.3
9.7

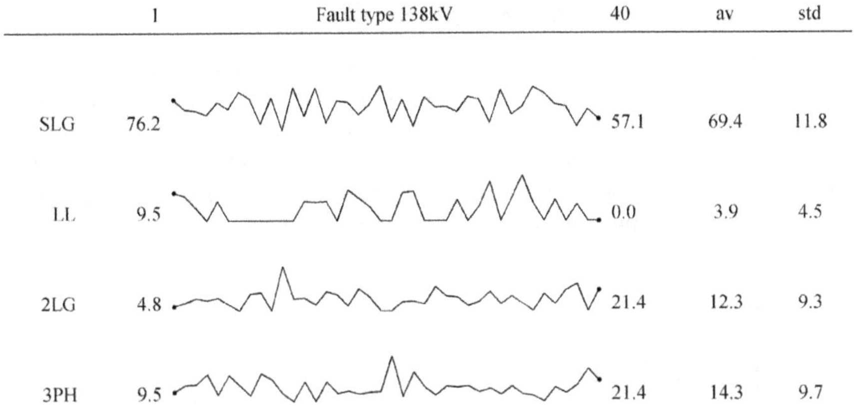

Fig. 14: Fault type in the 138 kV network of IEEE RTS – simulation results for 80 years of network operation.

distributions with the highest average values present the highest relative values of the standard deviations. The smaller average value corresponds to LL faults, which has the same order of magnitude as the standard deviation. As a result, some of the simulated years are characterized by the absence of such fault type.

Transmission line fault resistances, corresponding to SLG and 2LG, generated over the simulated time span were analyzed statistically considering 10 Ω resistance bins. Results per voltage level are presented in Fig. 15. For comparison with the predefined Weibull distribution, these are also plotted.

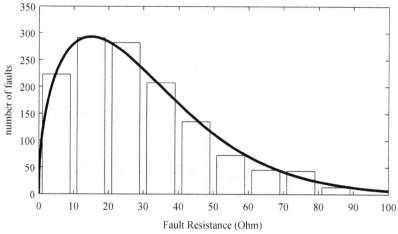

Fig. 15: Fault resistance in the 138 kV network of IEEE RTS – simulation results for 80 years of network operation. Comparison with predefined Weibull distribution.

12.6 Simulation Results—Equipment Outages

A convergence analysis is made by observing the *SARFI-SEMI$_{av}$* value in all network sites, during the progression of the Monte Carlo simulation. Results show that the *SARFI-SEMI$_{av}$* values rapidly stabilize as the simulation proceeds. For the tested network, results in Fig. 16 show that the index value starts to converge for a simulation time span longer than 500 years and, it completely stabilizes after 700 years.

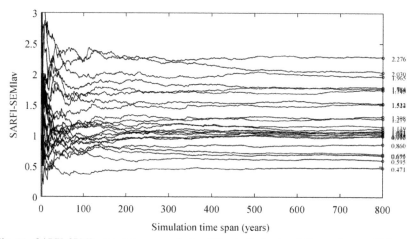

Fig. 16: *SARFI-SEMI$_{av}$* convergence analysis of Monte Carlo simulation outcomes. Values at all the IEEE RTS sites.

The chosen Monte Carlo simulation time span is 800 years, so to guarantee convergence of the $SARFI\text{-}SEMI_{av}$ value in all network sites.

The $SARFI\text{-}SEMI_{av}$ values for all network sites, corresponding to the 800 years simulation time span, are also shown in Fig. 16. Values range from 0.471 to 2.276 outages per year, highlighting the large differences, that can be expected, between network sites for a given equipment type.

In this example, two network sites are chosen for assessing the number of equipment outages, corresponding to a low and a highly exposed site. The chosen sites are Abel 101 and Agricola 104, characterized by $SARFI\text{-}SEMI_{av}$ equal to 0.695 and 2.276 respectively. In order to assess the adequacy between equipment characteristics and connecting network site, the equipment immunity level IEC 61000-4-11 Class 2 is also considered.

Accordingly, in the considered example, the prediction problem is defined as:

1. Assessing how many equipment outages are expected to occur in one year when a given equipment is connected to Abel 101 or Agricola 104 sites.

2. Assessing how many equipment outages are expected to occur in one year when the equipment immunity level is SEMI F47-0706 or IEC 61000-4-11 Class 2.

The computed density $f(n)$ and cumulative $F(n)$ distribution functions of *SARFI-SEMI* and *SARFI-IEC Class* 2 values for the network sites Abel 101 and Agricola 104 are shown in the histograms presented in Fig. 17.

In the case of connecting equipment type SEMI F47-0706 to Abel 101 site, the average number of outages is lower than 1 and the histogram is skewed to the right, showing the equipment robustness to the expected voltage sags. In the case of connecting equipment type IEC 61000-4-11 Class 2, the average number of equipment outages is between 3 and 4 and the histogram is more symmetrical in the vicinity of the 50% percentile.

For equipment connecting to Agricola 104 site, either SEMI F47-0706 or IEC 61000-4-11 Class 2 type, the histograms show symmetry in the vicinity of the 50% percentiles. The average number of equipment outages is close to 3 and 6 for SEMI F47-0706 and IEC 61000-4-11 Class 2 equipment type, respectively.

It can be concluded that the shape, skewed or symmetrical, indicates the degree of sensitivity of the equipment to the observed sags. Robust equipment for the expected sags has skew histograms, while sensitive equipment has symmetric histograms.

As the histograms in the Fig. 17 show, the number of equipment outages per year can range from 0 to a maximum of 14 in the case of IEC 61000-4-11 Class 2 equipment type connected to Agricola 104 site.

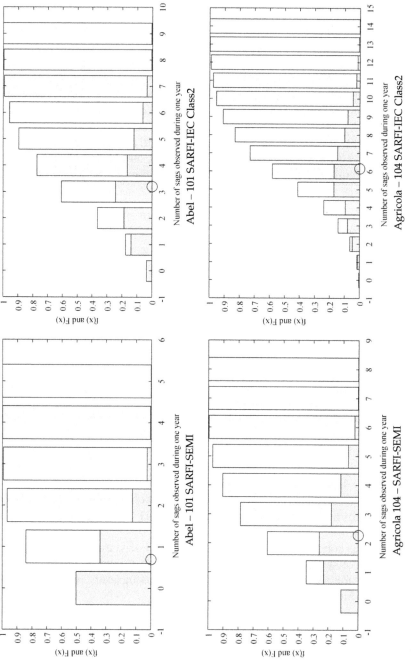

Fig. 17: *SARFI-SEMI and SARFI-IEC Class2 density f(n) and cumulative F(n) distribution functions. Average values (circle).*

Table 11: Number of equipment outages per year (*SARFI-SEMI*$_{95\%}$ and *SARFI-IE Cclass* $2_{95\%}$).

	SEMI F47-0706	IEC 61000-4-11 Class 2
Abel 101	2	6
Agricola 104	5	10

The computed 95% percentile values are summarized for the two equipment types and sites in Table 11.

Considering SEMI F47-0706 equipment type, in 95% of the simulated years, the number of outages is lower or equal to 2 if connected to Abel 101 site. This number is 5 if connected to Agricola 104 site. In case IEC 61000-4-11 Class 2 equipment type, the number of outages is 6 and 10 when connected to Abel 101 or Agricola 104 respectively.

When comparing equipment IEC 61000-4-11 Class 2 type and the SEMI F47-0706 type, both have equivalent performances, i.e., have practically the same number of outages per year, if connected to Agricola 104 site and Abel 101 site respectively.

The presented example shows the importance of combining the equipment performance, characterized by a voltage-tolerance curve, with the network site characterization, for assessing the number of expected equipment outages per year. In fact, the example illustrates that practically the same number of the outages can be expected for the two equipment types if connected to adequate network sites. Accordingly, the network site and the equipment performance should be chosen together to fulfill certain performance criteria defined by the user.

References

1999. Application Guide for AC High-Voltage Circuit Breakers Rated on a Symmetrical Current Basis, Std. C37.010.

Aucoin, M., J. Zeigler and B.D. Russell. 1985. Feeder protection and monitoring system, part II: stage fault test demonstration. IEEE Trans. Power App. and Syst. PAS-104(6): 2754–2758.

Aung, M.T. and J.V. Milanovic. 2006. Stochastic prediction of voltage sags by considering the probability of the failures of the protection system. IEEE Trans. Power Del. 21(1): 322–329.

Barnard, J. and A. Pahwa. 1993. Determination of the impacts of high impedance faults on protection of power distribution systems using a probabilistic model. Electric Power Systems Research 28: 11–18.

CEER. 2012. 5th CEER Benchmarking Report on the Quality of Electricity Supply 2011, CEER, Brussels.

Chambers, J.M., W.S. Cleveland, B. Kleiner and P.A. Tukey. 1983. Assessing distributional assumptions about data. In Graphical Methods for Data Analysis, Wadsworth International Group, Belmont, California.

CIGRE/CIRED/UIE JWG C4.110. 2010. Technical Brochure 412 - Voltage Dip Immunity of Equipment and Installations.

Conrad, L., K. Little and C. Grigg. 1999. Predicting and preventing problems associated with remote fault-clearing voltage dips. IEEE Trans. on Ind. App. 27(1): 167–171.

Dagenhart, J. 2000. The 40-Ω Ground-fault phenomenon. IEEE Trans. Ind. App. 36(1): 30–32.

de Andrade, V. and E. Sorrentino. 2010. Typical expected values of the fault resistance in power systems. In: IEEE/PES Transmission and Distribution Conference and Exposition: Latin America, Brazil.

dos Santos, A. and M.T. Correia de Barros. 2016. Comparative analysis of busbar protection architectures. IEEE Trans. Power Del. 31(1), DOI 10.1109/TPWRD.2015.2434415.

dos Santos, A., M.T. Correia de Barros and P.F. Correia. 2014. Reliability and availability analysis methodology for power system protection schemes. In: 18th Power Systems Computation Conference, Wroclaw, August 2014.

dos Santos, A. and M.T. Correia de Barros. 2015. Stochastic modeling of power system faults. Electr. Power Syst. Res. 126: 29–37.

dos Santos, A., M.T. Correia de Barros and P.F. Correia. 2015. Transmission line protection systems with aided communication channels – Part I: Performance analysis methodology. Electr. Power Syst. Res. 127: 332–338.

El-Kady, M.A. 1985. Probabilistic short-circuit analysis by Monte Carlo simulations. IEEE Trans. Power App. and Syst. PAS-102(5): 1308–1316.

ERSE. 2013. Energy Services Regulatory Authority [In Portuguese] Relatório da qualidade de serviço do setor elétrico.

European Committee for Electrotechnical Standardization. 2010. Voltage characteristics of electricity supplied by public electricity networks, EN 50160, CENELEC.

European Regulators' Group for Electricity and Gas. 2013. Towards Voltage Quality regulation in Europe - An ERGEG Conclusions Paper, ERGEG, Brussels.

Fonseca, V. and M. Alves. 2002. A dedicated software for voltage sag stochastic estimate. In: 10th Int. Conf. on Harmonics and Quality of Power, Rio de Janeiro, Brazil.

Ford, G.L. and S.S. Gengupta. 1981. Analytical methods for probabilistic short-circuit studies. Electric Power Systems Research 5: 13–20.

Gilkenson, C.L., P.A. Jeanne and J.C. Davenport, Jr. 1937. Power systems faults to ground Part I: Characteristics. Electrical Engineering AIEE, pp. 421–428.

Grigg, C., P. Wong, P. Albrecht et al. The IEEE Reliability Test System-1996. A report prepared by the Reliability Test System Task Force of the Application of Probability Methods Subcommittee. IEEE Trans. Power Syst. 14(3): 1010–1020.

Hanninen, S. and M. Lehtonen. 1998. Characteristics of earth faults in electrical distribution networks with high impedance earthing. Elect. Power. Syst. Research 44: 155–161.

Horowitz, S.H. and A.G. Phadke. 2006. Third zone revisited. IEEE Trans. Power Del. 21(1).

IEEE. 2014. IEEE Guide for Voltage Sag Indices, IEEE standard 1564.

International Electrotechnical Commission. 2001. Short-circuit currents in three-phase a.c. systems – Calculation of currents, IEC 60909-0.

International Electrotechnical Commission. 2004. Testing and measurement techniques – voltage dips, short interruptions and voltage variations immunity tests, IEC 61000-4-11.

International Electrotechnical Commission. 2005. Testing and measurement techniques – voltage dips, short interruptions and voltage variations immunity tests for equipment with input current more than 16 A per phase, IEC 61000-4-34.

International Electrotechnical Commission. 2008. Electromagnetic compatibility (EMC) – Part 4-30: Testing and measurement techniques – Power quality measurement methods, IEC 61000-4-30.

International Electrotechnical Commission. 2010. International Electrotechnical Vocabulary – Part 447: Measuring relays, IEC 60050-446.

Lee, R.E. and M.T. Bishop. 1985. A comparison of measured high impedance fault data to digital computer modeling results. IEEE Trans. Power App. and Syst. PAS-104(10): 2754–2758.

Martinez, J.A. and J. Martin-Arnedo. 2004. Voltage Sag stochastic prediction using an electromagnetic transients program. IEEE Trans. Power Del. 19(4).

Oliveira, T.C., J.M. Carvalho Filho, R.C. Leborgne and M. Bollen. 2009. Voltage sags: validating short-term monitoring by using long-term stochastic simulation. IEEE Trans. Power Del. 24(3).

Olguin, G., M. Aedo and M. Arias. 2005. A Monte Carlo simulation approach to the method of fault positions for stochastic assessment of voltage dips (Sags). In IEEE/ PES Transmission and Distribution Conference & Exhibition: Asia and Pacific, Dalian, China.

Papoulis, A. 1965. Probability, Random Variables, and Stochastic Processes, McGraw-Hill Kogakusha, Ltd. Tokyo, p. 280.

Qader, M.R., M.H.J. Bollen and R.N. Allan. 1999. Stochastic prediction of voltage sags in a large transmission system. IEEE Trans. Ind. Appl. 35(1): 152–162.

Regulatory Ordinance. 2010. Regulamento da rede de transporte, Diário da República, 1.ª série n.º 147, 30 de Julho de 2010. [In Portuguese].

SEMI. 2006. Semiconductor Equipment and Materials International (SEMI), Specification for semiconductor processing equipment voltage sag immunity, SEMI Standard F47-0706.

Shooman, M.L. 1968. Probabilistic Reliability: An Engineering Approch, McGraw-Hill Book Company.

Terzija, V. and H.J. Koglin. 2004. On the modeling of long arc in still air and arc resistance calculation. IEEE Trans. Power Del. 19(3): 1012–1017.

Vaahedi, E., W. Li, T. Chia and H. Dommel. 2000. Large scale probabilistic transient stability assessment using B.C. Hydro's on-Line tool. IEEE Trans. Power Syst. 15(2): 661–667.

Wamundsson, M. and M. Bollen. 2008. Predicting the number of voltage dips as function of both voltage and duration. In 13th Int. Conf. on Harmonics and Quality of Power, Wollongong, Australia.

Zhang, Y. 2008. Techno-economic assessment of voltage sag performance and mitigation. Ph.D. dissertation, Faculty of Engineering and Physical Sciences, The Univ. of Manchester.

CHAPTER 16

Domotics

Matteo Manganelli,[1,*] *Zbigniew Leonowicz*[2] and
Luigi Martirano[1]

1. Introduction

In industrialized countries, people spend most of their time in buildings—in workplaces, at home or in commercial buildings. Technological efforts have always been directed at improving comfort and quality of life within buildings. Now energy saving and environmental sustainability issues have also arisen. In the beginning, domotics had been particularly concerned, on the one hand, with aging or disabled persons and, on the other hand, with a niche of luxury buildings. In recent years, in a changing life style, it is becoming popular, due to the users' demand for comfort and energy saving.

Buildings account for a significant share of energy consumption. According to the most recently available data, reported by the European Commission in *EU Energy in Figures—Statistical Pocketbook 2016,* in the EU, 38.1% of all consumed energy is used in buildings (domestic or services). Analogously, as reported by the US EIA, in the USA, approximately 40% of total energy was consumed by the residential and commercial sectors

[1] Sapienza University of Rome, Department of Astronautical, Electrical and Energy Engineering, Via delle Sette Sale 12 B, 00184, Rome, Italy.
[2] Wrocław University of Science and Technology, Faculty of Electrical Engineering, Skr. poczt. 319, 50-950, Wrocław, Poland.
* Corresponding author: matteo.manganelli@uniroma1.it

(US EIA 2017); US buildings today consume 72% of electricity, as stated by the Lawrence Berkeley National Laboratory (2009). When most existing buildings were constructed, there was a much lesser concern about energy efficiency than there is today: Consequently, a huge amount of energy is used for heating, cooling and lighting. Designing new buildings and retrofitting the existing ones to make them as energy efficient as possible is a significant contribution to preserve energy resources and reduce environmental impact.

With the advancements in building technologies along with information and communication technologies, domotics has found its way to a rapid development and diffusion in a large market.

2. Historical Overview

Building automation is not a recent idea: The concept has been around for a long time and it just required the appropriate technologies to be implemented.

In 1883, Warren S. Johnson (U.S. college professor and founder of Johnson Controls) developed a thermostat (which he called an "electric tele-thermoscope") (Johnson 1883). Later, in 1895, he developed a system for automated multi-zone temperature control (Johnson 1885). In 1898, Nikola Tesla developed a wireless remote control (Tesla 1898), which he demonstrated by controlling a model boat at the Electrical Exposition in Madison Square Garden, New York City. In 1891, William Penn Powers (founder of Powers Regulator Company) developed an automated temperature regulator, which was later installed in a church and exhibited at the Chicago World Fair (Hevac Heritage 2017).

In 1907, a hotel in Chicago was the first to implement automated air conditioning in a large building. In those same years, the first skyscrapers had to deal with energy management and control (Domoticsweb.it 2015). In the beginning of the 20th century, the non-residential control industry rapidly evolved. The first control means and logics were pneumatic, i.e., operated via compressed air (up to the 1970's) (Hardcourt Brown and Carey 2015). In the 1930's, the World Fair presented the concepts of automated home and smart appliances, that fascinated the viewers (My Alarm Center 2015).

In the first two decades of the 20th century, appliances appeared in the homes, leading to the decline of full-time domestic workers and time-consuming housework.

The building-block of today's building automation came with microelectronics in the 1950's (My Alarm Center 2015).

In 1965, Westinghouse engineer James (Jim) Sutherland built the *Electronic Computing Home Operator* (ECHO IV), which automated many household chores previously performed by Mrs. Sutherland (control temperature and appliances, compute shopping lists, etc.). One of the most exciting features was its ability to act as a family message center (Infield 1968, Spicer 2000). In 1969, Neiman Marcus started advertising the Honeywell *Kitchen Computer* (a Honeywell 316 minicomputer in a fancy packaging). Purchasing and programming were prohibitive for the average user, no sales were reported and it was ultimately considered a fantasy product; nevertheless, it is regarded as the first time that a computer was presented as a consumer product (Spicer 2000).

In 1989, the *Maison Domotique Panorama* was built in Lyon.

In the 1990's, the term *gerontechnology* appeared (from *gerontology* and *technology*), describing technologies to facilitate the daily life of elderly persons.

3. What is Domotics?

Although the term *domotics* was coined to address automation in residential buildings (home automation), it is nowadays widely used for buildings in general and the term is overlapping with building automation. Thus, in this chapter, the terms will be used interchangeably. In addition, the wordings *smart home* and *smart buildings* appeared recently. According to interpretations, the name *domotics* comes from: Latin *domus* (home) and *robotics*, or from French *automatique* or *informatique*, via the French term *domotique* (coined in 1988 in the *Premiere Conference Europeenne sur l'Habitat Intelligent* in Paris). It is the use of information and communication technology, electrical engineering and electronics, to make a home or a building "smart".

A comparison of building automation and home automation characteristics is reported in Table 1.

Table 1: Comparison of building and home automation.

	Building automation	Home automation
Decision maker	Company	Inhabitants
User	Employees	Inhabitants
Manager	Building manager	Inhabitants
Extent	Building	House, apartment
Use of premises	Dynamic	Static
Motivation	Comfort, energy saving, management, productivity, safety, security	Accessibility, entertainment, quality of life, safety, security, status symbol

4. What is Domotics for?

The scope of building automation is: Energy saving and load management; users' comfort; quality of life for aging or disabled persons.

4.1 *Energy Saving and Load Management*

One aim of building automation is the reduction of the energy consumption of buildings and the manipulation of the load profile. Energy sustainability measures aim at reducing the amount of energy consumed, in order to reduce: (i) The depletion of exhaustible energy sources; (ii) The environmental impact of energy processes; (iii) The users' energy expenditures. Load management measures aim at attaining a more convenient load profile and can provide additional services with respect to energy saving (e.g., containing peak load demand). This, in turn, can be beneficial for energy sustainability itself. As an example, shifting a load can be beneficial for the power supply (reducing peak load and achieving a profile more compatible with power generation) and for energy sustainability (using different energy sources and decreasing the energy expenditures). Building automation systems can provide a significant contribution. The EN 15232 standard provides methods for the assessment of the impact of building automation on the energy performance of buildings.

The *efficient use of energy*, or *energy efficiency* in the broad sense, is the ensemble of means and actions to reduce the amount of energy used for given outcomes. These measures can be divided in *energy efficiency* measures (in a strict sense) and *energy efficacy* (or *effectiveness*) (Martirano 2011b).

The energy consumed by technical system, W, can be expressed as:

$$W = \int_0^T P(t)\, \mathrm{d}t \tag{1}$$

where $P(t)$ is the power consumption over time and T the operation time. Without control systems, (1) can be expressed as:

$$W = P_N \cdot T \tag{2}$$

where P_N is the installed power. With control system, (1) can be expressed as:

$$W \approx k_P P_N \cdot k_T T \tag{3}$$

where k_P and k_T are corrective factors that express the impact of control and management of power consumption and operation time. They are equal to 1 (no impact) if no control is used, between 0 and 1 (reduction) if a control system is used. They can also be greater than 1 (increase), depending on how the management is performed.

Energy efficiency measures (also called *passive* measures) aim at reducing the installed power P_N to achieve a given outcome, via the installation of high-efficiency components and the accurate design of systems to avoid unnecessary power demand. Passive measures in buildings usually involve building envelope, technical systems, components.

For instance, passive measures in a lighting system are the use of lighting fixture with higher lighting efficiency or the accurate choice and positioning of luminaires to avoid over-illumination.

Energy efficacy measures (also called *active* measures) aim at the reduction of the power demand over time $P(t)$ and the time of use T (expressed by $k_p P_N$ and $k_T T$). This is achieved via the implementation of control systems, regulating or turning off components as necessary. Active measures usually involve: (i) Control systems and building automation systems; (ii) Building management and maintenance. The definitions of *building automation control systems, building energy management* systems and *technical building management* apply, as defined in the nomenclature. In addition, it is possible to implement *demand side management* actions (e.g. control of appliances) and exploit *on-site generation* systems (e.g., PV plants).

For instance, active measures in a lighting system are the regulation of light output (*dimming*) to match a given illuminance value or the switching of luminaire groups (*control groups*) according to the occupancy

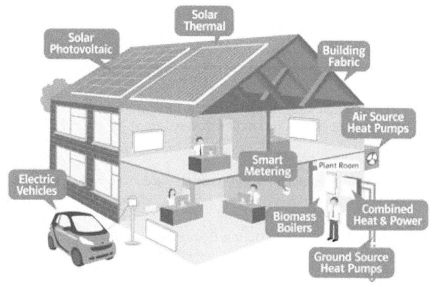

Fig. 1: Concept of smart building (E.On 2017).

in the premises. The control of technical systems is the core of building automation, which can greatly contribute to energy efficacy measures.

4.2 Users' Comfort

Another aim of building automation is the enhancement of users' comfort and the improvement of features in buildings. Ambient comfort usually involves temperature and humidity (thermohygrometric comfort), noise (acoustic comfort), illumination (visual comfort), indoor air quality, etc.

Thermohygrometric comfort and indoor air quality are also the scope of HVAC (heating, ventilation and air conditioning) systems. Control of temperature and humidity is already implemented on its own via HVAC systems. Indoor air quality is usually achieved, in large buildings, via a preset air circulation. Improvement opportunities given by building automation systems are:

- The automated control of ventilation, according to occupancy or a control variable (e.g., CO_2 concentration), see, e.g., Nielsen et al. (2011);
- The automated control of lighting systems (as illustrated in Section 7.1).
- The automated control of shading systems, for visual comfort (preventing glare) and psychophysical wellness (daylighting) (Section 7.2).

It is worth noting that energy efficiency measures and comfort measures can be synergic. For emxample, the adjustment of a lighting system both reduces energy consumption for lighting and provides visual comfort to the occupants.

4.3 Quality of Life of Aging or Disabled Persons

Building automation can provide useful elements to improve the quality of life of persons with special needs, e.g., aging persons or persons with disabilities. Thanks to the features of automation and communication systems, users with special needs can have greater autonomy and independence or overcome physical obstacles in usual technical systems. As an example, persons with motion impairments can control lighting and HVAC via a portable device and/or an automated system tailored to their needs, eliminating the obstacles of traditional command devices and a reduced reliance on assistance.

5. Concept and Key Elements

5.1 Architecture

The criteria behind a building automation system are to separate the communication system from the power system, and to have a common communication system. In this way, control is no longer performed via the power system, which considerably simplifies the installation. In addition, a common communication system allows for distributed command points and devices, making the system considerably flexible. Lastly, it is possible to programme devices to vary the functions according to needs. A schematic is reported in Fig. 2.

In the implementation of the building automation system, different media and protocols can coexist. Thus, in the overall system, it is possible to take advantage of the different characteristics of media (e.g., wired or wireless according to ease of installation) and protocols (e.g., a dedicated protocol for the lighting subsystem). This requires the use of devices (e.g., *gateways*) to interconnect different media and protocols, but makes the system flexible and suitable to the given installation. A schematic is reported in Fig. 3.

A key element of building automation system is the digitalization of commands. Command actions and inputs are translated into digital signals which are easy to elaborate. Thus, it is possible to consider the system and the interface separately. The system, e.g., a home and building electronic system (HBES), can be complex. The interface between users and the system, on the other hand, can be simple: Whichever device can be used, as also traditional devices, provided that inputs are then digitalized (Fig. 4).

The typical architecture of a building automation system is reported in Fig. 5. In addition to the core automation system, which processes inputs and outputs, there can be a supervisory subsystem or an integration system, which may communicate, e.g., with the Internet or with another BUS system.

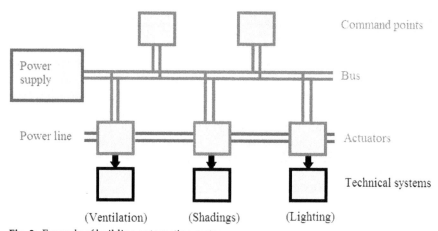

Fig. 2: Example of building automation system.

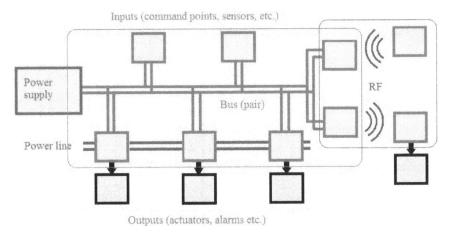

Fig. 3: Example of different media in a building automation system.

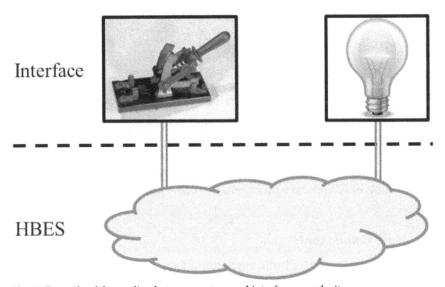

Fig. 4: Example of decoupling between system and interface complexity.

5.2 Functions

Functions performed by a building automation system can be divided into *elementary functions* or *complex functions*. These are also known as *first-level functions* and *second-level functions*, respectively. An elementary function can be the control of a single device, e.g., switching a lamp. A complex function, on the other hand, can be the control of a subsystem according to defined criteria, e.g., the control of the lighting system as a whole, according to a preset lighting scenario or an energy saving strategy. Elementary functions are usually performed via *distributed intelligence*,

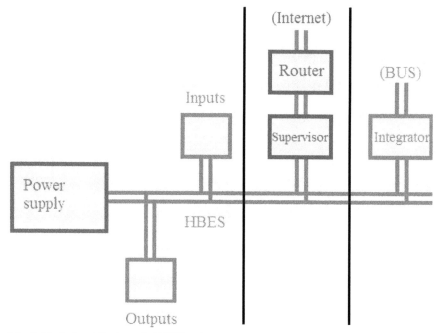

Fig. 5: Typical architecture of a HBES system.

i.e., a logic stored in single devices. On the other hand, complex functions are usually performed by central devices (e.g., a supervisory device). In general, the functions performed by a building automation system are a combination of elementary and complex function and not merely an aggregation of elementary functions. This also affects the reliability of the system.

5.3 Main Components

The main components of building automation systems are:

- *Router*: BUS device, connecting at least two subnetworks, circumscribing communications within the respective subnetwork and allowing communications between subnetworks according to defined criteria;
- *Bridge*: router with no filtering algorithm;
- *Signal repeater*;
- *Coupler*: device operating as a router, bridge or repeater;
- *Gateway*: device interconnecting different media or protocols;
- *Actuator*: device that receives a communication and executes defined actions (e.g., switch, dimmer, drive);

- *Sensor*: device sensing a physical quantity and communicate it to another device or into the network for further elaborations and actions (e.g., command button, occupancy detector, thermostat, light sensor).

5.4 *Communication Media and Protocols*

Media commonly used in building automation systems can be, e.g., Coaxial cable (CX); Infrared (IR); Fiber Optic (FO); Powerline communication (PLC); Radiofrequencies (RF); Twisted pair (TP). Communication protocols commonly used in building automation systems can be dedicated building automation protocols or, e.g., Ethernet, Bluetooth, WiFi, etc.

5.5 *Integration Levels*

The level of the integration among devices and system range from single non-interacting devices to the total integration into one seamless building automation system. The fundamental integration levels are reported in Fig. 6.

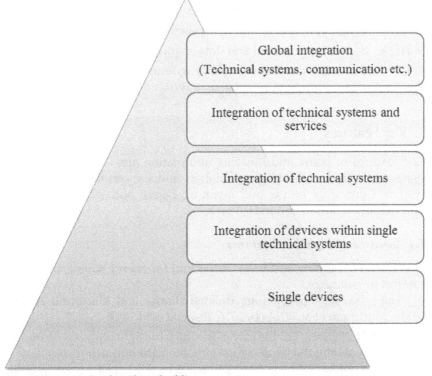

Fig. 6: Integration levels within a building.

6. Nomenclature

In the field of building automation and smart buildings, the main definitions are (Control Solutions Inc. 2015):

BACS	Building automation control system
BAS	Building automation system
BEMS	Building energy management system
BESS	Battery energy storage system
BMS	Building management system
EMCS	Energy management control system
BUS	Binary unit system (from Latin *omnibus* "for all")
CIB	Computer Integrated Building
ESS	Energy storage system
HBA	Home and building automation
HBES	Home and building electronic system
HVAC	Heating, ventilation and air conditioning
RESS	Residential energy storage system
SCADA	Supervisory control and data acquisition
PLC	Powerline communication, programmable logic controller
TBM	Technical building management

7. Key Features

Key features of home and building automation are: Control of lighting systems; Control of heating, ventilation and air conditioning (HVAC) system; Control of blinds and shading; Safety and security systems; Entertainment (audio and video) systems.

7.1 Control of Lighting Systems

The control of lighting systems is important for energy saving and visual comfort in buildings.

The following concepts are defined (Parise and Martirano 2011a,c, 2013a, Martirano et al. 2014a,b, 2016, Bisegna et al. 2016):

- *Control techniques*, i.e., the modes in which the light output is controlled (e.g., switching the lamps or varying the light output);
- *Control actions*, i.e., the means by which control techniques are implemented (e.g., switching the power supply or via an auxiliary device);

- *Control modes*, i.e., the modes in which control actions are operated (e.g., manually or automatically);
- *Control strategies*, i.e., the criteria based on which the control system operates (e.g., scheduling, occupancy or daylight).

7.1.1 Control Techniques

Control techniques are the modes in which the light output is controlled, i.e., *switching* or *dimming*. In switching control, lamps are turned on and off; in dimming control, the luminous flux emitted by the lamps is varied.

In switching control, the system can be arranged so that switching can be performed according to: single lamps (*switching in the lamp*); groups of lamps (*switching groups of lamps*); single luminaires (*switching in the luminaire*); groups of luminaires (*switching group of luminaires*). The more subdivided is the switching system (e.g., single lamps), the more control steps in the resulting illumination.

In either technique, luminaires can be arranged in groups, to allow for zone-based control strategies, as it will be illustrated in the following.

7.1.2 Control Actions

Control actions are the means by which control techniques are implemented. With traditional wiring, control operates via the switching of the supply circuit, or via an auxiliary control circuit (relay); with a building automation system, control operates via an auxiliary distributed BUS system, or via an auxiliary concentrated BUS system (PLC). A schematic of control actions is reported in Fig. 7 (Martirano 2011b).

With traditional wiring or PLC's, the wiring can be complex: Thus, switching group of luminaires is often preferred. With a distributed BUS system, the wiring is simplified and switching in the lamp is possible. An example of switching in the lamp is shown in Fig. 8 (Martirano 2011b), with traditional wiring (left panel) and distributed BUS system (right panel), for comparison.

7.1.3 Control Modes

Control modes are the modes in which control actions are operated:

- *Manual*: actions are triggered manually by the user;
- *Automated, stand-alone*: actions are triggered automatically by a local device which does not communicate with the rest of the control system;
- *Automated, centralized*: actions are triggered automatically by a central system; this can be either *concentrated intelligence* or *distributed intelligence*.

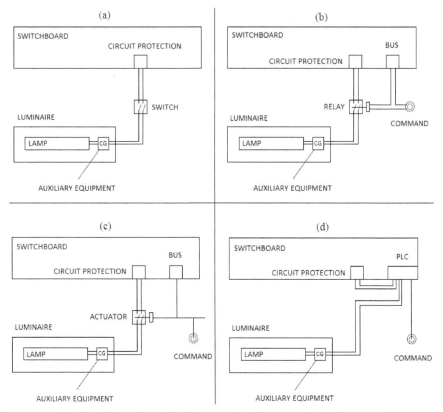

Fig. 7: Control actions: (a) traditional switch; (b) relay; (c) BUS actuator; (d) PLC.

7.1.4 Control Strategies

Control strategies are the criteria basing on which the control system operates:

- *Daylighting*: Integrating the contribution of daylight;
- *Integrating*: Controlling the light output to track a constant illuminance value;
- *Metering*: Measuring lighting energy consumption by areas;
- *Occupancy*: Controlling the lighting system basing it on occupancy in a room (*room occupancy*) or in a working area (*area occupancy*);
- *Remoting*: Remote control of the lighting system;
- *Scheduling*: Controlling the lighting system basing on a schedule, via a clock;
- *Zoning*: Arranging the lighting system according to areas, taking account of intended use and daylight penetration.

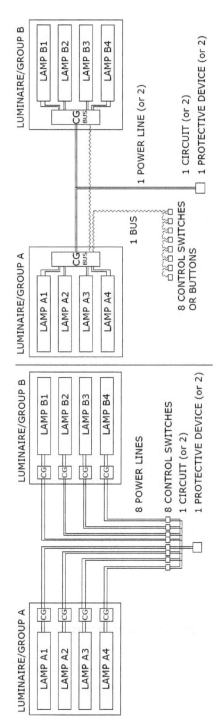

Fig. 8: Switching with traditional wiring (left) and with BUS system (right).

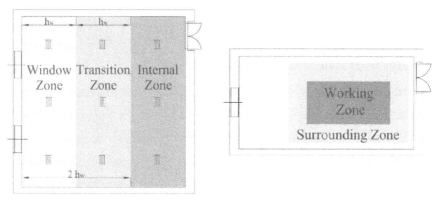

Fig. 9: Daylight-based (left) and task-based (right) zoning.

Methods for the evaluation of daylight penetration are available in literature (Martirano 2011a,b, Martirano et al. 2013, 2014a, 2016, Parise et al. 2009, 2011a,b,c, 2013a,b, 2014b, 2016).

An example of arrangements of luminaires are reported in Fig. 9, that take account of daylight penetration into a room (panel a) and working areas (panel b), to integrate available daylight and to match visual requirements of the area, respectively (Manganelli and Consalvi 2015).

7.1.5 Efficiency Indicators

Indicators are defined, to quantify the efficiency of a lighting system, as also taking account of control system:

- *Luminous efficiency of the lamp*
- *Luminous efficiency of the lamp and auxiliary equipment assembly*
- *Luminous efficiency of the luminaire (assembly of lamp, auxiliary equipment and optics)*
- *Luminous efficiency of the luminaire over time*
- *Efficiency of the lighting system (limit, effective and relative values)*
- *Building automation control factors*
- *Equivalent hours*
- *Lighting energy numerical indicator (LENI)*

Methods are available in literature for the calculation of indicators (Martirano and Di Ponio 2012, Parise and Martirano 2009, 2011b, 2013b, Parise et al. 2012, 2013 a,b, 2014a,b, 2016).

7.2 *Control of Shading Systems*

An increasing attention is being paid to shading systems. The building envelop is the interface between outdoor and indoor conditions. It is therefore a key element in maintaining comfortable indoor conditions while avoiding excessive energy consumption. In particular, window and shading systems channel solar energy (which causes daylight and heat gains). An accurate design of these elements is therefore essential for thermal and visual comfort. In this context, dynamic shading systems are gaining increasing attention, as they are a means to control the contribution of the variable solar energy (Nielsen 2011).

8. Advantages and Disadvantages

The advantages and disadvantages of building automation systems with respect to traditional systems can be expressed as reported in Table 2 (Scarpino 2010).

Table 2: Advantages and disadvantages of building automation systems with respect to traditional systems.

	Traditional system	Building automation system
Advantages	• Lower cost of devices • Well known to professionals and customers	• Flexibility • A single device can perform different functions • Much simpler wiring • Lower risk of direct contacts • Lower fire risk • Suitable for new buildings or retrofitting of historical buildings (due to limited wiring) • Lower cost of variations • Remote control • Efficient use of energy • Enhanced comfort for users • Accessibility for aging and disabled persons
Disadvantages	• Poor flexibility • More devices required • Higher costs for materials, wires, labour • More fire protection required • Higher risk of direct contact • Higher cost of variations • Poor remote control	• More expensive devices • Not well known to professionals and customers

9. Classification

Building automation systems can be classified based on the *functional level*, as per Table 3.

BAS's complying with standards EN 50090 are defined HBES's. Usually, they concern automation and control of electric systems in buildings. On the other hand, BAS's complying with standards EN ISO 16484 by CEN Technical Committee TC 247 are defined BACS's. Usually they concern automation and control of thermal (HVAC) systems. A unification process of the two series has been started jointly by the two Committees, in order to produce the standards CEN/CENELEC EN 50491 *General requirements for Home and Building Electronic Systems (HBES) and Building Automation and Control Systems (BACS)*, harmonizing the two types of systems. Standards EN 50090 classifies HBES's based on their features (Table 4).

Additionally, HBES's can be classified based on the concerned subsystem and its type (Table 5).

Table 3: Functional levels of building automation systems.

Level	Type of solutions	Type of products and technologies
0	Isolated	Stand-alone, non-interconnectible
1	Base	Interconnectible
2	Integrated	Interoperable

Table 4: Classification of HBES's as per standard EN 50090.

Class	Features (in addition to those of the previous class)	Example
1	Remote control features	Control, monitoring, measurement, alarm, low-speed data transfer
2	Commuted transfer of voice or other info with comparable bandwidth	Phone, fax
3	High-quality audio/video commuted transfer High-speed data transfer	Data network

10. Protocols

Building automation protocols can be *construction* protocols, concerning the manufacturing of devices, and *communication* protocols, concerning the communication among devices within a system. Protocols can be

Table 5: HBES's classification and main reference standards.

Subsystem	Type of system	HBES class	Main reference standards
Automation	Remote control	1	EN 50090, EN ISO 16484
Safety/security	Technical alarms	1	-
	Anti-intrusion	1	EN 50131
	Fire protection	1	EN 54
Audio/video	Sound system	2	EN 60728–11
	TV	3	EN 60728–11
	Audio/video	3	EN 60728–11
Telecommunications	Phone	2	EN 50173-3, EN 50174-2
	Data	3	EN 50310, EN 50346

proprietary of a manufacturer or open. The most common building automation protocols are illustrated in the following.

10.1 LonWorks

LonWorks (local operating network) protocol is based on a communication protocol developed by Echelon Corporation, once known as *LonTalk*, accepted as an ANSI standard in 1999 and as an international ISO/IEC standard in 2008. It is widely used in Europe and in the world, for industry, transportation, public lighting, remote metering and fire protection. Approximately 65% of building automation systems in Europe are based on this protocol. It is a comprehensive platform, open and independent from the media.

Each node can communicate with the other node and has a dedicated supply. The data rate is 2000–10000 baud. The media are pair, powerline (PLC), coaxial, fiber, wireless. As an example, the remote metering system of Enel (the main Italian DSO) is implemented in this protocol, via PLC, accounting for approximately 27 million nodes.

10.2 X10

X10 is a historical standard. It has been present for two decades in America and it has also had a widespread acceptance in Europe. It was developed in 1975 by Pico Electronics for the control of appliances. It is made up of a central unit that controls remote devices via PLC. The central unit can be controlled by a personal computer or via an infrared remote control. The data rate is 50 baud and the maximum number of addresses is 256 (it is suitable only for home automation). The media are PLC or radio frequencies (900 MHz, as GSM in Europe).

10.3 Batibus BCI

Developed on 1989 by Merlin Gerin, Airelec, Edf e Landis & Gyr (founders of Batibus Club International), it was later involved in the "convergence" process with EIB and EHS, merging into the KNX protocol. The topology is free, the data rate is 4800 baud, the media are twisted pair, infrared, radio frequency, and PLC.

10.4 EIB

EIB (European Installation Bus) protocol was developed by a pool of manufacturers of electric components, for the automation of electric system in buildings. It was later involved in the "convergence" (together with BatiBus and EHS) process into the KNX protocol.

It is an open standard. The topology is free, the data rate is 9600 baud, the maximum number of devices is 61455; the media are pair, PLC, Ethernet, radiofrequency, and infrared.

10.5 KNX

KNX was born following the foundation of the Konnex association in 1999, by EIBA, BCI and EHSA associations, with the purpose of implementing and promoting a unique standard for home and building automation.

It is based on the best features of the three starting protocols (EIB, BatiBus and EHS) on the EIB core, which is fully compatible with; and in addition includes BatiBus and EHS media and configuration modes. Devices are manufactured by different companies and certified as interoperable following a process by Konnex association. KNX has been also been conceived as a quality label.

The topology is free, the data rate is 9600 baud and the maximum number of devices is 61455. The media are twisted pair, PLC (110 and 132 kHz), radio frequency (868 MHz). It is the reference standard for CENELEC TC 205–CEI 83–EN 50090 standards.

10.6 ZigBee

ZigBee is a specification for a suite of communication protocols for the implementation of *Wireless Personal Area Networks* (*WPAN*) via low-power radio frequency transceiver. The specification has the purpose of being more simple and economic with respect to, e.g., Bluetooth or WiFi. A ZigBee device can be *ZigBee Coordinator* (*ZC*), *ZigBee Router* (*ZR*), *ZigBee End Device* (*ZED*). It is associated to IEEE standard 802.15.4. Development started in 1998 and it was accepted as a standard in 2003.

10.7 Other Protocols

Other noteworthy protocols are, e.g., *Modbus* (a common protocol for the remote control of electric systems), *Mbus* (a common protocol for metering), *Bacnet* (a common protocol for the control of HVAC systems; it is an ASHRAE, ANSI and ISO standard).

References

Bisegna, F., C. Burattini, M. Manganelli, L. Martirano, B. Mattoni and L. Parise. 2016. Adaptive control for lighting, shading and HVAC systems in near zero energy buildings. 2016 IEEE 16th International Conference on Environment and Electrical Engineering (EEEIC). Florence, 2016.

Control Solutions Inc. 2015. The Ultimate Guide to Building Automation. [Online] Available: http://controlyourbuilding.com/blog/entry/the-ultimate-guide-to-building-automation. Accessed: 10-7-2017.

Domoticsweb.it .2015. Storia della domotica [History of domotics]. [Online] Available: http://www.domoticsweb.it/index.php/domotica/2015-10-04-13-00-33/10-domotica/3-storia-della-domotica. Accessed: 10-7-2017. [In Italian].

E.On .2017. On Site Generation: Learn more about our onsite renewable energy generation technologies. [Online] Available: https://www.eonenergy.com/for-your-business/large-energy-users/manage-energy/on-site-generation. Accessed: 10-7-2017.

Hardcourt Brown and Carey. 2015. A brief history of building automation and controls. [Online] Available: http://www.harcourtbrown.com/a-brief-history-of-building-automation-and-controls Accessed: 10-7-2017.

Hevac Heritage. 2017. William Penn Powers. [Online] Available: http://www.hevac-heritage.org/built_environment/pioneers/37-Powers-Rayleigh.pdf. Accessed 10-7-2017.

Infield, G. 1968. A computer in the basement? Popular Mechanics April 1968: 77–79, 209, 229.

Johnson, W. 1883. United States Patent #281,884.

Johnson, W. 1885. Heat regulating apparatus. United States Patent #542,733.

Lawrence Berkeley National Laboratory. 2009. Working Toward the Very Low Energy Consumption Building of the Future [Online] Available: http://newscenter.lbl.gov/2009/06/02/working-toward-the-very-low-energy-consumption-building-of-the-future/ Accessed 10–7-2017.

Manganelli, M. and R. Consalvi. 2015. Design and energy performance assessment of high-efficiency lighting systems. IEEE 15th International Conference on Environment and Electrical Engineering (EEEIC), Rome, pp. 1035–1040.

Martirano, L. 2011a. A smart lighting control to save energy, Proceedings of the 6th IEEE International Conference on Intelligent Data Acquisition and Advanced Computing Systems. Prague, 2011.

Martirano, L. 2011b. Il controllo dei sistemi di illuminazione per il risparmio energetico [Control of lighting systems for energy saving]. [In Italian].

Martirano, L. and S. Di Ponio. 2012. Procedure to evaluate indoor lighting energy performance, 2012 11th International Conference on Environment and Electrical Engineering. Venice, 2012.

Martirano, L., M. Manganelli, L. Parise and D.A. Sbordone. 2014a. Design of a fuzzy-based control system for energy saving and users comfort. 2014 14th International Conference on Environment and Electrical Engineering, Krakow, 2014, pp. 142–147.

Martirano, L., G. Parise, L. Parise and M. Manganelli. 2014b. Simulation and sensitivity analysis of a fuzzy-based building automation control system. 2014 IEEE Industry Application Society Annual Meeting, Vancouver, BC, 2014.

Martirano, L., G. Parise, L. Parise and M. Manganelli. 2016. A fuzzy-based building automation control system: optimizing the level of energy performance and comfort in an office space by taking advantage of building automation systems and solar energy. IEEE Industry Applications Magazine 22: 10–17.

My Alarm Center. 2015. The history of home automation. [Online] Available: https://myalarmcenter.com/blog/the-history-of-home-automation/. Accessed: 10-7-2017.

Nielsen, M.V., S. Svendsen and L.B. Jensen. 2011. Quantifying the potential of automated dynamic solar shading in office buildings through integrated simulations of energy and daylight. Solar Energy 85: 757–768.

Parise, G. and L. Martirano. 2009. Impact of building automation, controls and building management on energy performance of lighting systems. Conference Record 2009 IEEE Industrial & Commercial Power Systems Technical Conference, Calgary, AB, 2009.

Parise, G. and L. Martirano. 2011a. Ecodesign of lighting systems. IEEE Industry Applications Magazine 17: 14–19.

Parise, G. and L. Martirano. 2011b Daylight impact on energy performance of internal lighting. 2011 IEEE Industry Applications Society Annual Meeting, Orlando, FL, 2011.

Parise, G. and L. Martirano. 2011c. Combined electric light and daylight systems ecodesign. 2011 IEEE Industry Applications Society Annual Meeting, Orlando, FL, 2011.

Parise, G., L. Martirano and S. Di Ponio. 2012. Energy performance of interior lighting systems. 2012 IEEE Industry Applications Society Annual Meeting, Las Vegas, NV, 2012.

Parise, G., L. Martirano and S. Di Ponio. 2013a. Energy performance of interior lighting systems. IEEE Transactions on Industry Applications 49: 2793–2801.

Parise, G., L. Martirano and G. Cecchini. 2013b. Design and energetic analysis of an advanced control upgrading existing lighting systems. 49th IEEE/IAS Industrial & Commercial Power Systems Technical Conference, Stone Mountain, GA, 2013.

Parise, G. and L. Martirano, 2013a. Combined electric light and daylight systems ecodesign. IEEE Transactions on Industry Applications 49: 1062–1070.

Parise, G. and L. Martirano. 2013b. Daylight impact on energy performance of internal lighting. In IEEE Transactions on Industry Applications 49: 242–249.

Parise, G., L. Martirano and G. Cecchini. 2014a. Design and energetic analysis of an advanced control upgrading existing lighting systems. IEEE Transactions on Industry Applications 50: 1338–1347.

Parise, G., L. Martirano and L. Parise. 2014b. Energy performance of buildings: An useful procedure to estimate the impact of the lighting control systems. 2014 IEEE/IAS 50th Industrial & Commercial Power Systems Technical Conference, Fort Worth, TX, 2014.

Parise, G., L. Martirano and L. Parise. 2016. A procedure to estimate the energy requirements for lighting. IEEE Transactions on Industry Applications 52: 34–41.

Scarpino, P.A. 2010. I Sistemi di Home e Building Automation [Home and Building Automation Systems]. Convegno GEWISS [GEWISS Convention]. Florence, 28 April 2010. [In Italian].

Spicer, D. 2000. If You Can't Stand the Coding, Stay Out of the Kitchen: Three Chapters in the History of Home Automation. [Online] Available: http://www.drdobbs.com/architecture-and-design/if-you-cant-stand-the-coding-stay-out-of/184404040. Accessed: 10-7-2017.

Tesla, N. 1898. Method of and apparatus for controlling mechanism of moving vessels or vehicles. United States Patent #613809.

US EIA. 2017. Frequently Asked Questions [Online] Available: https://www.eia.gov/tools/faqs/faq.php?id=86&t=1 Accessed 10-7-2017.

CHAPTER 17

Demand Response

Matti Lehtonen

1. Introduction

Demand Response (DR) is usually referred to as timely shifting of electrical load from high demand and expensive power price hours to lower demand and more affordable hours, or alternatively momentary changing of load in order to maintain the balance of generation and demand in a power system. The shifting can either be postponing or preponing of the demand, and in most cases the total energy usage mostly stays unchanged. Load curtailment is a special case of DR, where load is temporarily disconnected without any pay-back period, i.e., the function of the energy use is totally lost.

The Federal Energy Regulatory Commission (FERC) defines DR in a more general way as "Changes in electric usage by demand-side resources from their normal consumption patterns in response to changes in the price of electricity over time, or to incentive payments designed to induce lower electricity use at times of high wholesale market prices or when system reliability is jeopardized."

Demand Response has come to the focus in power systems due to the increased share of Variable Renewable Energy Sources (VRES), like wind and photovoltaic power. These are intermittent and forecasting of their generated power is prone to errors, especially in case of longer time spans. Huge and unexpected variation in VRES generation is a challenge to the power systems, and calls for new flexibility also from the load side, thus leading to the need of DR.

Aalto University, School of Electrical Engineering, PO Box 15500, FI-00076 Aalto, Maarintie 8, Espoo, Finland.
Email: matti.lehtonen@aalto.fi

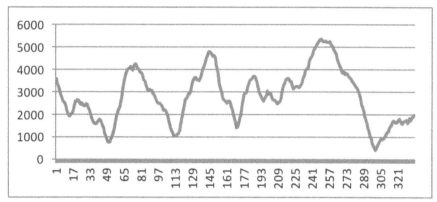

Fig. 1: Wind power production in Sweden (MW) during first two weeks in 2017 (source: Nordpool).

Figure 1 presents total wind power production in Sweden during first two weeks in 2017. When the share of wind power is increased, several new aspects emerge in power system balance management. First, steep generation ramps, upwards or downwards, make maintaining the instantaneous balance of a power system more challenging. The situation is made worse by the fact, that replacing conventional generation by wind turbines also reduces the inertia of the power system. Hence, the power system frequency becomes more sensitive and reacts more strongly to generation and demand power mismatch. Secondly, since wind production forecasting is prone to errors, the energy retailers having wind power in their generation portfolio tend occasionally to have large imbalances of their hourly traded power, which they have to buy from the balancing power market. Balancing power market is maintained (in Nordic Countries) by Transmission System Operators, and the power price in this market often is considerably higher than the regular hourly spot price in the power market.[2]

The third issue is the need of a power sink. If the VRES production is larger than the electrical load, there are two alternative ways to react: either to increase the demand (power sink), or to curtail the VRES production. When the share of VRES is increased in the power systems, the need of power sink becomes more and more critical. However, DR can only be used for shifting loads, thus leveling short term variation of the difference between VRES generation and demand, but it is not able to provide any longer term storages, not to speak about seasonal leveling of supply and

[2] http://www.fingrid.fi/en/electricity-market/imbalance-power/price-of-imbalance-power.

Table 1: Energy and power balancing in the Nordic power markets.

Physical market power transactions		Specific operating hour	Balancing in real time
ELSPOT	ELBAS	REGULATION POWER MARKET	Primary Reserves 1 s–3 min
12–36 hr	1–32 hr	15 min–1 hr	Real time
Day Ahead	Intra Day	Intra Hour	Secondary Reserves 15 min
Also Bilateral Transactions & Demand Response		Hourly Balance Management & Demand Response	Releasing Primary Reserves

demand. Hence, DR can extend into some degree the amount of VRES integrated into the power systems, but beyond a certain limit other options are needed, like conversion of power to gas, or perhaps electricity storages implemented as pumped hydro generation plants. At this point it is necessary to remind, that conventionally the best way of balancing the seasonal variations of VRES has been the hydro power generation with storages. In Nordic Countries, hydro generation covers about 50% of the total power generation capacity (Elovaara and Haarla 2011). However, when the Nordic power system is connected with stronger tie lines Central Europe, this hydro capacity is used to a higher degree for leveling the VRES generation also in Germany and other Central European countries. Hence, it is expected that the need of DR as a power sink is emerging also in the Nordic Countries in the future.

Before going to the details of DR, it is good to have a quick look at the operation of power market. The main properties of a power market (Nordic power market here as a case) are illustrated in Table 1. The generation scheduling and energy procurement starts at the previous day before the day of delivery. By noon of this day, the traders make bids of supply and demand, which are then summed into ascending and descending order of price, respectively. The point where these two summation curves intersect, defines the power price for all the deliveries at that specific hour. As a result of this trading is the fixed power price for each of the next day 24 hours, the Elspot price. Closer to the operation hour, the power balance can then be complemented by the Elbas market. Here the prices are set based on pay-as-bid basis for all transactions. Hence, in this intraday market the prices vary continuously. The Elbas market closes one hour prior to the hour of delivery. Those imbalances which have not been settled by Elbas trade are then subjected to the prices in balancing power (regulation power) market. Here the Transmission System Operator (TSO), in order to maintain the gross power system balance, receives bids of upregulation and downregulation, and the eventual price of these then defines the costs of the balance deviations of individual traders. This balancing power market then closes the hourly power market and the energy procurement of energy traders.

In order to maintain the power system security, the TSO also maintains markets for reserves. The primary reserves are frequency controlled and they should be activated in seconds. In case of disturbances, the secondary reserves are activated in 15 min to release the primary reserves for the next possible event.

DR can be utilized in all of these markets, either as a part of energy procurement in connection with Elspot and Elbas markets, for providing upregulation and downregulation in regulation power market, or providing reserves for real time balancing of the power system. In the following sections, the use of DR is first discussed for energy procurement and then for power system real time balance management. Hence, the focus in this chapter is in DR, in maintaining energy balance (procurement) or power balance (frequency controlled reserves).

2. Flexibility of Various Loads

Flexibility of demand can be found in all the main load categories. In the *industrial sector* the single loads are typically very large, and the necessary infrastructures for load control already exist. Also, industrial enterprises often already participate actively in power markets and they do have the necessary expertise, as well as communication and control systems required for advanced energy market operations. The possibilities of participating in the demand response however strongly depend on the nature of the industrial process. If the process allows timely altering of the energy use without causing losses in the planned production output, then DR could be feasible. A good example of industrial energy use allowing fast and accurate adjustments of demand is the aluminum smelter (Zhang and Hug 2014).

Another class of large customers having DR potential is *commercial buildings*. The possible flexible loads in this category are cold appliances, cold storages and heating, ventilation and air conditioning (HVAC) loads. Into a lesser degree also illumination may be controlled based on DR needs. In a study made in Florida USA by (Hao et al. 2014) the thermal storage potential of a commercial building and its capacity for DR was examined with the following results. It was found that about 15% of the rated power of HVAC load can be utilized for DR without having adverse effect on the building indoor temperature. It was also estimated that existing commercial buildings could provide about 70% of the required frequency controlled reserves in 8 s to 3 min time band. Lighting systems is the second largest power consumer in commercial buildings. The DR of illumination is relatively simple, since it can be implemented by dimming or switching off some lights. Reducing lighting may also result in reduced cooling need, since lighting systems produce heat (Shan et al. 2016).

In recent years, the biggest interest has however been in harnessing the DR potential of *residential loads*. Electrical appliances and energy uses in domestic housing can be divided into critical and non-critical loads, depending on their flexibility and importance. Critical loads that are assumed to have very little flexibility include consumer electronics, cooking and most part of illumination, for instance. Most other loads offer some degree of flexibility. According to (Stamminger et al. 2008) and (Lui et al. 2010) some rough estimates about the flexibility of various domestic appliances, in terms of maximum time that the use can be postponed, can be summarized as shown in Table 2.

Among the residential loads, thermostat controlled appliances (TCA) and HVAC loads are usually regarded having the highest DR potential. In the case of HVAC loads, including domestic hot water (DHW), the duration of timely shifting strongly depends on whether there is some heat or cool storage. In case of refrigerators and freezers, the potential of a single load is small, but these devices are made attractive for DR due to their large numbers and relative constant demand. These appliances however have relatively tight limits for how long a time their demand can be shifted. Among other white appliances, washing machines, clothes dryers and dish washers offer some degree of flexibility, depending on the willingness of people to optimize their time of use.

Table 2: Demand response potential of domestic appliances.

Appliance	DR potential (hr)
Washing Machine	4
Clothes Dryer	1
Dish Washer	5
Water Heater—Storage	3
Water Heater—Direct	-
Refrigerator and Freezer	1
Air Conditioner	1
Space Heating—Storage	5
Space Heating—Direct	1

In the future also electric vehicle (EV) charging systems are expected to offer a high potential of DR. EVs are parked most of the time and if plugged-in, they could easily participate both in energy and in power balancing. There is also huge interest in using EVs as a power source, that is, power flowing from vehicle-to-grid (V2G) instead of just being charged grid-to-vehicle (G2V). V2G capability would roughly double the DR capacity of EVs, compared to just G2V (Suryanarayanan et al. 2016).

The potential of HVAC loads not only depends on the degree of storages, but also on the comfort expectations and requirements of the residents. These can be expressed as the allowed indoor temperature variation band. Comfort is a subjective issue, and some people are more sensitive to temperature variations than the others. In the final end, the user comfort requirements together with the actual HVDC demand and thermal storage capability of house define the DR potential of space heating and space cooling loads.

One important factor that eventually affects the heating and cooling needs of buildings and hence also the available DR potential, is the heat gains. The heat gains can be divided into three classes: The energy used by electrical appliances and illumination that is converted to heat, the heat gain from people and the solar heat gain. For Nordic housing, for instance, these heat gains are well summarized in (Molina et al. 2015). There is a large discrepancy between different estimates of heat gains from different appliances. Depending on the appliance in question, the estimated heat gains vary from 50% to 100% of the electrical energy used. Anyway there is a common agreement that most part of the electrical energy being used eventually ends up as heat in the living space. The second class, the heat provided by people depends on their activity. A person in rest may supply around 100 W, whereas the heat gain from working people may reach some 570 W. The third class of heat gain is the solar heat gain. The yield of heat by solar insolation greatly depends on the shading of the house and especially the windows, and it is difficult to give any general figures.

The heat gains are into a high degree uncontrollable. They release thermal energy into the living space of the dwelling. In the heating season the heat gains may be useful, since they replace the heating energy that otherwise must be produced by the radiators. When planning the DR the interdependence of heat gains and space heating should be taken into account. If not so, power reduction in one appliance may lead to power increase in heating, which cancels the desired benefits of demand response application. In warm climates, and also in Nordic conditions in the summer time, the heat gains directly increase the cooling demand and thus power consumption. In the future energy efficient buildings the insulation level is expected to be higher than today, which will in turn emphasize the importance of considering the heat gains when planning DR applications.

3. Demand Response in Energy Procurement

From energy generators or traders point of view, the most important use case of DR is in maintaining the hourly power balance. The traders forecast their energy demand and VRES generation and prepare a generation and energy procurement plan for the next day correspondingly. In this phase the energy trading happens in the day-ahead Elspot-market. Closer to the hour of delivery, the forecasted and actual energies deviate for both demand and VRES generation, and the traders try to improve their energy balance by Elbas trade or by re-scheduling their generation plan. At the hour of delivery, the traders have to balance their procurement either by own flexible resources of by trading in TSO's balancing power market.

The main utilization of DR can be as a flexible energy resource in all the above three time spans, but since the price of power deviations tend to increase closer to the time of delivery, the best benefits are likely obtained in intra-day and intra-hour balancing. Hence we can consider the main purpose of DR to be reshaping the demand curve such that the deviations between procured and delivered energy is minimized, or that the cost of these deviations is minimized.

To utilize the flexibility of customer loads for DR requires some market mechanisms or agreements between the traders and the customers. In the past, various types of tariffs, like time of use tariffs (ToU) or critical peak pricing (CPP) have been used for demand side management. In ToU tariff, the customers are offered lowered prices between 10 pm and 7 am and during week end, for instance, while in CPP the prices are increased when peak demand hours. These methods can be used for obtaining more of less permanent changes in the load profiles. However, thinking of unpredictable hourly deviations in energy balance due to VRES variation, these means are infeasible.

More recently the power companies have developed dynamic pricing of power, mostly based on day-ahead, or Elspot-prices. These offer a transparent market to the customers, which enable them to gain benefit should they have the possibility of shifting demand to the cheap hours of the day in question.

An example of demand optimization based on day-ahead prices is shown in Fig. 2. The case illustrated is electric space heating with a heat storage based on water tank. The storage is charged during the cheap hours and discharged during the peak price period. In the case of highly variable prices, this kind of arrangement can offer savings to the customers without sacrificing their comfort, as the storage is able to maintain the unaltered room temperature for several hours. Assuming that the Elspot-prices reflect the shortage of generation (high prices) and the need of power sink (cheap hours), dynamic pricing like this helps in maintaining the gross power system energy balance.

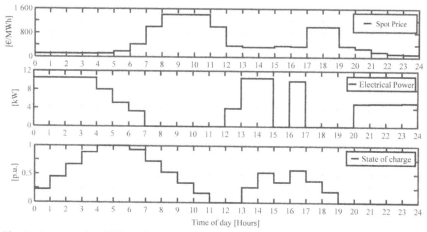

Fig. 2: An example of DR optimization based on day-ahead prices. Electric space heating with heat storage. Spot price, electrical power taken from grid, and storage level for each hour of the day. The size of storage covers 40% of the daily heat demand.

This kind of dynamic pricing is often referred as real time pricing (RTP). Because of intra-day forecasting errors and unexpected generation ramps, the RTP has a pressure to be developed towards reflecting better the intra-day or intra-hour energy costs. Examples of this are ComEd Hourly Pricing Program in Illinois USA and AEMO, Australian national energy market. In these markets the hourly prices are updated during the day, and the customers are provided forecasts of the coming prices in order to enable them to minimize their power costs.

The benefit of RTP is that it decouples the control systems of generator-trader and customer, since the sole intermediate mechanism is the power price. The customers can optimize their power use based on local demand forecasts and 24 hour rolling horizon of forecasted market prices. For the optimization various mathematical tools may be used, such as linear programming (LP), mixed-integer linear programming (MILP) or genetic algorithms (GA) (Mubbashir 2016).

More targeted DR can be obtained by direct controls. Here the trader sends a signal to shed, postpone or prepone the customers' loads. The loads controlled are specified in beforehand, and some bonus is usually paid to the customers, either for the actual happened controls or for the possibility of making the controls. In the case of larger customers, like industrial and commercial loads, the contract is directly between the customers and the energy company. In the case of residential loads, there can be an aggregator who collects a larger number of loads and sells their combined flexibility to the energy companies or to the energy markets.

In Nordic Countries, a typical case of direct load control has been shedding the electric space heating loads when the power price is high or generation adequacy is at risk. This kind of control has the problem of rebound effect. When the heating loads are reconnected, they start to recover the room temperature which causes increased demand at the respective hours. To better manage this pay-back period, some improved control schemes have been suggested, which are based on remote setting of space heating control thermostats. Smooth changes of thermostat setting allows leveling the rebound spike towards more favorable hours (Alahäivälä 2017).

In the future also various DR bidding schemes are possible between aggregators and customers. The flexibility of customers depends on their comfort requirements. The customers who are willing to accept larger temperature variations can offer higher DR potential from their heating or cooling systems. The real availability of DR also depends on the actual situation at the customers premises, like the state of charge in the EVs battery or the amount of heat stored in the water tank of electrical storage heating or domestic hot water. Real-time bidding schemes are able to consider customers as individuals and hence may offer higher potential of DR without sacrificing the customer comfort and preferences. The bidding schemes operate at the similar principles as power exchange or balancing power markets. In order to reduce the amount of communication, some hierarchical structure may be needed, including several aggregators. Some examples of bidding schemes are given in (Malik 2016).

Demand Response in Power System Balance Management

In a power system the generation and demand must be in balance practically in real time. Any deviation from this balance manifests itself as frequency deviation. If generation does not meet the demand, the frequency starts to fall. In the opposite case, the excess generation makes the system frequency to increase. Large frequency deviations put the system security under risk. The change of frequency in case that a major generation unit is lost is depicted in Fig. 3. The missing power is first taken from the inertia of the rotating masses in the power system, and as they discharge their kinetic energy, the frequency falls correspondingly. In some seconds, however, the frequency controlled primary reserves, also called spinning reserves, react and start to increase their power output. Hence the frequency decay starts to stabilize. The power-frequency control of primary reserves is usually implemented as a droop, as depicted in Fig. 4. After 15 min the primary reserves are released by secondary reserves and the power system is ready to manage a new disturbance. This kind of frequency controlled reserves are often referred as ancillary services, although ancillary services are a more general term which includes all the means needed to maintain the stability and security of a transmission grid, including also reactive power support and voltage control.

In Europe, the primary reserves are divided into Frequency Containment Reserves In Normal Operation (FCR-N) and Frequency Containment Reserves in Disturbances (FCR-D), see Table 3. The control of these reserves

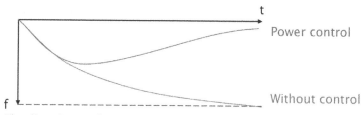

Fig. 3: The effect of power-frequency control in a power system.

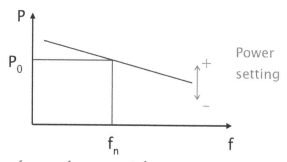

Fig. 4: The droop for power-frequency control.

Table 3: Types of reserves in the electricity markets maintained by Finnish TSO Fingrid.[3]

Reserve	Name	Minimum size MW	Activation times
Frequency containment reserve—normal	FCR-N	0,1	Continuously in frequency span 50.1–49.9 Hz. 0.1 Hz change 100% in 3 min
Frequency containment reserve—disturbance	FCR-D	1	Linearly in frequency span 49.9–49.5 Hz Below 49.5 Hz 50% in 5 s and 100% in 30 s
Frequency restoration reserve—automatic	FRR-A	5	After receiving a request, 100% in 2 min
Frequency restoration reserve—manual	FRR-M	10	According to the bids in balancing power market. 100% activated in 15 min

[3] http://www.fingrid.fi/en/electricity-market/reserves.

is based on frequency measurement. Secondary reserves are in turn called frequency restoration reserves, which can be either automatic (FRR-A) of manual (FRR-M). The latter one is traded in the TSO's balancing power market, and is activated within 15 min after the disturbance.

DR can participate also in primary or secondary reserves if it can satisfy the requirements presented in Table 3 for reaction time and minimum size. Some large customers may be able to meet the minimum power limits alone, but usually a combination of several loads is needed in order to form a large enough capacity. This gives rise to the idea of a virtual power plant (VPP), where an aggregator collects a large number of flexible loads and combines them into one large controllable unit to be offered to the market.

In the case that DR is used for primary reserves, a question rises about the suitable ratio of energy volume (E/MWh) and power offered (P/ MW). Since the primary reserves are activated for 15 minutes, the suitable dimensioning is $E = 0.5(hr)$ P, assuming that energy storage (battery of heat storage) is restored by secondary reserves back to the middle position.

4. Conclusions

This chapter has illustrated the use of DR in energy and power balance management of power systems. DR has come to the spotlight due to increased integration of variable renewable energy sources like wind and solar in the power systems. The VRES creates challenges due to unexpected generation ramps, which cause power balance problems and reflect in the system frequency. Another issue is the mismatch in hourly balance, which challenges the power procurement of the energy trader. And finally, in the case of excessive amount of VRES generation, a power sink is needed in order to avoid generation curtailment.

DR can be used for mitigating the power and energy balance problems in all the three cases mentioned above. Most promising DR potential is in thermostat controlled loads and in EV batteries.

In addition to the above use cases, DR can be used for managing distribution network congestion in the cases of fast EV charging or excessive photovoltaic production connected in the low voltage or medium voltage networks. To do so requires coordination between distribution network operator and energy trader or aggregator.

References

Alahäivälä, A. 2017. Harnessing Demand Response for Power System Flexibility, Residential Heating and Thermostatically Controlled Loads, PhD thesis, Aalto University School of Electrical Engineering, April 2017.

Elovaara, J. and L. Haarla. 2011. Power systems, Otatieto, Espoo, Finland.

Hao, H., Y. Lin, A.S. Kowli, P. Barooah and S. Meyn. 2014. Ancillary Service to the Grid Through Control of Fans in Commercial Building HVAC Systems. IEEE Transactions On Smart Grid 5:(4).

Lui, J., W. Stirling and H.O. Marcy. 2010. Get smart. IEEE Power and Energy Mag. 8(3): 66–78.

Malik, F. 2016. Flexible Loads in Smart Grids, Charging Solutions for Electric Vehicles and Storage Space Heating, PhD thesis. Aalto University School of Electrical Engineering, November 2016.

Molina, C., M. Lehtonen and M. Degefa. 2015. Heat gains, heating and cooling in Nordic housing with electrical space heating. International Review of Electrical Engineering 10(5): 599–606.

Mubbashir, A. 2016. Domestic Space Heating Load Management in Smart Grid, Potential Benefits and Realization, PhD thesis. Aalto University School of Electrical Engineering, October 2016.

Shan, K., S. Wang, C. Yan and F. Xiao. 2016. Building demand response and control methods for smart grids: A review. Science and Technology for the Built Environment 22: 692–704.

Stamminger, R., G. Broil, C. Pakula, H. Jungbecker, M. Braun, I. Rudenauer and C. Wendker. 2008. Synergy potential of smart appliances; D2. 3 of WP 2 from the Smart-A Project, Rep. no. D2.3.

Suryanarayanan, S., R. Roche and T.M. Hansen (eds.). 2016. Cyber-Physical-Social Systems and Constructs in Electric Power Engineering. IET Digital Library.

Zhang, X. and G. Hug. 2014. Optimal regulation provision by aluminum smelters. PES General Meeting, Washington D.C., 27–31 July 2014.

CHAPTER 18

Smart Agent and IoT Towards Smart City

George Cristian Lazaroiu[1] and
Mariacristina Roscia[2,*]

1. Introduction

Most urban areas want and like the label of Smart City, but often the impression is that the concept of a Smart City is only a digital or technological city and sometimes sustainable, but what does a Smart City really mean? A Smart City development is a question of "how" not "what"; a Smart City development is a question of "when" not "if".

The term Smart City is not a definition but a real life model to improve the quality of life and to adapt to the Human and Environment needs through advanced technologies. Cities are responsible for 67% of the total global energy consumption and more than 70% of greenhouse gas emissions, therefore the cities have a vital role to play in improving energy efficiency and reducing emissions.

There are many technical options for increased urban efficiency, such as more efficient street lighting, green buildings, increased use of public transportation, and reductions in traffic, but the concept of a "Smart City" model should include a certain harmony between Quality of Human Life, business activity, the exploitation of renewable resources, in other words a *'Smart Model of Economic, Social and Environmental Sustainability'*.

[1] University POLITEHNICA of Bucharest - Department of Power Systems, Splaiul Independentei 313 – 060042 Bucharest – Romania.
[2] Department of Engineering and Applied Sciences, University of Bergamo, Bergamo, Italy.
* Corresponding author: crisroscia@yahoo.it

The Smart City will be among the main objectives to achieve sustainability, environmental, economic and even the social inclusion of all citizens.

This can be done by investing more in (ICT) Information and Communication innovative technologies, and through the involvement of the majority of the population while taking public policy decisions, focusing increasingly on the participatory processes, such as online consultations and deliberations, as well as on 'activation of participatory creativity workshops'.

Sustainability, in all areas and sectors of the life of a city, is considered the most strategic component for Smart Cities. There are therefore, different aspects of sustainability, creativity, social inclusion and cultural development, used to determine the true notion of a "SmartCity".

We start with delinelation of the main dimensions for a Smart City (Odendaal 2003). The following characteristics represent the macro areas, which are then identified and according to which the related subsystems for a coordinated design, integrating all the parts of such a complex system (as will be seen in the following paragraphs) and for the identification of indicators, for the definition of the degree of the level reached in its realization.

1.1 Environment

➢ Reduction of greenhouse gas emissions, in according with the EU and International directives by:

  ✓ an intelligent distribution system and management of energy (possibly green), which is able, through the knowledge of production data and real-time consumption, to handle all energy carriers moment by moment;

  ✓ optimization of industrial emissions;

  ✓ improvement in the construction sector, so as to reduce the emissions from heating and air-conditioning;

  ✓ public lighting control;

  ✓ sustainable mobility, for each transport system as trains, planes, ships, bike, car, etc., through the restriction of private traffic, creating a model in which the movements are easy, ensuring effective innovative and sustainable public transport, which promotes the use of eco-friendly vehicles, such as electric vehicles, bicycles, which regulate access to town centers privileging the livability Limited Traffic Zone (LTZ); the adoption of advanced mobility management and info-mobility to manage the daily commute of citizens and exchanges with neighboring areas;

➢ reducing the amount of waste, and differentiation of their collection, their economic value;

➤ effective expansion of urban green management through the urban development based on "saving ground" and the remediation of brownfield sites.

➤ management of critical issues, such as natural disasters; health crises; terrorist events; environmental monitoring; accidents; in order to define priorities and methods of intervention, saving time and reducing human error.

1.2 Knowledge Economy

A Smart City is a continuous learning place, which will self-produce innovative training courses. It is a city that offers an atmosphere conducive to creativity and promotes it through encouraging innovation and experimentation in art, culture, and entertainment; which is perceived and represented as a laboratory for new ideas; and which focuses on the construction of a network of non-hierarchical networks, but inclusive, in which the various stakeholders and the community as a whole may have citizenship and a voice. It is a city that develops alliances with universities, research centers, institutions of education, and which gives space to free knowledge.

Therefore advanced solutions must be found in areas of public interest for the development of social integration models.

1.3 Urban Changes to Quality of Life

A Smart City has a strategic view of their development and can define, on the basis of this, decisions and courses of action. It considers the maintenance of property assets and its efficient management, using also advanced technologies for this purpose. It is, finally, a city that in his physical development, creates the conditions for promoting social cohesion and inclusion, eliminating barriers to full accessibility for all Smart Citizens.

1.4 Culture and Tourism

A Smart City promotes its tourist image with a Smart web presence; invests in the construction of a virtual dimension of its cultural heritage and traditions, creating network as a "common good" for use by all. It's a city that uses advanced techniques to create developments and "mapping" issues of the city and makes it easily accessible, promoting coordinated and intelligent offers of goods and services from its territory. It bases its growth on the respect of its history and its identity and emphasizes this in regard to the reuse and enhancement of the existing, in a renewal process that bases its assumptions on conservation.

It should also be noted that to design a Smart City from new, it is much easier than going to put new technology into an already built urban space. This means that for European cities the transition towards modernization will be more difficult and slow, compared to cities that can be realized in developing countries (Lazaroiu and Roscia 2012).

2. Systems for Smart City

Often the notions of what constitutes the systems of a Smart City are fragmented and it is a real difficulty to have an overview, showing the complexity of the key elements that should be outlined to get a realization of a sustainable city process. To this end, for the realization of a Smart City we have defined the macro sectors through which a Smart City is concretely realized, into which are inserted suits the keywords for the development and the design of a Smart City (Cosgrave et al. 2013) see Fig. 1, shown below:

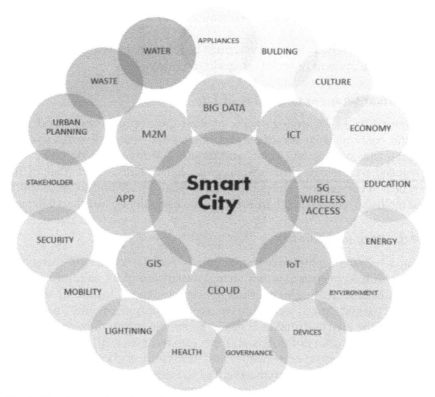

Fig. 1: Key elements for a Smart City.

2.1 Smart Building

➤ Urban planning
➤ Smart Building and Homes control, monitor and optimize building and home services
➤ System for offering greater comfort, security and reliability while reducing the costs normally associated with infrastructure management and maintenance needs
➤ Smart appliances and smart Metering Control
➤ Home and Building Energy Management System to reduce energy use

2.2 Smart Energy

➤ Smart and Micro Grid
➤ Renewable energy
➤ Utility services
➤ District heating/cooling management
➤ Gas distribution and management
➤ Energy efficiency

2.3 Smart Mobility

➤ EV infrastructure, charging, services
➤ Railway infrastructure and management
➤ Ship infrastructure and management
➤ Airport infrastructure and management
➤ E-bike infrastructure and management
➤ Public Intelligent Transport System (PITS)
➤ Smart logistic and Smart Parking Management
➤ Infomobility, Traffic management and Automatic drive
➤ Multimodality and sharing mobility

2.4 Smart Water

➤ Smart Water Management to improve quality and a reduction of water loss
➤ Water efficiency and smart metering
➤ Mapping and modelling of water distribution assets and network

- ➢ Real-time hydraulic modelling water distributions tool
- ➢ Water pressure optimiser
- ➢ Automated groundwater access and approval process
- ➢ Irrigation and Public use infrastructure and management

2.5 Smart Economy

- ➢ Energy, water, waste, mobility market
- ➢ Private assessment
- ➢ Business model and financing
- ➢ Emerging Economy
- ➢ Careers and employment
- ➢ E-commerce, retail sales

2.6 Smart Public Services

- ➢ Street lighting
- ➢ Public safety-emotion software and video surveillance
- ➢ Smart emergency disaster (environmental, health, terrorist) management
- ➢ Big data and privacy management
- ➢ E-Public Administration
- ➢ Smart governance
- ➢ Culture, tourism, education

2.7 Smart Health

- ➢ Smart hospitals
- ➢ Healthcare and wereable sensors and devices
- ➢ Diseases and health care professional management
- ➢ Systems for disabled and elderly people
- ➢ Food safety and quality
- ➢ Improvement quality of life

2.8 Smart Waste

- ➢ Energy from waste
- ➢ Disposing of waste in an environmentally-friendly manner

➤ Assess waste streams and develop the right recycling and disposal solution

➤ Making the garbage collection in the entire city more efficient and well-organized

➤ Monitoring and evaluating the performance of the waste and garbage trucks, in addition, contractor companies

➤ Keeping track of container services such as washing, disinfection, and fixing regularly or on request

2.9 Smart Data

➤ Infrastructure data enabled services
➤ Data centers
➤ Operation technology
➤ Internet of Things-IoT
➤ Information Processing and Communication Technology – IPCT
➤ Bio-Informatics

2.10 Smart Integration

➤ Smart city Cloud platform
➤ 5G wireless access
➤ Apps
➤ Gis and Weather management
➤ M2M
➤ Smart citizens
➤ Mesh topology
➤ Social network

3. Public Procurement for Smart City Solutions

Critical in an urban system, is the "governance", i.e., the ability to follow at the local level how the indicators are interpreted and translated for the citizen service.

Governance support is confined in all those systems that the intelligence motor can measure, interpret, plan and anticipate phenomena, new solutions are designed to help cities of all sizes, to obtain an overall view of the information located in the various departments and in the agencies working for the city itself. Through the application of advanced analytical models to the local administrative activities processes, and managed through a single, central command—the city will be able to anticipate problems, respond to crises and better manage its resources.

The Public Administration (PA) in the realization of the Smart City, has a key role, that being: promoting and financing of Act and services; coordinating the development of infrastructure in the cities; keeper of relevant Data and information, in quantity and quality.

By linking many administrations, particularly local authorities and having their realities linked, they have expressed an interest in a better understanding of technological, economic and social issues related to the adoption of Models for Smart Communities and Smart City. Fundamentals for the development of intelligent solutions are public procurement, only choices coming from politicians can implement the process of transformation through innovative solutions.

4. The Smart City Value: Revenue Opportunities

The cities have a limited access to financial resources, therefore for the sustainable transformation of the city it is possible only when it is done in a Smart way. In order to realize the Smart City projects or subsets that in the future may be "assembled" for the establishment of Smart Cities, requires major funds, which are allocated around the world both by governments and by the private sector.

This is because from the point of view of the public it is not only opportunistic policy choice but is also a long-term economic return, and by stakeholders because it opens up opportunities for business offers.

The following is the data globally available (Bellifemine et al. 2007) until 2025 about possible investments along the main lines of action for the establishment of the Smart City

1. Smart governance and smart security,

   market value $1230.26 billion

   CAGR: 11,6% (2012–2025)

   More projects coming up in North American, Middle East and European regions

2. Smart Energy

   market value: $ 781,66 billion

   CAGR: 28,7% (2012–2025)

   Highest growth in Nord America

3. Smart Infrastructure

   market value: $ 381.53 billion

CAGR: 12,0% (2012–2025)

Highest growth in Nord America and in Asia

4. Smart transportation

market value: $ 351.13 billion

CAGR: 19,6% (2012–2025)

Market growth in Latin America, Middle East & Africa, and CIS + Eastern Europe

5. Smart healthcare

market value: $ 348.51 billion

CAGR: 8,8% (2012–2025)

North America to dominate the smart healthcare market with more 50% share by 2025, followed by Europe

6. Smart building

market value: $ 248.65 billion

CAGR: 4.1% (2012–2025)

China and India are the fastest growing markets followed by Japan and Korea.

5. Data Collection for Smart City

A further key element for the realization of a Smart City is to collect, manage and preserve the privacy of big data which must be a complete and not partial management of the city.

Here are the basic steps to be taken for Big Data, in the path towards moving to the Smart City:

✓ Integration of operational systems with respect to the territory ("Geographic Information System"). This is the only way get timely updates and sure information regarding: population, business and trade, construction, property, planning instruments, taxes, green, education, roads and traffic, etc.

✓ Integration with all public subjects for sharing through the exchange of information relating to the use of the land: Land Registry (classification, properties and texture of immovable property), Chambers of Commerce (changes relating to companies), water services, energy (electricity and gas) and environmental services (waste collection and recycling, street cleaning, etc.). Databases become sensors that detect changes in the daily life of buildings, businesses, citizens and consumption.

✓ Involvement in the integration of private external actors, in this way the geographical information system would be complete and able to collect and process, georeferencing, a very substantial amount of information on people, businesses, building transformations, heritage, schools, public parks, streets, shops, distribution of energy and gas, variations of urban planning, management/SIT control, etc.

An information system full of updated data is a necessary, but not sufficient to turn data into actionable information and be able to guide decisions. It is necessary that data and information are made available to *"changemakers"* and decoding and predictive models are necessary organs that can synthesize and actually explain the information.

✓ Sharing and circularity data at all levels of decision making in various areas: decision makers must have the information fully available in its own operating environment. For this purpose they are essential to decode and for predictive models to access data and the simulation of the effects of decisions.

✓ Open data: free data available to everyone for further use. If the Data can be made available to all those who request it, the prospects for development and the further future changes are endless and hard to conceive. At the same time, the Data information must be treated and managed very carefully to privacy and undue, wrong and illegal use.

The Smart City will be advanced solutions in the areas of public interest and social integration models, using technologies, applications, models of integration and inclusion. Through the use of the Internet of Things (IoT), it will be possible build a network of information concerning not only people but also objects and processes, thus obtaining a large quantity of data to be managed by advanced and autonomous decision software systems.

6. Combination of Technology for Smart City Solution

The Smart City is often considered as a simple definition, but the goal is to provide a model, which can adapt to the singularity of each city, modulating it as a living structure, through the definition of the necessary physical systems that make it feasible. The Smart City concept is founded on a set of solutions which are a combination of today's stand alone technologies.

The components that physically allow the construction of a Smart City system are:

6.1 Sensors, Actuators, Controller

Sensors must be located in different areas (temperature, pressure, light flow, etc.) and with these variable data is involved and managed by a controller with adaptative logic (fuzzy logic, neural networks, etc.), these are processed with variables provided by external systems (Cloud, smartphones, laptops, tablets, APP) and are used for weather information, irrigation management, transmitting severe weather warnings, and load control in relation to use as part of a micro or smart grids, etc. In Fig. 2 the control scheme for the generic system or subsystem inside a Smart City is illustrated. The controller will send orders for the actuators, with the intention to accomplish to the system, the action and events desired, in real time. These interactions can also be handled by Cloud, Human Interface, and TCP/IP. In addition to the controller remote monitoring will be provided in order to make the most reliable controls.

The proposed scheme will highlight the connections in both wired and wireless modes, the last in particular will be more useful for their ease in deployment and because new wireless devices may be fitted in a Plug and Play use.

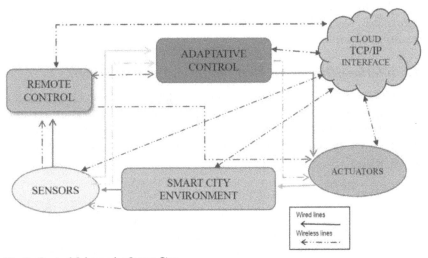

Fig. 2: Control Scheme for Smart City.

6.2 Topology of Network

The network topology is the connection geometry between the different nodes (device) of a system, and it is critical because it affects the rules of the infrastructure, such as the connections and communications systems (wired or wireless).

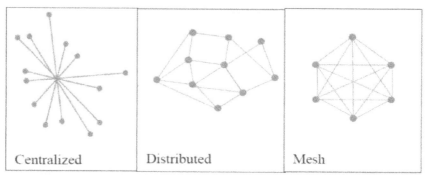

Fig. 3: Different Topology of Network.

The topology of a network (see Fig. 3) can be of different types:

✓ Centralized, where intelligence resides in a central node

✓ Distributed: where intelligence is distributed across all devices (node)

✓ Mesh: intelligence is distributed and where each of the devices (node) can communicate with all other devices on the network.

The possibility to connect one device to another, makes the network highly fast, safe, reliable, and easy, but also economically unfeasible with the only wired logic.

6.3 Interface/Display for User Interfaces (User Friendly)

The UI (User Interface) is an information device with which a user can interact, for example, the most popular and well used UIs are screens, smartphones and tablets. In addition to other technologies such as the touch screen, systems are being used that use as an interface the touching of the skin.

6.4 ICT

While it is possible to identify the elements that make up ICT, on the other hand it is not easy to provide one single definition, because there is no generally agreed upon definition. By its nature the field of ICT is a very dynamic field and evolving with time in relatively confined areas.

ICT has finished with the increasingly tying the components of Information Technology (IT) with the issue of Communication Technology (CT). Particularly when the latter has taken on new clothes, that is, with the advent of network technology, information that ended up losing represented characteristics as an elaboration of stand-alone machines to become a shared component with other machines in a network (both LAN and to the global Internet).

6.5 Smart Meter

The spread of the Smart Meter installation, makes it possible to adopt flexible billing and pricing patterns, the management of smart appliances, some of them interoperable, allowing the user to proactively manage power consumption with a comfortable mode, in a affordable and environmentally conscious environment.

An additional layer is the constitution within the city of so-called "balanced energetically campus" or "Micro Smart Grid" which consist of areas interconnected by a point of view of production and energy consumption with the aim to achieve and/or exceed the balance of power. In a city there may be many examples of these balanced islands such as hospitals, college campuses, shopping centers, etc. In this context it is necessary to support the monitoring systems and analysis of production systems and consumption points.

6.6 Memory for Parameterization and Data Acquisition Events Memory

With the integration with the mobile network or, in general, the Internet allows, for example, the automatic sending of status and alarm messages, centralized storage of data packets on a central database server (cloud) or remote access for system analysis or re-parameterization from anywhere in the world. If the data recorder can not be integrated with a network, the replacement of the flash memory card is offered as an on-site service.

6.6.1 Connection Systems

The different physical means for communication between the different devices are represented by:

✓ twister pair (TP)
✓ BUS
✓ power line PL
✓ radiofrequency (RF) or infrared
✓ wireless

6.6.2 Standards for Communication

The following are the most developed and widely used communication systems, and protocols that are used by different devices to communicate with each other, draw information and implement actions on the system on which operate:

✓ X10
✓ INSTEON

- ✓ Z-WAVE
- ✓ ZIGBEE
- ✓ 6LOWPAN
- ✓ WIFI HIGH SPEED WIRELESS
- ✓ Desigo
- ✓ Enocean
- ✓ DALI
- ✓ KONNEX
- ✓ BACNET
- ✓ GOOGLE PLATFORM
- ✓ RFID
- ✓ NFC
- ✓ Bluetooth
- ✓ LON
- ✓ TCP/IP
- ✓ Z-Wave

6.6.3 UltraBroadband Internet Service (100 Mbps)

This is the most used form of Internet access because of its high access speeds; for example DSL (or Digital Subscriber Line), fiber-optics, cable, and satellite. Internet users are moving towards a faster broadband Internet connection.

6.6.4 Cloud Computing

The term cloud computing refers to a set of technologies that allow the form of a service offered by a provider to the customer, store/store and/or process data (via CPU or software) through the use of distributed hardware/software resources and virtualized on the Net.

Cloud computing may be defined as a new approach to the use of existing technologies: Internet, virtualization, Web, etc., But it has been remodeled in order to create a computing platform that is independent from a physical location and that abstracts the hardware resources and used software.

Cloud computing is consolidated for economies of scale, to make it globally available, hide the complexity, make it easily consumable, and tailor the experience to the customer. The benefits to the cloud are: it is easily upgraded, no IT maintenance costs, off site data storage, productivity anywhere, always up, ower cost of ownership, disaster assistance.

6.6.5 M2M

Machine to machine (M2M) can be used to describe any technology that enables networked devices to exchange information and perform actions without the manual assistance of humans.

M2M is the basic element that has led to the creation of ICT through the use of increasingly popular technologies such as RFID, WiFi, and software that can make decisions such as those based on smart agents.

6.6.6 Internet of Things (IoT): The Evolution of Connection

All these devices must be able to communicate and collaborate with each other, even remotely, and this involves a large amount of data that must be managed, this is now made possible by the multiplicity of systems connected to the network, and used by its users in a very widespread, simple and integration through the Iot. In fact the App is the preferred interface by users (70%), demonstrating how smartphones and tablets have a key role in bringing technology to the consumer, for example WiFi is the most well-known standard and is the ideal solution for all wireless Internet network and wireless broadband access points. With technology's transformational impact everything is connected, in fact today 400 Million of devices are connected, and in Table 1 the future of the net and the connections at 2020 and it is estimable about 4$ trillion revenue opportunity is shown.

The development of a Smart City comes from the possibility of being able to use highly distributed devices and connect between them, without necessarily using a wired system. For example the great spread of wireless systems have given a strong impetus for the spread of home automation systems, in fact, the spread of this technology over the past two years has been exponential, and this will drive the construction of Smart Cities.

For the development of Smart City, is the Wireless Mesh Network (WMN): in it each node of the network, can be both HOST and Router, this way a very flessbile and robust network can be achieved, since each node is connected to another node. Since the WMN have no fixed infrastructure, how can be in mobile networks via the base station or access point. It presents a high degree of connectivity, it can work with an auto configuration and topology control algorithms, however, they require a careful design. It must be compatible with standards such as

Table 1: The Future of the IoT.

2020				
4 billion of connected people	25 Million of APPS	50 Trillion of GBs of DATA	25+Billion Embedded and Intelligent systems	5G Wireless access

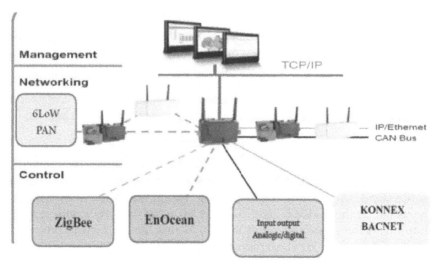

Fig. 4: Example of architecture for a Wireless Mesh Network (WMN).

Zigbee Enocean, or HSDPA (high speed downlink packet access) this way we will come to an integration with what is basically Internet of Things (IoT).

In Fig. 4 an example is shown of the possible configuration of a WMN, in which there coexists different types of connections, for different devices, and with different protocols. The simplification is in the wireless systems, which automatically achieve a mesh network, in which each node can communicate with another node.

Multiple integrated wireless access solutions have improved the way to connect both for people and for the machines, with an automatic transfer of data between different devices for a remote control.

In the near future—cars, buildings, hospitals, schools, government, etc., will be possible to be connected. In fact, as shown in Fig. 5, the illustrated example of how the IoT can allow the connection of people, objects, and processes, contributing to the realization of a Smart City, in much faster times compared to previous wired systems.

7. Adaptive Technology in the Smart Cities

The idea is to build within the city intelligent networks that combine various electronic devices and services that can be inserted and removed. In other words, the organizations in charge of the hardware and software systems can reconfigure them with new components when requirements change.

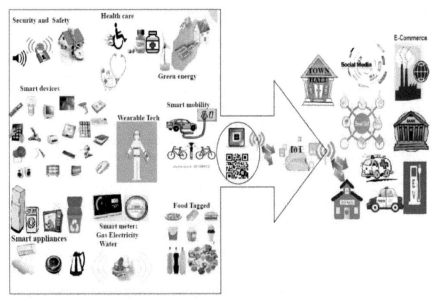

Fig. 5: Example of IoT inside Smart City.

In this manner, it is not technology to control the urban environment; is the environment reshaping the technology.

In this context, the new network and virtualization technologies (the virtual desktop, virtual teller, the cloud) manage to break free from the physical limits and ensure the management of the complexity of optimizing the resources available. With this "oriented architecture" services, you will be able to achieve a goal, which in this case is to extend the life of the project, the software and hardware rather than the single realization, that it is a repeatable, exportable, deployable model.

The challenge for Smart Cities is to urbanize the technologies they use, to make them responsive and available for people whose lives they will affect. This is because you can not build a Smart City and simply "plug in" new technologies in an urban space.

In fact, the technological systems that may work well in one city may not work in another city or should be completely modified to be useful elsewhere. This would neutralize all attempts to define a Smart City model.

With Smart urban spaces, we plan to create places that can react in real-time to the behavior of people present. Urban space adapts to the needs and intentions of the people to provide the right services at the right time in the best possible way, all without the direct control of the services by the users.

8. Smart City as a Complex System

In the past and also in part the trend today is to entrust the control operations to a 'single central unit equipped with a lot of computing power, this in order to simplify the system architecture, however, this is not the best solution, a fault unit or a malfunctioning of the software would cripple the entire system. The computational complexity of new technologies and communication interfaces in electronic applications is growing rapidly due to the accessibility of these technologies from a large number of users, making it dangerous to entrust a large computational load to a single unit.

The optimal solution is the Distributed Control Systems by Multi Agent, to better manage the complexity level of the network devices and to reach a state of the collective intelligence of a Smart City.

Besides the spread of interfaces means that you can equip each device with its own computing power and memory, they can collaborate with each other, reducing the intrinsic complexity of a system such as the Smart City.

A distributed network of intelligent devices will create a very high performance system and with a high level of reliability, replace a component of the system or install new ones that would require a minimum effort as it would be enough to connect to the network and integrate our device through a software to edge of a more complex device such as a network hub, a smartphone or a laptop.

The integration of a new device will be of plug-and-play, type able to configure itself by acquiring the necessary parameters themselves.

A complex system, such as the Smart City, is a system composed of a multitude of elements capable of being subjected to continuous changes in the manner of the future, performance of the individual elements is not linear, predictable, but it is not the global behavior of the system.

Some examples of these systems are ecosystems, economic and social systems, the nervous and the climate system.

The greater the variety and number of relations between elements of the system, the greater is its complexity. In Fig. 6 we can see a vision of a complex system, where each individual system of a Smart City (mobility, energy, health), it has as many subsystems. All these elements can have continuous interactions (within the same system) or interactions that alternate or are infrequent. The variables at this, also point to having linear behavior that can be subjected to "disturbances" by all other external elements which increases the complexity exponentially.

It will be necessary therefore to provide structures and techniques that make it easier to manage these systems due to their complex nature.

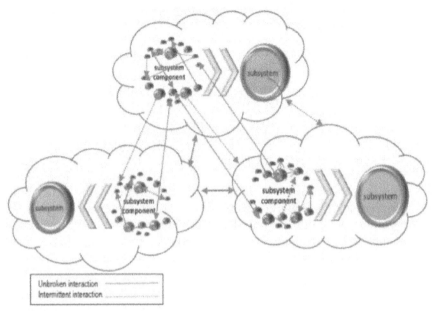

Fig. 6: Vision of Complex System.

Although it is difficult if not impossible to predict future states, complex systems often show recurring behavior, regularities:

- The complexity takes a hierarchical form, or to a system composed of subsystems that interact with each other, each of whom in turn assumes a hierarchical structure, up to the lowest level of elementary subsystem. The type of these organizational relationships varies between subsystems, however, some forms of the generic relationship such as client-server, peer to peer and teams can be found. These relationships are dynamic and can change over time.

- Definition of the components in the system is relatively arbitrary and defined by the observer's goal.

- Hierarchical systems evolve more quickly than non-hierarchical systems of comparable size, the complex systems evolve from simple systems more rapidly if there are stable and identifiable intermediate forms.

- It is possible to distinguish the interaction between subsystems from those inside of the subsystems, the latter are much more frequent than the first, approximately an order of magnitude at most, as well as more predictable. This supports the idea that if complex systems are dismantled, they can be divided into subsystems that may be considered almost completely independent, given the interactions between them. Only some of these interactions can be envisaged in the design phase.

Based on these observations it will be possible manage a complex system such as the Smart City, through:

✓ Breakdown: The basic technique to deal with the big problems is to divide them into smaller parts, thus more easily managed.

✓ Abstraction: The process of defining a simplified model that emphasizes some details and property while it hides others.

✓ Hierarchy (organization): Is the process of defining and managing the interrelationships between the various components. The ability to specify and implement organizational relationships helps to deal with the complexity, allowing a number of basic components to be grouped together and treated as a unit of high-level analysis and providing a means to describe the relationships between the various units.

9. Smart Agent System

In order to manage a complex system such as the Smart City, there should be a suitable software and hardware.

An "agent" software is a piece of code that acts on behalf of a program or a user, to achieve a goal established in the design phase, an agent must also be able to act independently and if appropriate, where they do not act as strictly invoked for a task, but can take action on their own.

Characteristics of a Smart Agent is that they must be able to:

• manifest aspects of artificial intelligence, such as learning or reasoning

• change the path by which they reach their goal

• distributed over separate physical machines

• relocate their execution on another processor

In practice an "agent" software describes an abstraction, a concept, as the paradigm of GOAL-oriented programming in terms of methods and functions. The concept of agent provides a way to describe an entity, complex software capable of operating with a certain degree of autonomy to achieve a goal. Unlike the object-oriented programming an agent is not defined through the methods or attributes, but in terms of behavior.

In Fig. 7 a New Smart Agent Model is shown, that must have the following characteristics:

Autonomous, means that agents are independent and make their own decisions. Another key property is that they pursue goals over time, that is, they are **proactive**. Smart Agents almost always need to interact with

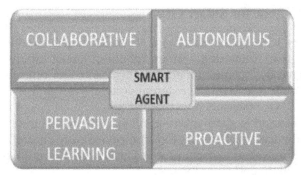

Fig. 7: Attributes of Smart Agent.

other agents, that is, agents are collaborative and finally a Smart Agent must have a **Pervasive Learning**, that is learning at the speed of need through formal, informal and social learning modalities (Pontefract 2013).

Pervasive learning is a social process that connects learners to communities of devices, people, and situations, including other pervasive leaning situations (Thomas 2005). In addition to increasing accessibility, mobile devices have the potential to increase connections between everything and decrease isolation in distance (Hulsmann 2004). The proliferation of wireless and mobile technologies has increased flexibility for learners and the boundaries of anytime, anywhere learning (McGreal 2005).

The concept of the Smart Agent System is very suitable when we are dealing with complex distributed systems. An agent is an entity of problem-solving clearly identifiable, with well-defined boundaries and interfaces; located (or embedded) in a particular environment where it has partial control and observability, receives input state of the environment through sensors and can act on it through actuators; it is designed to play a specific role and is autonomous, and is able to control its internal state and its behavior. It is able to perform problem-solving flexible operations in order to achieve the goal for which it was designed.

Smart Agents tend to be used where the environment is challenging, dynamic in that they change 'rapidly', that is the agent cannot assume that the environment will remain static while it is trying to achieve a goal. The environments, are in fact unpredictable, in that it is not possible to predict the future changes of the environment; often because it is not possible for an Smart Agent to have perfect and complete information about their environment, and because the environment is being modified beyond the agent's knowledge and influence.

10. Smart Agent for Smart City

A Smart City is to be regarded as a set of intelligent modules (sensors, actuators, smart metering, smart phones, smart appliances, etc.). Interconnected by a common network they can communicate with every part of the same and with the external system. Each unit is composed of a hardware device (sensors or actuators) and a Smart Agent associated is able to process and to share information with the rest of the system.

Other devices such as laptop computers can control the overall behavior of the system in order to avoid critical decisions are taken by the system that might endanger the user.

To achieve the goal of designing a Smart Agent System that serves a stable infrastructure, accessible, extensible and widely used.

Standards for various aspects of an Agent's to infrastructure are being developed, as a language of communication between agents (for example ACL agent communication language). In a system of this type agents must be able to coordinate the sharing of information and services by following trading protocols which are also very complex and to perform socially complex operations.

A Smart Agent infrastructure (shown in Fig. 8) is essential for the development of a Smart City, built without it, even if through the use of intelligent and interoperable devices, is unfeasible.

The Smart Agent system infrastructure shall not be defined or rigid, but a structure in which agents do not know their task, they do not know what actions should be implemented to achieve their goal, but they have to learn it in a manner not supervised by the rest of the system, trying to achieve the common goal using techniques of collection and analysis of data by advanced algorithms.

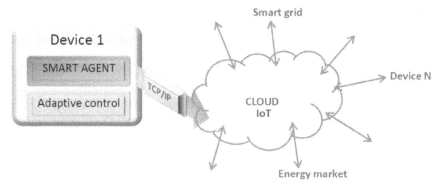

Fig. 8: Proposed architecture of Smart Cities through Smart Agent.

Each device will be equipped with a Smart Agent, which through an evolutionary algorithm or adaptive control technique, will succeed in real time to connect via IoT to Cloud, where Big data will be interchanged, so as to make the system act as a neural system intelligent.

The intelligence, the initiative, the information, the resources and the control can be fully distributed on mobile terminals but not in a fixed network (Karnouskos and de Holanda 2009).

To exchange data with the Smart City platform, an S-function (The MathWork 2010) is programmed to create a TCP/IP connection between the Simulink models on the RTT PC and the SPA agents.

The principal PR agents of each cell receive from the communication SPA agents the state variables which are collected from the physical setup. This will allow you to get stand-alone controls, as with intelligent agents you can report with a star-level agent in relation to network condition and goals can be coordinated in real time (Nguyen et al. 2011). This allows to develop forms of adaptation and learning on the part of the system that improves over time thanks to:

- Introduction of knowledge in the system: re-use of data collected previously in special situations, facts, behaviors and rules.
- Through processing of **pervasive learning**.
- Concepts are generalized to multiple instances.
- Information is reorganized in the more efficiently the system.
- It is used to the experience.

Smart hardware devices with enough information and a decision-making capacity can show a functional and autonomous behavior, developing a knowledge that can not reside in the device in advance but must be developed in a learning phase by soft-computing algorithms that allow the unit to interpret the information running on the network and decide which are useful.

To make all this possible, we must use techniques such as: Evolutionary Algorithms, Neural Networks, Fuzzy Systems which are techniques of Soft-Computing and Artificial Intelligence to integrate with a 'Smart Agent architecture', achieving a better solution than to centralized intelligence, being able to learn better and more quickly. The increase in complexity associated with collaborative organization is warranted by a much more efficient system control.

This approach requires the use of very sophisticated software and a micro controller with sufficient hardware resources to allow the execution.

11. Planning of Smart Agent Architecture in a Smart City Environment

In planning of system based on Smart Agent it must have a decentralized approach, in order to have exchanges of messages between all agents, without them becoming ineffective and/or obsolete.

So instead of exchanging many messages to a central manger, that should check each agent, the Smart Agent must be designed so that their decisions on information that is "perceived" by evolutionary algorithms, is automatically up to date information, reducing the overall level of communication.

The decentralized system (see Fig. 9) implies that the Smart Agent must make decisions in an autonomous way, also through pervasive learning. This could cause confusion in the attempt by the Smart Agent to perform the same tasks, within that connection they must be equipped with skills to predict what other agents will do. This is obtained through a definition of a particular goal.

A **goal** is a target that the agent that must reach and often, it is defined as states of the environment that the smart agent wants to make; Goals give the agent proactiveness and it is very important because they represent the persistence of goals, for example if the system for achieving a given goal is not valid, the action plan is modified, it fits as long as the goal is not achieved or considered impossible.

The **event** is a significant occurrence that the agent should respond to in some way.

An **action** is something that an agent does, it is an agent's ability to affect its environment and actions are instantaneous. Actions can be behaviour over time and can produce partial effect.

In addition to actions that relate to the outside environment of the Agent, there are actions that affect the capacity, level of the same Agent, which is not defined through goals. The execution cycle then has to relate events, objectives, learning that determines the decision making process and is performed by the Agent. In this way, succession of Events will change learning, which will change the Agent's plans, so that the goals set for that Agent can be achieved. This according to a cycle of perception-thinking-action, in which the part devoted to thought that defines the decisions.

The structure of the planes to reach a goal is generally a tree, consisting of sub-objectives, so as to reach the final goal with a step-by-step system (see Fig. 10).

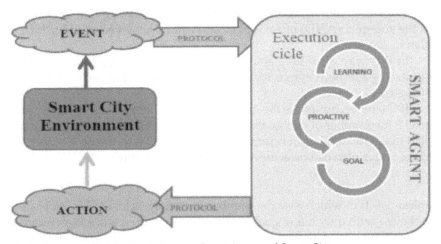

Fig. 9: Diagram of interaction between a Smart Agent and Smart City.

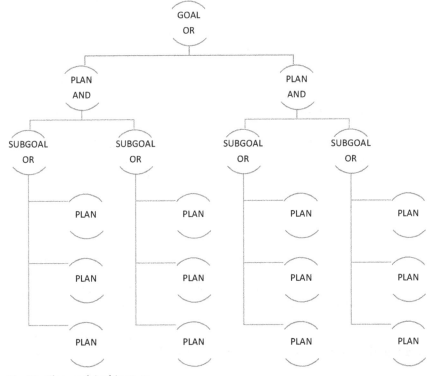

Fig. 10: Plan-goal Architecture.

The sub-goal (OR) are alternative ways of achieving that goal where as the sub-goal (AND) of each plan must all be achieved in order for the plan to succeed.

In this way even by a simple example it is possible to have a large number of alternatives to achieving a top-level goal that can be generated by a goal-plan tree.

If C (choice, option) is the number of plans that are applicable for each goal, S the number of sub-goals for each plan, and D the depth of the goal-plan tree. The number of actions where the target is at the base of the tree of the target plan can be achieved is following equation:

$$C^{((S^D-1)/(S-1))}$$

unless $S = 1$, in which case the number of options is just C^D and it is clear that even with only a few goals, sub-way and plan, it has a high number of possible ways of and achieving the top-level goal for a goal-plan tree.

Wherefore is really important that the program and protocol used are appropriate for the decision-lock, so that performance is carried out in Real Time.

This program execution provides a large number of action plans to reach the target, so this means that the environment agent the environment explored is running. Even if there are different communication languages and protocols, when using the Agent platform, not all of them support Plan-Agent.

The range and status of available agent platforms most latest can be found online, for example: PRS, UMPRS, JAM, JACK, DECAF, Zeus, AgentBuilder and JADEX (which is an extension of JADE), JADE, Zeus, OAA and mobile agents: Grasshopper, D'Agents, Aglets (Bellifemine et al. 2007). JADE-LEAP Java Agent Development Environment—Lightweight Extensible Agent Platform chose to develop MAS in mobile devices because JADE has a run-time for J2ME-CLDC (Connected Limited Device Configuration) and CDC (Connected Device Configuration) platforms. (http://jade.tilab.com).

MAS-based architecture has a high degree of scalability and therefore flexibility, through simple and powerful software that can develop a pervasive learning environment, while, in the same time, running multiple stationary and mobile agents equipped resource-limited mobile device (Huber, AOS Group, Agent Builder 2018).

JADELEAP, it serves as an agent platform: to facilitate real-world learning, through prediction or knowledge of the choices a user is making or interested.

PBL: Problem Based Learning is an SAL Smart Agent Learner centered approach.

When implementing PBL the SAL would have interactions with contexts and other agents. Learning environments have the potential to provide multiple ways to support different interactions at anytime.

Adaptive learning environments can help agents learn based on their individual goals, so integrating learning environments are adaptable and with the ability to continuously improve learning.

Of course, as with all learning methods, there are different models depending on learning objectives and the characteristics of learning environments. Ubiquitous and adaptive learning environments are new areas of research (Graf et al. 2008, Padgham and Winikoff 2004).

In this chapter, we aim at introducing and evaluating learning models that integrate PBL in adaptive learning environments and learning environments.

At the start of each reasoning cycle of a Goal agent, all percepts are collected and processed and the Smart Agents (SA) start a new cycle after receiving information by the Smart City environment and decide what to do next using condition-action rules. The choice is done when the selected action is sent to the SA Environment and Goal Agents run acycle of Observe-Decide-Act (ODA).

The goal is a logical-based agent programming language and one of its main strengths is to facilitate the development of agent strategies, since Goal-Agents can extract information from the environment to reach goals. Goal agents result their choice of action by their beliefs and goals by means of condition action rules. Every agent has a similar structure and each can handle all different roles. Each agent uses specific modules during module sharing.

The main module function is to connect all the smaller submodels and jump back and forth between them. Once all common tasks have been performed, the primary module controls the role of the agent in question and selects the role-specific modules that handle all the tasks and decisions that affect the specific role (Hurtado et al. 2014).

The agents who are in the same node are classified and assigned a unique number called Agent-Status, which is used when a different number of agents are designed to perform the same action; for which the agent that will perform the action depend on the type of rank, while other agents will perform other actions. This rank-based mechanism allows agents to divide the tasks between themselves without the need of communication, while at the same time ensuring that each agent performs an action only when possible (Hurtado et al. 2014, Dekker et al. 2011).

12. Critical Point of Smart City

As the technology is highly prompt, the most critical in the implementation, development and design of a sc are varied and mostly economic, political, managerial, for example:

> ➤ policy administrators often make choices on the basis of an immediate return, and they do not plan long-term;
> ➤ absence of interdisciplinary strategic vision: very often the different sectors of administration know little or nothing of what others are doing, is not structured interdisciplinary collaboration;
> ➤ excessive bureaucracy: the number needed to consult, involve, can be really excessive and often characterized hostility to change;
> ➤ absence of common objectives between public and private;
> ➤ widespread dissemination of smart systems;
> ➤ management of Big Data, regardless of the software producers and hardware;
> ➤ capacity to manage the privacy of all information interchanged;
> ➤ security in relation to possible hacker attacks;
> ➤ ability of citizens to understand the technology, changing the role by just consumers to prosumers;
> ➤ open standards to ensure interoperability and modularity;
> ➤ critical infrastructures can be made more Resilient by discovering the vulnerabilities and reducing them with enhanced failure analysis tools (Angelidou 2014).

13. Conclusion

The technology available today is adequate to enable the creation of a Smart City, in fact, thanks distributed Embedded Systems, their hardware and software building blocks, wireless Internet connectivity, interoperable machine to machine (M2M) services, to the widespread use of smart phones, the Internet of Things and the application of the Smart Agent is possible the design and implementation of a Sustainable city. This is also due to the fact that it is easier to use the technology, because without the ability of citizens to be able to use it does not make the city accessible.

In the near future, engaging citizens in research and innovation is becoming ever more important and Social/human capital (HC) is essential for urban development, as a knowledge-based city implies that its population grows intellectually by knowing how to adapt to the environment.

The quality of the infrastructure for the transmission and data management increasingly affects the effectiveness and efficiency of services offered to citizens in urban areas. However, it seems essential to act also on the quality of the information and knowledge they generate.

It appears essential to have an interdisciplinary strategic vision to create a city that is really smart. With this work we wanted to outline a vision of the overall model that has in itself all the elements constituting a Smart City, in order to define the guidelines for change-making. It is in this perspective that the Intelligent Community paradigm and in particular of Smart City were introduced as a conceptual tool to promote and implement, in a unified manner, some of the qualifying factors and competitiveness of urban areas.

In fact, the *change-makers* have difficulty in evaluating and justifying technology investments because the benefits are difficult to estimate. Change-makers responsible for planning and investments on smart solutions through new technologies find themselves confronted with numerous stakeholders having multiple needs and value perception. Therefore getting information and alternatives are crucial for decision-change-making in order to increase the transparency and accountability of the decisions.

References

Agent Builder. 2018. http://www.agentbuilder.com.

Angelidou, M. 2014. Smart city policies: A spatial approach. Cities: Volume 41, Supplement 1, July 2014, Pages S3-S11. Elsevier.

AOS Group. http://www.agent-software.com.

Bellifemine, F., G. Caire and D. Greenwood. 2007. Developing Multi-Agent Systems with JADE, John Wiley & Sons, Ltd.

Cosgrave, E., K. Arbuthnot and T. Tryfonas. 2013. Living labs, innovation districts and information marketplaces: A systems approach for smart cities. Elsevier.

Dekker, M., P. Hameete, M. Hegemans, S. Leysen, J. van den Oever, J. Smits and K.V. Hindriks. 2011. HactarV2: An agent team strategy based on implicit coordination. *In*: Dennis, L., O. Boissier and R.H. Bordini (eds.). Programming Multi-Agent Systems. ProMAS 2011. Lecture Notes in Computer Science, vol. 7217. Springer, Berlin, Heidelberg.

European Smart Cities. 2018. http://www.smart-cities.eu/model.html.

Graf, S., K. MacCallum, T.-C. Liu, M. Chang, D. Wen, Q. Tan, J. Dron, F. Lin, N.-S. Chen and R. McGreal. 2008. An Infrastructure for Developing Pervasive Learning Environments. IEEE XPlore. Kinshuk.

http://jade.tilab.com/.

Huber, M.J. 2018. Intelligent Reasoning Systems. www.marcush.net/IRS/irs_downloads.html.

Hulsmann, T. 2004. The Two-Pronged Attack on Learner Support: Costs and Centrifugal Forces of Convergence. Proceedings of the Third EDEN Research Workshop (electronic version), 2004. Retrieved March 31, 2005.

Hurtado, L.A., P.H. Nguyen, W.L. Kling. 2014. Agent-based control for building energy management in the smart grid framework. pp. 1–6. *In*: Innovative Smart Grid

Technologies Conference Europe (ISGT-Europe), 2014 IEEE PES, Istanbul, 12–15 Oct. 2014.

JadeX. Sourceforge. http://sourceforge.net/projects/jadex.

Karnouskos, S. and T. de Holanda. 2009. Simulation of a Smart Grid City with Software Agents. IEEE XPlore.

Lazaroiu, G.C. and M. Roscia. 2012. Definition methodology for the smart cities model. Energy 47: 326–332.

McGreal, R. 2005. Mobile technologies and the future of global education. Proceedings of ICDE International Conference on Open Learning and Distance Education, New Delhi, India, Nov. 19–23, 2005.

Nguyen, P.H., W.L. Kling and P.F. Ribeiro. 2011. Smart Power Router: A Flexible Agent-Based Converter Interface in Active Distribution Networks.

Odendaal, N. 2003. Information and communication technology and local governance: understanding the difference between cities in developed and emerging economies - Computers, Environment and Urban Systems – Elsevier.

Padgham, L. and M. Winikoff. 2004. Developing Intelligent Agent Systems. John Wiley & Sons, Ltd. ISBN: 0-470-86120-7 (HB).

Pontefract, D. 2013. Book, Flat Army: Creating a Connected and Engaged Organization, Wiley.

The MathWork. 2010. In Real-time Workshop for Use With Simulink. Available: http://www.mathworks.com/products/rtw/.

Thomas, S. 2005. Pervasive, persuasive elearning: Modeling the pervasive learning space. Proceedings of the Pervasive Computing and Communications Workshops, pp. 332–336.

Index

Index